Engineering Computational Technology

Saxe-Coburg Publications on Computational Engineering

Computational Structures Technology
Edited by: B.H.V. Topping and Z. Bittnar

Object-Oriented Methods and Finite Element Analysis
R.I. Mackie

Computational Modelling of Masonry, Brickwork and Blockwork Structures
Edited by: J.W. Bull

Innovative Computational Methods for Structural Mechanics
Edited by: M. Papadrakakis and B.H.V. Topping

Parallel and Distributed Processing for Computational Mechanics: Systems and Tools
Edited by: B.H.V. Topping

High Performance Computing for Computational Mechanics
Edited by: B.H.V. Topping and L. Lämmer

Computational Mechanics for the Twenty-First Century
Edited by: B.H.V. Topping

Parallel Finite Element Computations
B.H.V. Topping and A.I. Khan

Engineering Computational Technology

Edited by

B.H.V. Topping and Z. Bittnar

SAXE-COBURG
PUBLICATIONS

© Civil-Comp Ltd., Stirling, Scotland

published 2002 by
Saxe-Coburg Publications
Dun Eaglais, Kippen
Stirling, FK8 3DY, Scotland

Saxe-Coburg Publications is an imprint of Civil-Comp Ltd

ISBN 1-874672-17-2

British Library Cataloguing in Publication Data
A catalogue record for this book is available from the British Library

Cover Images: Test section of a wind tunnel simulator
*Developed by the ODU Center of Advanced Engineering Environments
at NASA Langley Research Center, Hampton, U.S.A.*

Printed in Great Britain by Bell & Bain Ltd, Glasgow

Contents

Preface

This volume comprises the invited lectures presented at The Third International Conference on Engineering Computational Technology (ECT 2002). The conference was held concurrently with The Sixth International Conference on Computational Structures Technology (CST 2002). Both conferences were organised in conjuction with, and held at, the Faculty of Civil Engineering, Czech Technical University in Prague, Czech Republic, from 4 to 6 September 2002. These conferences are part of the CST–ECT series that commenced in 1991.

The First Computational Structures Technology Conference was held in Edinburgh in 1991. From pure structural engineering the theme of this conference was expanded in The First Engineering Computational Technology Conference held in Edinburgh in 1998. The additional themes of the ECT series include: geotechnical engineering, fluid flow problems; electromagnetic problems; mechanical engineering, aeronautics and aerospace engineering. The themes also cover all aspects of computational technology including computing hardware and software developments.

Hardware and software developments provide engineers with opportunities to develop new algorthims and techniques which could not be implemented using old technology. In the first chapter of this book, Professor Noor reviews recent developments in computer technology and shows us new and emerging technologies for engineering analysis and design. Recent developments discussed include virtual reality, distributed collaboration and tele-immersion. Among the emerging technologies reviewed are grid, pervasive and autonomic computing techniques.

One of the recent developments, virtual reality, is further discussed in Chapter 2 by Professor Thabet and his colleagues. Their review takes a detailed look at how virtual reality technology is contributing to construction engineering. In Chapter 3, Dr McCarthy reviews developments in knowledge-based systems (KBS) for the design of structural steelwork over the last twenty years concluding that their age is just begining!

In Chapter 4, Professor Armstrong's group look at integrating CAE concepts with CAD geometry. This technology is key to the efficient processing and simulation of many engineering problems. In Chapter 5, Professor Morgan and his colleagues describe techniques for finite element analysis of electromagnetic scattering problems. In Chapter 6, Professor Crouch reviews finite element techniques for nonlinear dynamic fluid-soil-structure interaction of reinforced concrete structures. Professor Meyer considers, in Chapter 7, projection techniques used within the preconditioned conjugate gradient method for solving contact problems with hanging and contact nodes.

In Chapter 8, Professors Nikishkov and Atluri describe a method of combining the Galerkin boundary element method and the finite element method for modelling crack growth. The alternating procedure is implemented using an object-oriented (O-O) code. In Chapter 9, Dr Jimack reviews domain decomposition techniques for parallel solution of partial differential equations using iterative solvers. The theme of parallel computing is also discussed in Chapter 10 by Professors Sotelino and Dere. Again, an object-oriented approach is used by the authors to develop the software for parallel and distributed finite element dynamic analysis. In Chapter 11, Professor Benim reviews techniques for the equilibrium Euler-Euler modelling of pulverized coal combustion.

In Chapter 12, Dr Zimmermann discusses how intelligent object-oriented techniques may be used to solve engineering problems. This chapter links themes that have also been discussed in Chapters 3 (KBS), 8 (O-O) and 10 (O-O).

We are grateful to the authors and co-authors of the invited lectures included in this volume. Their contribution both to the ECT 2002 conference and this book is greatly appreciated.

We are indebted to Professor A.K. Noor and J. Peters at the Center for Advanced Engineering Environments, Old Dominion University, NASA Langley Research Center, Hampton VA for the two computer generated images shown on the cover of this book.

Other papers presented at the conferences in 2002 are published as follows:

- *The Invited Lectures from CST 2002 are published in:* Computational Structures Technology, B.H.V. Topping and Z. Bittnar (Editors), Saxe-Coburg Publications, Stirling, Scotland, 2002.

- *The Contributed Papers from CST 2002 are published in:* Proceedings of the Sixth International Conference on Computational Structures Technology, B.H.V. Topping and Z. Bittnar (Editors), (Book of Abstracts and CD-ROM), Civil-Comp Press, Stirling, Scotland, 2002.

- *The Contributed Papers from ECT 2002 are published in:* Proceedings of the Third International Conference on Engineering Computational Technology, B.H. V. Topping and Z. Bittnar (Editors), Saxe-Coburg Publications, Stirling, Scotland, 2002.

These Conferences could not have been organised without the contribution of many who helped in their planning, organisation and execution. We are particularly grateful to Jelle Muylle who once again, so expertly guided the design of this volume of lectures. We are also grateful to the following staff and students of the Faculty of Civil Engineering at the Czech Technical University: Alexandra Kurfürstová, secretary of the Department of Structural Mechanics and PhD students Jitka Poděbradská, Richard Vondráček and Matěj Lepš.

Barry H.V. Topping and Zdeněk Bittnar

Chapter 1

©2002, Saxe-Coburg Publications, Stirling, Scotland
Engineering Computational Technology
B.H.V. Topping and Z. Bittnar, (Editors)
Saxe-Coburg Publications, Stirling, Scotland, 1-23.

Computing Technology: Frontiers and Beyond

A.K. Noor
Center for Advanced Engineering Environments
Old Dominion University, NASA Langley Research Center
Hampton, Virginia, United States of America

Abstract

A brief history is given of the status and some recent developments in computing technology. Discussion focuses on recent advances in microelectronics, high-performance computing, adaptive/configurable computing, and user-interface hardware and software. Moore's law and the fundamental physical limits of computation are outlined. The emerging paradigms of grid computing, ubiquitous/pervasive computing and autonomic computing are described. Future computing environment and novel non-silicon computing paradigms are outlined.

Keywords: computing technology, emerging and future computing paradigms, future computing environment

1 Introduction

The field of computing is less than sixty years old. The first electronic computers were built in the 1940s as part of the war effort. By 1950s, IBM and Univac built business computers, intended for scientific and mathematical calculations to determine ballistic trajectories and break ciphers. Soon other companies joined the effort – names like RCA, Burroughs, ICL and General Electric – most of whom disappeared or left the computer business. The first programming languages – Algol, FORTRAN, Cobol, and Lisp – were designed in the late 1950s, and the first operating systems in the early 1960s. The first computer chip appeared in the late 1970s, the personal computer around the same time, and the IBM PC in 1981. Ethernet was invented in 1973 and did not appear in the market until 1980. It operated at 10 megabits per second (Mb/s) and increased to 1 Gb/s (10^9 bits/s) in 1997. The internet, which descended from the ARPANET in the 1970s, and the World Wide Web was created in 1989.

The notion of a hierarchical system was introduced with the NSF-sponsored backbone network NSFNET. The development started in 1986, and the network was retired in April 1995. The current internet has a multiplicity of backbone, regional and metropolitan networks connected in a myriad of ways. A number of government network projects provided experiments in advanced technology and applications, including NSF's very high speed Backbone Network Service (vBNS), DOE's Energy Services Network (ESnet), Advanced Technology Development Network (ATDnet), and National Research and Education Network (NREN).

Progress in computer technology has driven the evolution of computers. The mainframe computer became obsolete except as a computing engine for large applications in business, science and engineering. Computers are now everywhere from desktops and laptops to handhelds, from the telephone networks to embedded and wearable computers.

A number of papers and monographs have been written on the evolution of computers as well as on various aspects of current computing technologies (see, for example, [1] to [4]). Also, a number of special issues of journals [5], and workshops and conferences have been devoted to the subject, and proceedings have been published. The objectives of this paper are to summarize some of the recent developments and trends in computing technologies, and identify future computing paradigms and environment.

2 Evolution of Microprocessors and Microprocessor-based computers

2.1 Microprocessors

Although the first computers used relays and vacuum tubes for the switching elements, the age of digital electronics is usually said to have begun in 1947, when a research team at Bell Laboratories designed the first transistor. The transistor soon displaced the vacuum tube as the basic switching element in digital design. The nerve center for a computer, or a computing device, is its integrated circuit (IC or chip), the small electronic device made out of a semiconductor material. Integrated circuits, which appeared in the mid-1960's and allowed mass fabrication of transistors on silicon substrates are often classified by the number of transistors and other electronic components they contain. The ever-increasing number of devices packaged on a chip has given rise to the acronyms SSI, MSI, LSI, VLSI, ULSI, and GSI, which stand for small scale (1960s - with up to 20 gates per chip), medium-scale (late 1960s - 20-200 gates), large-scale (1970s - 200-5000 gates per chip), very large- scale (1980s - over 5000 gates per chip), ultra large-scale (1990s - over million transistors per chip), and giga-scale integration (over billion transistors per chip), respectively (Figure 1). These generations of microprocessors have increased the speed of computers by more than trillion times during the last five decades, while dramatically reducing the cost (see Figure 2).

A number of technologies have been used to achieve ultrafast logic circuits. These include use of: new material systems such as gallium arsenide (Ga As); multichip modules (MCM); monolithic and hybrid wafer scale integration (WSI); new transistor structures such as the quantum-coupled devices using hetero-junction-based super lattices; and optical interconnections and integrated optical circuits. More recently, the use of carbon nanotubes as transistors in chips; clockless (asynchronous) chips and; hyperthreading, which makes a single CPU act in some ways like two chips, have been demonstrated.

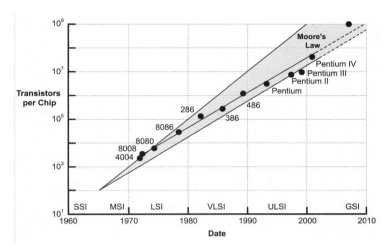

Figure 1: Evolution of Intel Processors

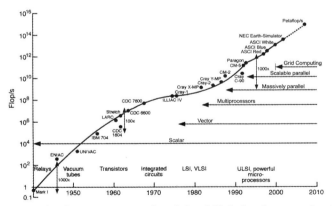

Figure 2: Growth of computer speed and the shift in hardware technology and computer architecture

The massive demand for higher processor performance resulted in two development paths. The first is based on the successive introduction of temporal, issue, and intra-instruction parallelism into the processor operation. The sequence was basically determined by hardware efficiency aspects. The subsequent evolutionary phases of microprocessors in this development path are: traditional sequential processors, pipelined processors, superscalar processors, and superscalar processors with multimedia and 3-D support. The second development path is based on the introduction of temporal parallelism followed by intra-instruction parallelism in the form of a Very Long Instruction Word (VLIW) – instructions, instead of issue parallelism. However, this approach requires compilers that can resolve the dependencies, and optimize parallel operations in VLIW processors. Recently, the Explicitly Parallel Instruction Computing (EPIC) style of architecture was developed to enable high levels of instruction – level parallelism. The EPIC model has the potential of achieving unparalleled levels of performance in future computer systems.

2.2 Memory and Storage Devices

Memory is the most rapidly advancing technology in microelectronics. Recent progress includes development of an entire hierarchy of addressable memories, and of high-speed, random-access memory chips with many bits of data. Each level in the hierarchy represents an order-of-magnitude increase in capacity for the same cost.

Memory can be categorized in dozens of different groups, types and classes. Two important types are static and dynamic random access memories (SRAMs and DRAMs). Over the years DRAM has improved in reliability, cost and speed. This lead to the development of several high-performance memory systems ([6], [7]), such as magnetic random access memories (MRAMs), which store data in magnetic tunnel junctions, and retain their states even when the power is off.

The significant increase in computing power is associated with the need for a significant increase in the storage requirements, as well as new approaches for creating storage devices. Among these approaches are holographic, polymer-based, and protein-based memories (Figure 3).

Figure 3: 3-D optical memory developed by using photo synthetic protein bacteriorhodopsin

The current trend is towards the merging of computing and communications in every device; and the creation of Microsystems, system-on-a-chip computers for special applications.

3 Moore's Law and Fundamental Limits of Computation

In 1965 Gordon Moore hypothesized that processing power (number of transistors and computing speed) of computer chips was doubling every 18 – 24 months or so. For nearly four decades the chip industry has marched in lock step to this pattern or rule of thumb, which is referred to as Moore's law (see Figure 1).

This remarkable rate of advancement has resulted in smaller feature sizes; improved manufacturing techniques, which allowed making larger chips and wafers; more efficient circuit designs and materials; and brought large speed and functionality increases with a substantial drop in costs. A silicon substrate of just a few square centimetres can hold 100 million transistors, with key features measuring 0.18 micron.

Predictions by the semiconductor industry association is for a processor clock speed of 3.5 Ghz by 2006 and 6.0 Ghz by 2009.

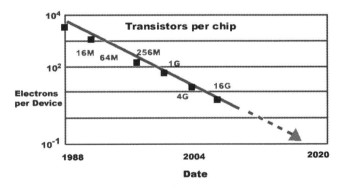

Figure 4: Limits of Moore's Law - Vanishing Electrons

Although the computer industry will continue to improve the performance of processors and overcome the associated technological limitations, at some future point, Moore's law will saturate. There are fundamental, hard physical limits to computing (Figure 4). These include thermodynamic and quantum limits of information storage, communication limits, and computation rate limits [8]. Specifically:

- *Feature Size* – cannot be smaller than 0.1 nm. It is anticipated that the storage density of approximately 1 bit/Å^3 will be reached in 40 years.

- *Thermodynamic limit* of 0.7 KT for heat generation involved in storing a bit of information in memory and destroying the information already there (energy dissipation of irresivible computation). However, this limit can be avoided by developing storage techniques that do not involve destroying previously stored information (e.g., reversible chips).

• *Processor Clock Rate* – cannot exceed the limit set by the maximum rate of atomic transitions (about 10^{15} Hz – million times above present rate). This limit is expected to be reached in 30 years.

4   High-Performance Computing

The incessant demand for computing power to enable accurate simulation of complex phenomena in science and engineering has resulted in the development of a class of general-purpose supersystems designed for extremely high-performance throughput, and new paradigms for achieving the high-performance. These include:

• Vector/pipeline processing

• Parallel processing on multiple (hundreds or thousands) CPUs, and

• Multitasking with cache memory microprocessors

The first generation of vector/pipeline supersystems included the array of processors ILLIAC IV, the pipeline computer CDC STAR-100, and the Texas Instruments Advanced Scientific computer (ASC). The second generation included the popular CRAY-1 computer, which featured the pipelined vector instructions introduced in the first generation machines, but carried out in a clever register-to-register mode. The introduction of vector/pipeline computing provided an order of magnitude increase to computing performance. However, the increase came at the cost of a restrictive programming model. Large numbers of identical operations had to be performed in sequence.

In the early 1980s, the move towards parallel processing supersystems started. The number of processors on the third generation supersystems grew from one to eight and to 16 a year or two later. Other machines developed during the 1980s used a hybrid combination of pipeline and array processors to achieve high performance. Examples included CRAY 932, CRAY-3, ETA-10 and the Japanese machines Fujitsu VP-400, Hitachi S810/20 and S820/80, NEC SX2-400.

Scalable parallel systems (SPPs) are massively parallel processor systems which comprise hundreds or thousands of VLSI processors integrated through high bandwidth networks. The challenge for SPPs is to increase not so much the number of processors but the logical and physical interconnections, allowing a large number to be used effectively as one computing system. Both single instruction multiple data (SIMD) and multiple instruction multiple data (MIMD) architectures were developed, which exploit parallelism at an unprecedented scale while leveraging the rapid advances in semiconductor technology.

An MIMD machine consists of a number of processors with some form of interconnection between them. The interconnection can be either through a common memory, the simplest form of which is the shared memory, or through message passing. Examples of parallel processing machines are the connection machines CM-2 and CM-5, and the CRAY T3D and T3E. The CRAY T3D and T3E demonstrated the technical feasibility of interconnecting 512 or even 1024 processors in such a way that the latency of one processor obtaining a word from the memory of any other was as little as 2 – 3 microseconds.

More recent developments in supersystems included the introduction of microprocessors with cache memory, Very Long Instruction Word (VLIW) processors, which have more than one functional unit for such tasks as load/store, floating-point add/multiply, and integer arithmetic/logic operations; and superscalar processors.

Although the peak performance of the first generation supersystems was less than 100 Mflop/s, the gigaflop barrier (1 Gflop/s) was passed in 1988/89, and the teraflop barrier (1 Tflop/s) in 1996/97. In 1995, the US Department of Energy supported the development of three terascale machines through its Accelerated Strategic Computing Initiative (ASCI). The three machines are: ASCI Red, with 9,472 Intel Pentium II Xeon processors – 2.379 Tflop/s at Sandia National Labs; ASCI Blue Mountain with 5,856 IBM PowerPC 604E processors – 1.608 Tflop/s at Los Alamos National Lab; and ASCI White with 8, 192 IBM Power 3-II processors – 7.226 Tflop/s at Lawrence Livermore National Lab.

To date, there are over 17 terascale machines worldwide. The maximum performance reported today is 35.6 Tflop/s of the Earth Simulator at Kanazawa, Japan, which consists of 5,104 vector processors (with peak performance of 40 Tflop/s).

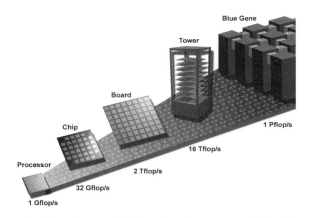

Figure 5: Five Steps to the Petaflop Supersystem Blue Gene/P

In December 1999, IBM announced a five year effort to build a petaflop (10^{15} Flop/s) supersystem – The Blue Gene Project. The project has the two primary goals of advancing the state of the art of biomolecular simulation, and computer design of extremely large-scale systems. Two systems are planned: Blue Gene/L, in collaboration with Lawrence Livermore National Lab, which leverages high speed interconnect and system-on-a-chip technologies and has a peak performance of 200 Tflop/s; and Blue Gene/P, the petaflop-scale system. The system will consist of more than one million processors, each capable of one billion operations per second. Thirty-two of these ultra-fast processors will be placed on a single chip (32 Gflop/s). A compact two-foot by two-foot board containing 64 of these chips will be capable of 2 Tflop/s. Eight of the boards will be placed in 6-foot-high racks (16 Tflop/s) and the final system will consist of 64 racks linked together to achieve the one Pflop/s performance.

The dramatic increase in performance delivered by the supersystems to applications follows from a combination of advances, each making the system more scalable:

- The microprocessor building block became faster (especially for floating-point operations) through greater internal parallelism, use of huge multilevel caches, and faster clock speeds.

- Schemes were devised for effectively connecting large numbers of processors and memories.

- Researchers learned to use large numbers of processors and deep memory hierarchy more effectively.

- System software and tools improved through experience and the leveraging of mainstream advances.

5   Adaptive/Configurable Computing

The software-programmed microprocessor and the Application Specific Integrated Circuit (ASIC) are the two primary methods in conventional computing for the execution of algorithms. The first method has the advantage of flexibility, the functionality of the system can be altered by changing the software instructions, without changing the hardware. But its performance can suffer, if not in clock speed, then in work rate. The second method is very fast and efficient when executing the exact computation for which ASIC was designed. However, ASIC are inflexible – the circuit cannot be altered after fabrication. Changes in application require a board redesign and replacement.

Adaptive/configurable computing is intended to fill the gap between hardware and software, achieving potentially much higher performance than software, while maintaining a higher level of flexibility than ASIC hardware (Figure 6). Reconfigurable devices, including field programmable gate arrays (FPGAs) contain an array of computational elements whose functionality is determined through multiple programmable configuration bits. They provide fine-grained logic and interconnection elements whose function and structure users can program to suit an application's needs [9].

Early FPGA-based computing machines were essentially customized and domain specific. They were configurable to an extent, but had a fixed arrangement of many FPGAs, memories, and interconnections. Their board-level architecture limited the type and range of applications that programmers could effectively map and execute on these machines.

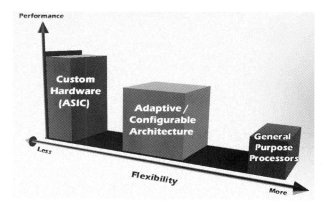

Figure 6: Adaptive/Configurable Computing

In the mid-1990s, a new form of configurable element emerged – the reconfigurable data paths, which have coarser-grained reconfigurable units than their fine-grained FPGA predecessors. They can accommodate reconfigurable nibble, byte, or wider arithmetic logic units (ALUs). These, along with other improved features, efficient programming abstractions and compiler techniques, let a reconfigurable data path exceed a CPU's performance in a range of applications.

Recently, coherent application-development environments have emerged, where a program's small but time-consuming kernels can map to a reconfigurable coprocessor while the program's remaining parts execute on a traditional instruction – set processor.

Another recent trend of adaptive computing is to use very large scale integration (VLSI) to develop specialized, domain-specific architectures that are configurable for a few applications in a specialized domain (e.g., 2-D image processing or wireless signal processing).

6   User Interface Hardware and Software

6.1   Multimodal human-computer interfaces

As the computer's role has evolved over the past 50 years, so have the user's expectations of the human-computer interfaces. Although the WIMP (windows, icons, menus, pointers) paradigm has provided a stable and global interface, it will not scale to match the myriad form factors and uses of platforms in the future collaborative distributed environment.

Recent work has focused on intelligent multimodal human-computer interfaces, which synthetically combine human senses, enable new sensory-motor control of computer systems; and improve dexterity and naturalness of human-computer interaction ([10], [11]). An interface is referred to as intelligent if it can:

- Communicate with the user in human language,
- Perform intelligent functions, or
- Adapt to a specific task and user.

Intelligent interfaces make the interaction with the computer easier, more intuitive, and more flexible.

Among the interface technologies which have high potential for meeting future needs are Perceptual user interfaces (PUIs), and brain computer (or neural) interfaces (BCIs). PUIs integrate perceptive, multimodal and multimedia interfaces to bring human capabilities to bear on creating more natural and intuitive interfaces. They enable multiple styles of interactions, such as speech only, speech and gesture, haptic, vision, and synthetic sound, each of which may be appropriate in different applications. These new technologies can enable broad uses of computers as assistants, or agents, that interact in more human-like ways.

A BCI interface refers to a direct data link between a computer and the human nervous system. It enables biocontrol of the computer – the user can control the activities of the computer directly from nerve or muscle signals. To date, biocontrol systems have utilized two different types of bioelectric signals: electroencephalogram (EEG), and electromyogram (EMG). The EEG measures the brain's electrical activity and is composed of four frequency ranges, or states – alpha, beta, delta, and theta. The EMG is the bioelectric potential associated with muscle movement. BCI technologies can add completely new modes of interaction, which operate in parallel with the conventional modes, thereby increasing the bandwidth of human computer interactions. They will also allow computers to comprehend and respond to the user's changing emotional states – an important aspect of affective computing, which received increasing attention in recent years. Among the activities in this area are creating systems which sense human affect signals, recognize patterns of affective expression, and respond in an emotionally-aware way to the user; and systems which modify their behaviour in response to affective cues. IBM is developing sensors in the mouse that sense, physiological attributes (e.g., skin temperature and hand sweatiness), and special software to correlate those factors to previous measurements to gauge the user's emotional state moment to moment as he/she use the mouse.

Future well-designed multimodal interfaces will integrate complementary input modes to create a synergistic blend, permitting the strengths of each mode to overcome weaknesses in the other modes and to support mutual compensation of recognition errors.

6.2   Multimedia, Virtual Reality, and Interfaces for Volumetric Displays

Multimedia and virtual reality facilities, which embrace different forms of human communication and provide elaborate graphics, video, animation and visualization capabilities, have been developed. Virtual reality places the analyst in the space with the information being visualized. It provides multisensory experience – through dimensional sight, sound, touch, forced back or forced resistance and motion.

Commercially available virtual environments span two generations. The first generation include the Helmet Mounted Displays (HMDs) and the Binocular Omni Orientation Monitor (BOOM). The second generation systems include the Cave Automatic Virtual Environment (CAVE) – an immersive, multiperson, 10 x 10 x 10 – ft^3 room – sized, high resolution 3-D video and audio; the rigid Powerwalk screen system (workwall) developed by Fakespace Systems; the Vision Dome and the Vision Station of Elumens Corporation, and the desktop display from Panoram Technologies (Figure 7).

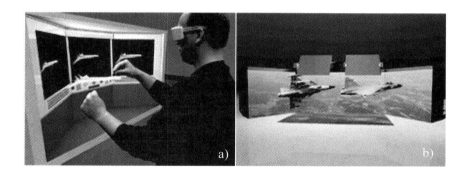

Figure 7: New Virtual Reality facilities a) Desktop display from Panoram Technologies, b) The RAVE II - fully reconfigurable display modules from Fakespace Systems, Inc.

In contrast to virtual reality, there are conceptual and technological developments emphasizing that the objects of the real environment should, and can be augmented instead of immersing into virtual reality. *Augmented and mixed reality* are the results of overlaying and adding digital information (e.g., virtual objects), to real objects, which become, in a sense, an equal part of the environment.

Recently, a number of volumetric displays, have been developed which enable true 3-D image visualization, and hold the promise of enhancing rendered 3-D graphics' sense of realism by providing all depth cues humans require. Volumetric displays generate true 3-D volumetric images. Holographic displays, which produce a 3-D image by reproducing the diffraction of light from a 3-D scene is an example of volumetric displays (Figure 8).

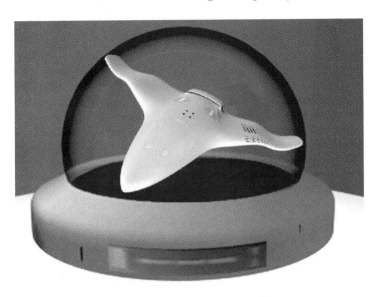

Figure 8: The Perspecta Spatial 3-D visualization platform from
Actuality Systems, Inc.

6.3   Distributed Collaboration and Tele-immersion

The accelerating pace of technology development has made it incumbent on high-tech organizations to efficiently and effectively collaborate and share knowledge across geographically dispersed, multi-disciplined teams.   An explosion of distributed collaboration facilities, technologies and infrastructure have been developed including, decision support systems, computer-supported cooperative work, and virtual immersive workspaces. Herein, only tele-immersive environments will be described.

Tele-immersion creates coordinated, partially simulated environments at geographically distributed sites so that users can collaborate as if they were in the same physical room. The computers track the participants as well as the physical and virtual objects at all locations, and project them onto stereo-immersive surfaces. The collaborators at remote sites share details of virtual world that can autonomously control computations, query databases and gather results.

In early tele-immersive environments, collaborators entered the environment as avatars – lifelike computer-generated representations. The environment transmitted gestures, as well as audio and video, so users had a greater sense of presence in the shared space than they would with other collaborative media.

The significant increase in network bandwidth and capacity will enable the creation of new paradigms in distributed collaboration, including virtual laboratories, and tele-presence – long-distance transmission of life-size, 3-D synthesized scenes, accurately sampled and rendered in real time using advanced computer graphics and vision technologies. This should lead to more natural teleconferencing work environment computer-generated 3-D stereoscopic image (Figure 9).

Figure 9: Applications of immersive telepresence a) collaborative design project, b) coupling of experiments and numerical simulations

7   Emerging Computing Paradigms

7.1   Grid Computing

The rapidly increasing power of computers and networks in the 1990s, led the new paradigm of distributed computing. A flurry of experiments were conducted on "peer-to-peer" computing, all devoted to harnessing the computer power and storage capacity of idle desktop machines. These included cluster computing using networks of standard single-processor workstations to solve single problems. At the same time, the high-performance computer community began the more ambitious experiments in *metacomputing*. The objective of Metacomputing was to make many distributed computers function like one giant computer – metasystem (e.g., the virtual national machine). Metasystems give users the illusion that the files, databases, computers and external devices they can reach over a network constitute one giant transparent computational environment. However, no standard software and no infrastructure were available to support it.

In 1996 the U.S. Defense Advanced Research Projects Agency funded the Globus Project to provide a solid foundation for large scale coordinated sharing of computer resources. A grid with some 80 sites worldwide running Globus software was demonstrated in 1997. The name "grid" computing was coined by analogy to the electrical power grid. Since then "grid" computing has emerged as an important new field, distinguished from conventional distributed computing by its focus on large scale coordinated resource sharing, innovative applications and problem solving in dynamic, multi-institutional virtual organizations, and, in some cases, high-performance orientation.

The term grid computing is now used to refer to massive integration of computer systems to offer performance unattainable by a single machine. It provides pervasive, dependable, consistent, and inexpensive access to facilities and services that live in cyberspace, assembling and reassembling them on the fly to meet specified needs ([12], [13], [14], [15]).

The Globus Project software development has resulted in the Globus Toolkit, a set of services and software libraries to support Grids and Grid applications. The toolkit includes software for security, information infrastructure, resource management, data management, communication, fault detection, and portability. Current Globus Project research targets technical challenges that arise from building grids and developing grid applications. Typical research areas include resource management, data management and access, application development environments, information services, and security.

Although grid technologies are currently distinct from other major technology trends, such as internet, enterprise, distributed, and peer-to-peer computing, these other trends can benefit significantly from growing into the problem spaces addressed by grid technologies.

Grid technologies enable the clustering of a wide variety of geographically distributed resources, such as high-performance computers, storage systems, data sources, special devices and services that can be used as a unified resource.

The essential building blocks of grid computing are: Fast processors, parallel computer architectures, advanced optical networks, communication protocols, distributed software structures and security mechanisms. Once the concept of grid computing was introduced, several grid computing projects were launched all over the world. A sampling of grid computing projects are listed in Table 1.

In the future, grids of every size will be interlinked. The "supernodes" like TeraGrid will be networked clusters of supersystems serving users on a national or international scale. Still more numerous will be the millions of individual nodes: personal machines that users plug into the grid to tap its power as needed. With wireless networks and miniaturization of components, that can evolve into billions of sensors, actuators and embedded processors as micronodes.

Project	Date	Sponsor	Main Purpose
Information Power Grid http://www.ipg.nasa.gov	1999	NASA	Computational support for aerospace development, planetary science and other NASA research
Access Grid http://www-fp.mcs.anl.gov/ fl/accessgrid/	1999	U.S. Department of Energy, National Science Foundation	Internet based collaboration, including lectures and meetings – among scientists at facilities around the world.
Grid Physics Network (GriPhyN) http://www.griphyn.org	2000	NSF	Data analysis for four physics projects
Unicore http://www.unicore.de	2000	German Federal Ministry for Education and Research	A seamless interface to high-performance computer centers at nine government, industry and academic labs.
Network for Earthquake Engineering and Simulation (NEESgrid) http://www.neesgrid.org	2001	NSF	Integrated computing environment for 20 earthquake engineering labs
European Data Grid http://www.eu-datagrid.org	2001	European Union	Data analysis in high-energy physics, environmental science and bioinformatics
U.K. National Grid http://www.grid-support.ac.uk	2001	U.K. Office of Science and Technology	Support for grid projects within Britain
TeraGrid http://www.teragrid.org	2002	NSF	General-purpose infrastructure for U.S. science – will link four sites at 40 gigabits per second and compute up to 13.6 Teraflop/s
International Virtual Data Grid Laboratory (iVDGL) http://www.ivdgl.org	2002	NSF and counterparts in Europe, Australia and Japan	World's first truly global grid – will link high-performance computer centers in Europe, Australia, Japan and the United States

Table 1: A sampling of grid computing projects 1999-2002 [15]

7.2   Ubiquitous/Pervasive Computing

The trend of computers getting smaller is likely to lead to an environment with computing functionality embedded in physical devices that are widely distributed and connected in a wireless web.

In a seminal article written in 1991, Mark Weiser described a hypothetical world in which humans and computers were seamlessly united [16]. This vision was referred to as ubiquitous computing. Its essence was the creation of environments saturated with computing and communication, yet gracefully integrated with human users.

In the mid-1990s, the term pervasive computing came to represent the same vision as that described by Weiser. More recently, other related visions were proposed, such as *proactive computing* and *autonomic computing*, described subsequently. Proactive computing focuses on improving performance and user experience through speculative or anticipatory actions.

The key components of ubiquitous/pervasive computing are:

- *Pervasive devices*, including
  - Small, low-powered hardware (CPU, storage, display devices, sensors),
  - Devices that come in different sizes for different purposes, and
  - Devices that are aware of their environment, their users, and their locations,

- *Pervasive communication* – a high degree of communication among devices, sensors and users provided by ubiquitous and secure network infrastructure (wireless and wired) and mobile computing,

- *Pervasive interaction* – more natural and human modes of interacting with information technology, and

- *Flexible, adaptable distributed systems* – dynamic configuration, functionality on demand, mobile agents, mobile resources

7.3 Autonomic Computing

The increasing capacity and complexity of the emerging computing systems, and the associated cost to manage them, combined with a shortage of skilled workforce are providing the motivation for a paradigm shift to systems that are self-managing, self-optimizing, and do not require the expensive management services needed today. A useful biological metaphor is found in the autonomic nervous system of the human body – it tells the heart how many times to beat, monitors the body temperature, and adjusts the blood flow, but most significantly, it does all this without any conscious recognition or effort on the part of the person - hence the name *autonomic computing* was coined.

Autonomic computing is a new research area led by IBM focusing on making computing systems smarter and easier to administer. Many of its concepts are modelled on self-regulating biological systems.

Autonomic computing is envisioned to include the ability of the system to respond to problems, repair faults and recover from system outages without the need for human intervention. An autonomic computing system consists of a large collection of computing engines, storage devices, visualization facilities, operating systems, middleware and application software. It combines the following seven characteristics:

1. *Self-defining* - Has detailed knowledge of its components, current status, ultimate capacity and performance, and all connections to other systems.

2. *Self-configuring* – can configure and reconfigure itself under varying and unpredictable conditions. System configuration or setup must occur

automatically, as must dynamic adjustments to that configuration to handle changing environments.

3. *Self-optimizing* – never settles for status quo. Always looks for ways to optimize its performance. Monitors constituent parts, and metrics, using advanced feedback control mechanisms and makes changes (e.g., fine-tune workflow) to achieve predetermined system goals.

4. *Self-healing* – able to recover from routine and extraordinary events that might cause some components to malfunction or damage. It must be able to discover problems, reconfigure the system to keep functioning smoothly.

5. *Self-protecting* – detect, identify and protect itself against various types of failure. Maintains overall system security and integrity.

6. *Contextually Aware* – This is almost self-optimization turned outward. The system must know the environment and the context of the surrounding activity, and adapts itself (in real-time) accordingly.

7. *Anticipatory* – anticipates the optimized resources, configuration, and components needed.

IBM is already working on several interdisciplinary autonomic computing projects, some involving artificial intelligence; complex adaptive systems; and self-healing technology, which lets systems keep running even if certain components fail. Current research projects at labs and universities include self-evolving systems that can monitor themselves and adjust to some changes; cellular chips capable of recovering from failures to keep long-term applications running; heterogeneous work-load management that can balance and adjust work loads of many applications over various servers; and traditional control theory applied to the realm of computer science.

Self-managing capabilities have been demonstrated by the IBM project eLiza. Its objective is to create servers that respond to unexpected capacity demands and system glitches without human intervention. Some aspects of autonomic computing are not entirely new to industry. However, this innovation needs to be taken to an entirely new level. Among the experimental autonomic systems being built are *Oceano, Regetta* and *Planetary computing systems.* The first is under construction at the IBM's Thomas J. Watson Research Center. Its goal is to enable pools of servers to share software workloads. It combines the first three of the aforementioned characteristics of autonomic systems. Regetta entails the building of self-healing memory resources, which are able to repartition themselves, on the fly, to correct faults. The planetary computing system is being developed at Hewlett-Packard Labs in Palo Alto, CA, and is envisioned as a 50,000-node computing fabric that can be built, reconfigured, and managed automatically. All system changes will occur in the software. Autonomic functionality is expected to begin appearing in servers and software by the middle of the decade.

8   Future Computing Environment and Novel Computing Paradigms

8.1   Future Computing Environment

Significant advances continue to be made in the entire spectrum of computing and communication technologies.   Speculations about the future of computers and computing have been attempted in several monographs ([17], [18], [19], [20]). Herein, only the emerging trends are identified, which include:

- Ubiquitous / Pervasive computing
- Computer-supported distributed collaboration
- Augmented reality and tele-immersion facilities
- Novel computing paradigms
- Smarter, self regulating computing systems
- Hierarchical knowledge nets
- Optical networks supplemented by wireless communication

Figure 10: Wind Tunnel Simulator

The combination of these will affect every facet of our life. It will enable unprecedented levels of sophistication in the modelling and simulation of complex engineering systems (Figure 10).

Realizing the full potential of the new environment, however, requires novel approaches for human-machine interfaces. The full power of creative artificial intelligence research should be brought to bear on the development of intelligent, adaptive interfaces, which should effectively link to as many of the human sensing and communication mechanisms as possible. The interfaces should be tailored to specific user needs, and be able to know how to best interact with a user.

8.2 Novel Computing Paradigms

Silicon-based technology is expected to reach its physical limits in the next decades. But silicon and computing are not inextricably linked, although they often seem to be. For example, when silicon microelectronics reaches ultimate physical limits a number of new approaches and technologies have already been proposed. These include ([21], [22]);

- Molecular computing,
- Quantum computing,
- Chemical and biochemical computing,
- DNA computing, and
- Optical and optoelectronic computing

None of these approaches is ready to serve as an all-purpose replacement for silicon. In fact, some approaches may be only appropriate as specialized methods in particular niches, such as high-level cryptography.

8.2.1 Molecular Computing

When silicon microelectronics reaches ultimate physical limits to further miniaturization, the smallest silicon transistor will still contain over a million atoms.

Commercial fabrication methods cannot economically make silicon transistors much smaller than 100 nm. Even if they can be etched onto a chip, ultra small silicon components may not work. At transistor dimensions of around 50 nanometers, the electrons begin to obey odd quantum laws, wandering where they are not supposed to be.

Molecules, on the other hand, are only a few nanometers in size, making possible chips containing billions – even trillions of switches and components.

In the last few years molecules and nanoparticles were integrated into scaleable electronic logic and memory devices. Researchers at IBM modified and positioned a single carbon nanotube to form two types of transistors, each a few nanometers in diameter - one hundred times smaller than the transistors now found on computer chips.

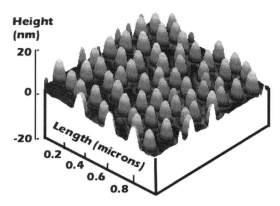

Figure 11: Quantum dots – Germanium on 15 nanometers silicon substrate

8.2.2   Quantum Computing

Conventional computer circuits are based on solid state devices which represent binary digits, or bits. By contrast, quantum computers encode information as quantum bits, or qubits (Figure 11). A qubit can be a one or a zero, or it can exist in a superposition that is simultaneously one or zero or somewhere in between. Qubits represent atoms that are working together to serve as a computer memory and microprocessor. Because a quantum computer can contain these multiple states simultaneously, it has the potential to be millions of times more powerful than today's most powerful supersystems ([23], [24]). A 30-qubit quantum computer would equal the processing power of a conventional computer capable of running at ten trillion operations per second.

8.2.3   Chemical and Biochemical Computing

A chemical computer processes information by making and breaking chemical bonds, and it stores logic states or information in the resulting chemical (i.e., molecular) structures. It is a process-control computer. A chemical nanocomputer would perform such operations selectively among molecules taken just a few at a time in volumes only a few nanometers on a side. A biochemically based computer is a variant of chemical computers which mimics the activities of humans and other animals with multicellular nervous systems.

8.2.4   DNA Computing

DNA computing refers to the use of biological molecules such as DNA (Deoxyribose Nucleic Acid) and RNA (Ribonucleic Acid) to solve mathematical problems. It was first proposed by L. Adleman in 1994.

A key advantage of DNA is that they can make computers smaller, while at the same time significantly increasing their storage capacity. More than 10^{12} DNA molecules can fit into a volume of one cubic centimeter. With this small amount of

DNA, a computer would be able to hold ten terabytes of data and perform 10^{12} calculations at a time. By adding more DNA, more calculations could be performed.

8.2.5 Optical and Optoelectronic Computing

After three decades of research on optical computing, it is now realized that optical computing cannot supplant electronics in the foreseeable future. An all-optical computer cannot import, modify or export information – such a device would serve no useful purpose. However, the coupling of electronics and optical technologies can improve the performance of future computing systems ([25], [26], [27]). High-end electronic devices can efficiently perform data processing on a single chip, and optics can provide high-bandwidth, high connectivity and high density communication channels, thereby overcoming the inter-chip interconnection bottleneck on multi-chip modules, and also between printed circuit boards. In addition, the use of intra-chip (reconfigurable) optical interconnections can reduce wiring congestion and provide flexibility to the interconnection topology on SIMD processor arrays.

9   Concluding Remarks

The accelerating pace of the computing technology development shows no signs of abating. Computing power reaching 100 Tflop/s is likely to be reached by 2004 and Pflop/s (10^{15} Flop/s) by 2007. The fundamental physical limits of computation, including information storage limits, communication limits and computation rate limits will likely be reached by the middle of the present millennium. To overcome these limits, novel technologies and new computing paradigms will be developed.

As a result of convergence of wireless communication and portable computers, pervasive autonomic computing will become the prevalent computing paradigm. The environment will be saturated with computing and communication devices, yet gracefully integrated with human users. The devices will be small and powerful, light and durable, able to perform multiple functions, easy to use and have embedded intelligence. The key elements of the environment are: handheld, embedded and wearable computers; wireless and optical networks; advanced modelling, simulation and multisensory immersive visualization tools and facilities; devices to sense, influence and control the physical world; and natural cooperative human-machine collaboration, with mixed initiatives between human and computer-generated interactions. Intelligent, affective technologies will be developed to allow computers to comprehend and respond to the user's changing emotional states.

Customization (adaptively) and intelligence will become two of the major characteristics of future networks, collaboration infrastructure, modelling, simulation and visualization facilities.

As the Internet's functional architecture and orientation evolve from data service to multimedia, application-specific functional hierarchical networks will be developed. These will be networks of networks, configured as neural networks. Some or all of the nodes are themselves regional wide area networks (WANs) consisting of metropolitan area networks (MANs); these, in turn, consisting of local area networks (LANs), that ultimately link individual processors (computing devices and human/computer pairs). Through multimedia and multiple sensors, humans and computing devices will be seamlessly united in the environment.

References

[1]   Noor, A.K., *"New Computing Systems and Future High-Performance Computing Environment and Their Impact on Structural Analysis and Design"*, Computers and Structures, Vol. 64, Nos. 1-4, July-August 1997, pp. 1-30.

[2]   Williams, M.R., *"A History of Computing Technology"*, IEEE Computer Society; 2nd edition, 1997.

[3]   Ceruzzi, P.E., *"A History of Modern Computing"*, Massachusetts Institute of Technology, 1999.

[4]   Greenia, M.W. *"History of Computing: An Encyclopaedia of the People and Machines that Made Computer History"*. CD-ROM, Lexikon Services, 2000.

[5]   *"50 Years of Computing"*, Special issue of IEEE Computer, Vol. 29, No. 10, October 1996.

[6]   Prince, B., *"High Performance Memories: New Architecture Drams and Srams – Evolution and Function"*, Wiley, John & Sons, Incorporated, 1996.

[7]   Luryi, S., Xu, J., and Zaslavsky, A., *"Future Trends in Microelectronics: The Road Ahead"*, John Wiley and Sons, Inc., 2002.

[8]   *"Limits of Computation"*, Special issue of Computing in Science and Engineering, May/June 2002.

[9]   Compton, K., and Hauck, S., *"Reconfigurable Computing: A Survey of Systems and Software"*, ACM Computing Surveys, Vol. 34, No. 2, June 2002, pp. 171-210.

[10] Maybury, M.T., and Wahlster, W., (editors), *"Readings in Intelligent User Interfaces"*, Morgan Kaufmann Publishers, Inc., 1998.

[11] Carroll, J.M. (editor), *"Human-Computer Interaction in the New Millennium"*, ACM Press, New York, 2002.

[12] Foster, I., and Kesselman, C., (editors), *"The Grid: Blueprint for a New Computing Infrastructure"*, Morgan Kaufmann Publishers, Inc., San Francisco, CA, 1999.

[13] Kim, D., and Hariri, S., *"Virtual Computing: Concept, Design, and Evaluation"*, Kluwer Academic Publishers, August 2001.

[14] Grigoras, D., Nicolau, A., Toursel, B. and Folliot, B., (editors), *"Advanced Environments, Tools, and Applications for Cluster Computing"*, NATO Advanced Research Workshop, Iwcc 2001, Mangalis, Romania, September 2001: Revised Papers, Springer-Verlag New York, Inc., May 2002.

[15] Waldrop, M.M., *"Grid Computing"*, Technology Review, May 2002, pp. 31-37.

[16] Weiser, M., "The Computer of the 21st Century", Scientific American, Vol. 265, No. 3, September 1991, pp. 66-75.

[17] Shurkin, J., *"Engines of the Mind"*, W.W. Norton & Company, New York. 1996.

[18] Denning, P.J., and Metcalfe, R.M., (editors), *"Beyond Calculation: The Next Fifty Years of Computing"*, Springer-Verlag, New York, 1997.

[19] Denning, P.J., (editor), *"Talking Back to the Machine: Computers and Human Aspiration"*, Springer-Verlag, New York, 1999.

[20] Kurzweil, R., *"The Age of Spiritual Machines: When Computers Exceed Human Intelligence"*, Penguin Putnam, Inc., New York, 1999.

[21] "The Future of Computing – Beyond Silicon", Special Issue of Technology Review, MIT's Magazine of Innovation, May/June 2000.

[22] Calude, C.S., Paun, G., *"Computing with Cells and Atoms: An Introduction to Quantum, DNA and Membrane Computing"*, Taylor & Francis, Inc., January 2001.

[23] Williams, C.P., and Clearwater, S.H., *"Explorations in Quantum Computing"*, Springer-Verlag, New York, 1998.

[24] Hirvensalo, M., *"Quantum Computing"*, Springer-Verlag TELOS, June 2001.

[25] AcAulay, A.D., *"Optical Computer Architectures: The Application of Optical Concepts t Next Generation Computers"*, Wiley, John & Sons, Inc., 1991.

[26] Wherrett, B.S., and Chavel, P. (editors), *"Optical Computing"*, Proceedings of the International Conference, Heriot-Watt University, Edinburgh, U.K., August 22 – 25, 1994, Iop Publishers, March 1995.

[27] Caulfield, H.J., *"Perspectives in Optical Computing"*, IEEE Computer, Vol. 31, No. 2, February 1998, pp. 22-25.

Chapter 2

©2002, Saxe-Coburg Publications, Stirling, Scotland
Engineering Computational Technology
B.H.V. Topping and Z. Bittnar, (Editors)
Saxe-Coburg Publications, Stirling, Scotland, 25-52.

Virtual Reality in Construction: A Review

W. Thabet†, M.F. Shiratuddin† and D. Bowman‡
†Department of Building Construction
‡Department of Computer Science
Virginia Tech, Blacksburg, Virginia, USA

Abstract

Virtual Reality (VR) and the development of Virtual Environments (VEs) can make a significant impact on how construction project stakeholders can perceive and successfully complete their projects. VR techniques have the potential to enhance the efficiency and effectiveness of all stages of a project, from initial conceptual design through detailed design, planning and preparation, to construction completion. The ability to review the design and rehearse the construction of the facility in a 3D interactive and immersive environment can increase the understanding of the design intent, improve the constructability of the project, and minimize changes and abortive work that can be detected prior to the start of construction. Unlimited virtual walkthroughs of the facility can be performed to allow for experiencing, in near-reality sense, what to expect when construction is complete.

Various efforts in the industry and academia are underway to explore these possible benefits of VR in construction. This paper provides an overview of recent examples of successful adoption of VR technology as applications in construction. The paper also provides an overview of what Virtual Reality (VR) is, and discusses some of the hardware and software that can be used for VR applications.

Keywords: virtual reality, virtual environments, immersive, interactive, design, construction planning, visualization, CAVE, 3D modeling.

1 What is Virtual Reality (VR)?

During the last two decades, the word *virtual* became one of the most exposed words in the English language [1]. The Webster dictionary defines it as "being such practical or in effect, although not in actual fact or name." What we refer to as reality is based upon something we call the external physical world [2]. Therefore, a virtual reality seems to suggest a reality that is believable, and yet does not physically exist. Isdale [3] indicated that the term virtual reality (VR) is also interpreted by many different people with many meanings. To some people, VR is a specific collection of technologies and others stretch the term to include conventional books, motion pictures, radio, etc. - any medium that can present an environment that draws the receiver into its world.

In this paper, the term VR is restricted to being computer-mediated systems. This involves the use of computers to create and visualize 3D scenes supported by auditory or other sensual outputs, with which one can navigate through and interact. Navigation includes the ability to move around and explore features of the 3D scene, whilst the interaction implies the ability to select and manipulate objects in the scene. Interaction with the virtual world, at least with near real time control of its 3D scenes, is a critical test for a virtual reality.

Various definitions of virtual reality have been given by many researchers and practitioners. VR has been defined as a way for humans to visualize, manipulate and interact with computers and extremely complex data [4]. Barfield and Furness [5] described VR as "the representation of a computer model or database, which can be interactively experienced and manipulated by the virtual environment participant(s)". Boman [6] characterizes VR as interactive, virtual image displays enhanced by special processing and by non-visual display modalities, such as auditory, tactile (touch), and haptic (force), to convince users that they are immersed in a synthetic space.

Halfawy *et al.* [7] classified VR as computer generated models of real environments in which users can visualize, navigate through, and interact with these models in an intuitive way. Sherman and Craig [8] defined virtual reality as a medium composed of highly interactive computer simulations that senses the user position, and replaces and augments the feedback of one or more senses - giving the feeling of being immersed or being present in the simulation.

Since the early 90's, VR has matured considerably, and has begun to offer many powerful solutions to very difficult problems. New hardware platforms coupled with new complex graphics cards capable of generating high-end computer images have appeared. Other devices that provide tactile (touch) and haptic (force) images to complete the sense of illusion have been greatly improved. New more flexible and powerful commercial software systems geared towards generating real time virtual environments have also become available, with better features and capabilities. VR is no longer a technology looking for applications, but rather, it is a solution to many problems that involve real-time visualization of complex data.

Simultaneous with these developments, the term 'Virtual Reality' has been gradually ignored by some in preference for the term 'Virtual Environments' (VE). The use of the VE term avoids any possible implication that there is any ambition to remodel the universe [2]. Other terms that have been used include visually coupled systems, synthetic environments, cyberspace, artificial reality, virtual presence, and simulator technology. For the purpose of this paper, the terms virtual reality (VR) and virtual environments (VE) will be used interchangeably.

The following sections of the paper present a general overview of the different categories of VR, the current hardware and software technologies used for VR implementations, and provide a summary of current applications of VR in construction. The intention here is not to provide an exhaustive list or a complete review, but rather provide some example representations of what is available to make the reader aware of current technology and applications of VR in construction, and point the reader in the direction of more detailed references.

2 Types of VR Systems

As noted above, there is still a debate about what constitutes "virtual reality." Thus, there are many different types of systems that have been called VR. This section explores some of those categories of systems. The differences between these categories are due to several factors, including display hardware, graphics rendering algorithms, level of user involvement, and level of integration with the physical world.

2.1   Immersive VR

The most common popular image of VR is that of an *immersive* VR system – typically a user wearing a head-mounted display (HMD) or standing inside a spatially immersive display (SID). The concept of *immersion* is that the virtual environment surrounds the user, at least partially. Immersion is not a binary quantity – there are various levels of immersion. For example, a standard CAVE (section 3.1.1.2) has screens on four of the six sides of a cube; the user is only partially surrounded by the display.

A related, but separate concept is *presence* – the subjective feeling of "being there." For a user experiencing high levels of presence, the virtual world replaces the physical world as his "reality" [9]. An immersive VR system may produce higher levels of presence, but presence may also be achieved in non-immersive systems, or even in non-VR systems (e.g. a reader may become so engrossed in a novel that she feels present in the "world" of the book).

Immersive VR systems also generally use 3D interaction techniques based on whole-body input. In other words, to see what is to my left, I turn my head/body to the left; to move forward I take a step forward; to grab an object I reach out my hand and grasp the object with my fingers. These natural techniques may also be enhanced to allow for greater efficiency or usability. For example, the user might be given a virtual arm that can reach much further than his physical arm [10]. All of these techniques require the use of a tracking system.

2.2   Desktop VR

Non-immersive VR systems normally run on standard desktop computer workstations, thus the term "desktop VR." Desktop VR systems use the same 3D computer graphics as immersive VR systems, but there are two key differences. First, the display of the virtual environment does not surround the user – it is seen only on a single screen in front of the user. Second, the user typically navigates through and interacts with the environment using traditional desktop input devices such as a mouse and keyboard (although specialized 3D input devices may also be used).

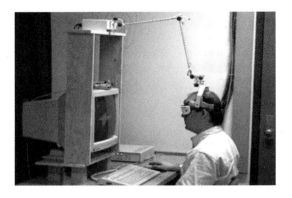

Figure 1: Fish tank VR setup using a mechanical head tracker

2.3   Fish Tank VR

It is possible to use a desktop computer combined with a head tracker (Figure 1) to provide a "window" into a small virtual world, so that users can obtain different views of the world using natural head motions. This has been called "fish tank VR" [11] because the effect is similar to looking into various parts of a fish tank by moving one's head relative to the tank. The head tracking provides motion parallax information to allow the user to more easily comprehend the depth of 3D objects in the scene. Often, stereo graphics are used to enhance the 3D depth; this requires stereo glasses.

2.4   Image-based VR

Most VR systems display completely synthetic (computer-generated) virtual environments. The objects in these environments are made up of geometric primitives (usually triangles) along with colors and textures. However, it is also possible to display a realistic virtual world using only images. This is called *image-based rendering*, and its basic approach is to manipulate the pixels in images to produce the illusion of a 3D scene, rather than to build the 3D scene explicitly.

The simplest type of image-based VR is a panorama – a series of images taken with a camera at a single position pointed in multiple directions. These images can then be "stitched" together so that the seams are not visible, allowing the user to look in any direction and see the appropriate part of the scene. Apple's QuickTime VR is one of the most common types of panoramic VR. Of course, the limitation of panoramas is that the view is only correct when the viewpoint is at the exact position from which the images were taken. Thus, researchers have been studying the more general image-based rendering problem – that is, given a set of images taken from many positions and in many directions within an environment, how can we produce the correct perspective view from an arbitrary viewpoint in that environment [12, 13]. Image-based rendering has not yet made a large impact in VR, but it is clearly a way to increase the realism of virtual environments – an important consideration for the construction industry.

2.4   Highly Interactive VR

All VR applications involve *interaction*; that is, they allow the user some degree of control over what is happening. This is the characteristic that distinguishes VR from static 3D images or pre-computed 3D animations. Many VR systems, however, are simply walkthroughs or flythroughs – they display a static environment and allow the user to navigate (position and orient the viewpoint) through that environment. Highly interactive VR systems allow the user to perform other tasks in the VE, such as selection, manipulation, system control, and symbolic input.

Highly interactive VR systems can allow users to perform work in a VE that until now has only been available in 2D desktop systems (e.g. designing a building in a VE rather than in a CAD package). This takes VR from being simply a visualization tool to being a tool for producing real-world results. The key challenge for highly interactive VR is the design of usable and efficient interaction techniques and user interfaces.

2.5   Telepresence

Telepresence is related to VR, and involves interacting with real environments that are remote from the user [3]. Teleoperated systems are developed as a result of the need to interact with environments from a distance. This technology links remote sensors and actuators in the real world (teleoperators) with human operators who are at a remote location from that environment (Figure 2). This link provides the operator with a remote view of and limited control over the teleoperator's environment. This leads to a sense of telepresence. Current applications of telepresence involve the use of remotely operated vehicles (e.g. robots) to handle dangerous conditions (e.g. nuclear accident sites) or for deep sea and space exploration.

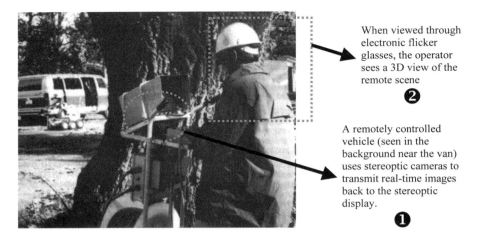

When viewed through electronic flicker glasses, the operator sees a 3D view of the remote scene ❷

A remotely controlled vehicle (seen in the background near the van) uses stereoptic cameras to transmit real-time images back to the stereoptic display. ❶

Figure 2: An example of a Telepresence system, adapted from [14]

2.6   Augmented/Mixed Reality

While VR provides a view of a completely virtual world, and telepresence a view of the physical world, augmented reality (AR) and mixed reality (MR) combine both virtual and physical environments into one. In AR systems, the user views the local real-world environment, but the system augments that view with virtual objects. For example, someone on a walking tour of a college campus might see labels that named all the buildings in his field of view (Figure 3) [15]. Augmented reality systems often use see-through head-mounted displays so that virtual and real images can be combined. Mixed reality (MR) refers to a whole continuum of systems that merge virtual and real environments, all the way from AR to mostly virtual environments that include some real-world objects.

Figure 3: Mobile augmented reality, adapted from [16]

3   VR Hardware

A wide range of hardware technologies is used to realize VR systems. We provide a high-level overview of some of the most important VR devices here, beginning with output devices, including visual, auditory, and haptic displays, and concluding with input devices, including discrete, continuous, and hybrid devices.

3.1   Output devices

The term display, in general, refers to a device that presents perceptual information. In VR systems, one goal is to involve as many of the user's senses as possible, so displays have been researched for almost all of the senses (visual, auditory, haptic (touch), olfactory (smell), and vestibular (motion) – we do not know of any research on a gustatory (taste) display). Here we present the three most common types of displays: visual, auditory, and haptic.

3.1.1   Visual Displays

The visual display is the one indispensable piece of hardware in a VR system, and often the visual display is the defining characteristic of the system as well. There are four general categories of VR visual displays.

3.1.1.1 Desktop Displays

The most common VR visual display is a simple desktop monitor, used for Desktop VR (section 2.2) or Fish Tank VR (section 2.3). Sometimes such displays are enhanced by using stereo graphics and stereo glasses to allow better depth perception, but for the most part they are identical to the displays used by all computer users.

3.1.1.2 Spatially Immersive Displays (SIDs)

To achieve immersive VR, a more complex visual display device is needed that allows the VE to appear to surround the user, or that fills much of the user's field of view with the computer graphics. One approach, called a spatially immersive display (SID) is to physically surround the user with the display(s). For example, the CAVE [17] uses 4-6 large projection screens in the shape of a cube into which the user can walk (Figure 4). Stereo graphics are projected onto the screens, so that the user sees nothing but the VE and their own body. The CAVE allows multiple viewers, unlike some other VR technologies, but normally only one viewer has the correct perspective view.

Another type of SID, called a dome or hemispherical display, uses only a single curved screen that wraps around the user to provide immersion. The Hemispherium™ (Figure 5) is an example of this type of display [18]. It uses a six-meter dome mounted vertically. The user sits in front of the dome so that the display fills his entire field of view. Smaller domes include the elumens VisionStation™.

SIDs often provide excellent sensations of presence and high-resolution graphics, but can be quite costly, and are usually not fully-immersive (except in the case of the 6-sided CAVE).

3.1.1.3 Head-Mounted Displays (HMDs)

The head-mounted display (HMD) was the earliest VR display [19], and represents another method for achieving immersion. HMDs use two small LCD or CRT screens mounted inside a helmet worn on the user's head (Figure 6). The two screens can be controlled separately to allow for stereo graphics. The HMD blocks out the user's view of the physical world so that only the virtual world can be seen. When the HMD is combined with a head tracker, the user can view the VE in any direction simply by turning his head.

HMDs can be much less costly and more portable than SIDs, and provide complete physical immersion. However, they generally have a narrow field of view (30-90 degrees horizontal), a low display resolution (typically 640x480 or 800x600), and can be heavy and cumbersome.

Figure 4: A CAVE system

Figure 5: User seated in the flight chair of the Hemispherium™

Figure 6: User wearing a head-mounted display

3.1.1.4 Workbenches and Walls

The final approach to providing some level of immersion and a wide field of view is to simply use a single, flat, very large display surface. Typically, this takes the form of a horizontal screen upon which the user looks down (a workbench) or a vertical screen mounted in front of the user (a wall). Most workbench and wall displays use rear-projected stereo graphics and stereo glasses for the user. Head tracking is also common, although not required. Some displays, such as Fakespace's Immersive Workbench (Figure 7) allow the screen to be rotated to produce a workbench configuration, a wall configuration, or anything in between. Workbenches and walls provide only limited immersion, but they do allow multiple users to share the display, and afford direct manipulation of the displayed data.

Figure 7: Immersive Workbench

3.1.2 Auditory Displays

In this case, the display hardware is not generally specific to VR – standard speakers or headphones can be used. The challenge is in creating sound that is realistic for a given environment. Many advanced VR systems use *spatial audio*, which presents auditory information that sounds as if its source is at a particular location in 3D space.

3.1.3 Haptic Displays

Haptics refers to all of the "skin senses" (force, pressure, texture, heat, pain, etc.). In VR systems, haptic devices allow the user to feel the virtual world. There are many different types of haptic displays that have been used in virtual environments. The most common type is a mechanical linkage device that displays force to the user at a single point, such as the Sensable Technologies Phantom device. This device allows users to probe virtual objects with a fingertip or stylus in order to feel the surface. Another type of device, called an exoskeleton (Figure 8, left), allows users to grasp virtual objects and feel the force on all five fingers. These two types can be combined to allow both grasping and probing (Figure 8, right). A third type of haptic display is the tactile display, which presents surface features, such as texture.

Near-field haptics [20] is a different approach to providing haptic sensations to the user, which uses real-world objects, called *props*, which match virtual objects in the VE. For example, a physical railing can be placed around the user in a virtual elevator, so that when she reaches out her hand to the virtual railing, she feels the physical railing at the same location [21].

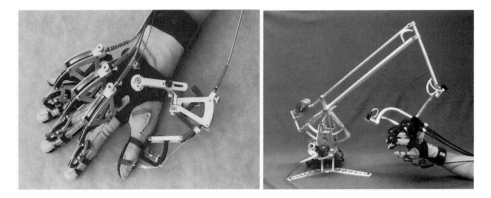

Figure 8: Exoskeleton (left) and combination exoskeleton and mechanical linkage (right) haptic devices from Virtual Technologies, Inc.

3.2 Input Devices

Input devices are used to control a VR application. VR systems typically use novel 3D input devices. The most common 3D input device is the position tracker (section 3.2.2.1), but there is a wide range of other discrete, continuous, and hybrid devices for VR systems.

3.2.1 Discrete Input Devices

A discrete input device produces individual events when the user performs some action. In desktop environments, the keyboard is a typical discrete device – each key press signals the system to do something, and no input is produced unless a key is pressed. In VR systems, keyboards may also be used, but often they are non-traditional keyboards, such as chord keyboards or soft keyboards. A chord keyboard [22] is a one-handed device with a small number of keys that allows the user to input any symbol by pressing "chords" (multiple keys held down at once). Soft or virtual keyboards do not use a specialized input device at all, but instead use virtual keys presented in the environment. The user selects these virtual keys to input symbols.

Another typical discrete input device for VR systems is the Pinch Glove™ (Figure 9). Pinch gloves are gloves with conductive cloth sewn into the fingertips, so that when two or more fingers touch one another, a signal is generated. Pinch gloves have been used for a variety of different tasks in VEs, including navigation, manipulation, menus, and symbolic input [23].

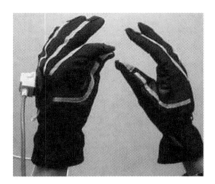

Figure 9: Pinch Gloves™, from Fakespace Labs

3.2.2 Continuous Input Devices

A continuous input device sends signals back to the host computer constantly, with or without any action by the user. The positional component of a mouse is a continuous input device used in desktop systems – it reports the X and Y position of the mouse constantly. In immersive VR systems, trackers and data gloves are the two most common continuous input devices.

3.2.2.1 Tracking Devices

As noted above, head tracking is an important component of many types of VR systems. A head tracker allows users to view the virtual world naturally using head motions, rather than indirect movements using a mouse or some other device. Many VR applications also track users' hands, bodies, or feet, or other physical objects (props). There are many different technologies used in tracking devices, including electromagnetic, inertial, acoustic, optical, mechanical, and vision-based. A complete survey of these technologies is beyond the scope of this paper. However, it is important to use tracking technologies that provide high precision, high update rates, and low levels of latency, or lag. Latency is especially crucial for head-tracked immersive VR. If the tracking system has too much latency, the virtual world will lag behind the user's head movements, possibly causing simulator sickness.

3.2.2.2 Data Gloves

Unlike pinch gloves, data gloves continuously report on the posture of the user's hand by measuring joint angles. When combined with a tracker, data gloves allow the recognition of gestures (postures + orientation or motion). This allows users to gesture naturally to interact with the VE, such as grabbing an object, waving to someone else, or giving a "thumbs-up" signal.

3.2.3 Hybrid Input Devices

Finally, we need to mention hybrid input devices – those which combine both discrete and continuous inputs. On the desktop, the mouse is a hybrid device, combining discrete buttons and continuous positional input. A large number of VR input devices are also hybrid in nature. The most common example is a tracking device that also includes discrete buttons, so that it can be used something like a 3D mouse.

4 VR Software

Current VR development software supports a wide range of VR implementations. VR software is not only concerned with 3D object generation, but also needs to allow for navigation and interaction within the 3D world. VR software also includes other features that need to be considered when selecting VR software. This include features to support importing 3D models from other systems, 3D libraries, optimization of level of detail, object scaling, rotating and translating, stereo viewing, animation, collision detection, and multi-user (avatars) networking. A good review of these features can be found in [1, 3].

There are two major categories of VR implementation software: toolkits or software development kits (SDKs), and authoring systems [3]. SDKs are programming libraries (generally written in C or C++) that provide a set of common functions with which a skilled programmer can quickly create a basic layout of the VR application. Authoring systems are mostly icon-based programs with graphical user interfaces (GUIs) to create virtual worlds without going through detailed programming. This section examines some of the available software for VR implementation.

4.1 Sense 8: World Toolkit (SDK) and WorldUp (authoring tool)

The World Toolkit (WTK) [24] is a commercial package that consists of a library of over 1000 functions written in C that enable users to rapidly develop new virtual reality applications. Using WTK, programmers build virtual worlds by writing codes to call the WTK functions.

WorldUp, a component of the WTK, is a 3D content authoring tool for real-time graphical simulations. It provides an easy-to-use graphical user interface from which users create objects and properties and design simulations. It can create or import 3D scenes, make them interactive with an easy-to-use drag and drop assembly, and can also integrate them with the industry standard tools that are already available. In adding behaviors to the objects, users can author customer behaviors or change a property of an existing behavior by writing scripts using the BasicScript language, or use property change events to trigger behaviors.

4.2   Paradigm: Multigen (authoring tool)

Paradigm [25] provides modular based commercial VR and 3D content creation. They offer an industry-leading range of fully integrated, highly automated real-time 3D database development and visual and sensor simulation tools for the IRIX™ and Microsoft Windows NT® operating systems. Some of the products available are MultiGen Creator for modeling, TerrainPro for Large Area Terrain generation, RoadPro for creating roads that meet real-world engineering standards and Vega for the creation of real-time visual and audio simulation, virtual reality, and general visualization applications.

4.3   Perilith Industrielle - Unrealty (authoring tool)

Unrealty [26] is a new innovative commercial VR creation kit. The VR kit was developed utilizing the game engine of a currently popular 3D game i.e. Unreal Tournament from Epic games. Due to ease of use and highly realistic and believable real-time images it can produce, Unrealty was used to generate a 3D walkthrough environment of the Virtual Norte Dame Cathedral of France project (Figure 10). NASA has also used Unrealty to create their Virtual International Space Station (Figure 11).

Figure 10: The VRND project

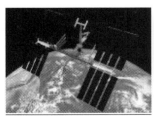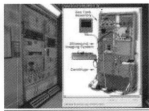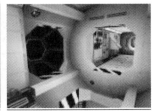

Figure 11: NASA's Virtual International Space Station project

4.4   VRML

VRML [27, 28] (an acronym for the Virtual Reality Modeling Language) allows to store descriptions of 3D worlds on the web and provide ability to transform (i.e. scale, translate, rotate, etc.) graphical objects in real time. VRML contains a number of features such as hierarchical transformations, light sources, viewpoints, geometry, animation, fog, material properties, and texture mapping (see Figure 12).

Unlike HTML, VRML also provides the technology that combines 3D, 2D, text, and multimedia into a logical model. When these media types are integrated with scripting languages and Internet utilities, an entirely new generation of interactive applications is possible.

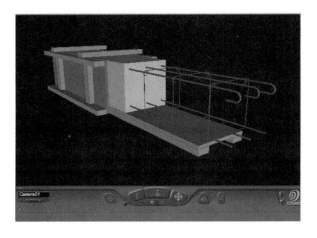

Figure-12: A VRML model of a Construction Assembly

4.5   Wild Tangent: Web Driver™

Wild Tangent has developed new 3D content creation software for the web. The whole technology evolves from Genesis 3D, which is an Open Source 3D Game Engine and it is available for public at no cost [29]. Wild Tangent's Web Driver™ has the ability to create powerful, interactive applications for the Web in super-small-sized files. The Web Driver™ 2.2 SDK when combined with the environment editor WTStudio, will provide everything that the user needs to develop interactive media applications. The best part of all, it's FREE to use.

4.6 3DState

3DState [30] specializes in 3D technology for the Internet. Their technology is suited for 3D virtual malls, 3D virtual expos and 3D web-sites. Their development tools include: WorldBuilder to create professional interactive 3D environments, TerrainBuilder to convert standard 2D topographical map into an interactive 3D world. Users can flythrough and explore this newly created terrain, or export it to WorldBuilder for further development, and 3D Webmaker Web Sites can be built in minutes using drag and drop functionality.

4.7 3D Game Engines (3DGEs)

The use of 3D Game Engines (3DGE) to create real-world VR applications is a promising new alternative to currently available commercial VR development platforms. Recent research efforts have proven that the use of 3D Game Engine, which is also known as Game Development Kits (GDKs), is a viable solution to creating visually engaging virtual environments while still maintaining a low-cost development and execution platform [31, 32]. Two of the most widely used 3D Game Engines to develop VR applications (mainly walkthroughs) are the Unreal Tournament 3DGE from Epicgames [33] and Quake 3DGE from id Software [34]. Table 1 lists examples of VR projects that utilized 3DGE for their development [35].

Recently, a group of game enthusiasts released a version of Quake for the PocketPC known as PocketQuake (see Figure 13). Using Pocket Quake, users are able to use the PocketPC to navigate through a 3D virtual environment that is fully lighted and textured. Since id Software has made it Open Source, users can create 3D models using various third party 3D level editors for Quake [32]. More information on how to create these levels is available at [36, 37, 38].

Figure 13: Pocket Quake on the Pocket PC

Year	VE Project	GDK	Developer	Description/Comments
1998	Virtual Florida Everglades National Park	Unreal	Project leader: Victor DeLeon	A project to educate the public and also promote ecological awareness
1998	Notre Dame Cathedral of France	Unreal	Digitalo Studio	Funded by UNESCO. Demo can download at http://www.vrndproject.com.
1999	Long Island Technology Center	Unreal	Perillith Industrielle for Rudin Management	Demo can be downloaded at http://www.unrealty.net.
1999	Heartland Business Center	Unreal	Perillith Industrielle	An office complex in New York.
1999	HypoVereins Bank	Unreal	Perillith Industrielle for Turbo D3	Virtual bank in Germany. Demo can be downloaded at http://www.unrealty.net.
2000	Virtual Graz of Austria	Unreal	Bongfish	Graz is the second largest city in Austria. Funded by UNESCO.
2000	Virtual International Space Station - VISS	Unreal	NASA Langley Research Center Spacecraft & Sensors Branch An International Virtual Space Station	Demo can be downloaded at http://www.unrealty.net.
2000	Cambridge University and Microsoft Science and Technology site in West Cambridge	Quake 2	Martin Centre for Architectural and Urban Studies, Cambridge University	Part of a project on using electronic communication between buildings' architects and their eventual users.
2000	CAVE Quake 3	Quake 3	Quake 3 Visualization and Virtual Environments Group, NCSA.	A CAVE system based on the Quake 3 Arena engine. Web-site at: http://www.visbox.com/cq3a/.
2001	CAVE UT	Unreal Tournament	Medical Virtual Reality Center, Department of Otolaryngology, University of Pittsburgh	A CAVE system based on the Unreal Tournament engine. Web-site at: http://www2.sis.pitt.edu/~jacobson/ut/CaveUT.html.

Table 1: Major VE projects utilizing 3D Game Engines

5   VR in the Construction – Current Research Work

What can construction project stakeholders do in a 3D virtual environment ? They can actually "climb inside" a building and visualize its elements and components from any visual perspective to evaluate the design and make modifications. They can virtually "disassemble" the components and "reassemble" them repeatedly to rehearse the construction process, develop a construction sequence, assess the constructability of the design and identify potential interference problems. They can take unlimited virtual walkthroughs of the facility and experience, in near-reality sense, what to expect when construction is complete.

This section explores some applications of VR in construction. The list of work reviewed is not intended to be exhaustive, but rather to provide an opportunity to discuss and review a range of applications associated with VR and the way it is applied in this sector.

5.1   VR in House Building

In a survey that investigated the IT systems used for product design, development modeling and sales in the British house building industry [39], over half of the questioner respondents had previously seen a demonstration of VR in the construction industry, 82% thought that it was potentially useful for their company, and 76% of the respondents indicated that it would take less than five years to see VR techniques used in their company. Specific results included:

- Use of 3D and VR can be beneficial if used early at the conceptual design process - detailed construction drawings could then be automatically generated.
- Housing developers had mixed feelings about giving users total freedom of movement around models in a VR environment - they had preservation on exposing users to views from sensitive unrealistic angles.
- VR and advanced visualization must be used with care when dealing with local authorities to avoid exposing the design/construction to other issues.

Whyte and Bourchlaghem [40] reported a study visit to Japan where three case studies were conducted to investigate Japanese house builders' use of VR. Findings indicated that the Japanese house builders have recognized the benefits of VR in construction and have been using VR technologies for over three to four years.

5.2   VR in Construction Safety and Training

Neville [41] suggested that training is important for rehearsal purposes and to prevent accidents and injuries. Barsoum *et al* [42] developed an interactive virtual training model, SAfety in construction using Virtual Reality (SAVR), to train construction workers on avoiding falls from platform-metal scaffolding. Using HMDs, users are able to interact with the VE and detect hazardous conditions (e.g. missing guardrails, loose, weak or inadequately spaced planks, inadequate connections between scaffolding components, and defective components) and attempt to eliminate it. A scoring system is used to evaluate performance of participants. SAVR comprised of two main modules; an erect module, and an inspection module. The erection module is used to demonstrate appropriate procedures to erect scaffolding. The inspection module is used to detect and correct the potential causes of falls. Sense8 WTK on an SGI platform and 3D Studio were used for creating the virtual environment and engine. An Onyx Reality Engine 2 (ORE2) from SGI was used to allow real time rendering.

Soedarmono *et. al* [43] also developed a prototype VR model for training personnel on avoiding falls during construction. Occupational Safety and Health Administration (OSHA) regulations were integrated into the model as 2D text and audio information. Warning messages (i.e. required safety standards) are displayed or announced when a user approaches a working platform in the VE from which they could fall.

5.3   VR for Project Planning and Monitoring

Although modern project management software aids the development and assessment of project schedules, their 2D symbolic representation communication among the parties involved difficult and mistake prone. Retik [44] developed an approach to tackle these issues through the use of a computer based system for visual planning and monitoring of the construction processes. The system allows the creation of a 'virtual construction project' (Figure 14) from a schedule and subsequent visual monitoring of and interaction with the progress of the simulated project.

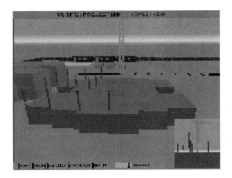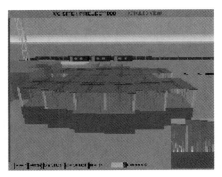

Figure 14: Images from the Virtual Construction Project system

Research in progress at the University of Teesside, UK [45], is developing the Virtual Construction Site (VIRCON), a prototype application for evaluation, visualization, and optimization of construction schedules within a virtual reality interface. The structure of the VIRCON system is designed with three main components: Project database, analysis tools and decision support components to process time critical and space critical tasks, and a visualization component to communicate the project database and analysis results through a range of interconnected graphic windows.

Mahachi *et al* [46] reported on the development of an integrated construction management tools in a virtual reality environment. This was possible through the integration of project management scheduler with cost forecasting techniques and using genetic algorithms for site-layout planning. Primavera P3™ is integrated with VR to enable the simulation of construction scheduling in real-time, allowing the project planner to see the progress of construction activities on site.

Kim *et al* [47] proposed a Construction Visualization System (4D-VR, 4 Dimension-VR). The system was intended to be applied to large and complicated projects that require milestone schedule management and detailed activity control. The system has a software structure with five modules i.e. 3D CAD modeling, VR modeling, schedule data processing, linking graphic data with schedule data and visualization output modules.

5.4   Augmented VR

Webster *et al* [48] developed two experimental augmented reality systems. One prototype called "Architecture Anatomy," allows viewers to overlap graphical images of columns and rebar on top of the user's view of portions of a building architectural or structural finishes. The system was developed using C, C++, and CLIPS running on a Unix operating system. The second augmented reality prototype developed guide workers through the assembly of a space frame structure. Using bar coding technology coupled with sound files containing verbal instructions, the system guides the user to select the correct component by reading its barcode, and utilizing textual installation instructions along with sound explaining how to complete installation. The prototype was developed using Modula-3 running on an assortment of hardware under Unix, WinNT and Win95.

5.5   VR for Site layout Planning

Boussabaine *et al* [49] implemented a working VR prototype system to simulate the layout of construction site facilities. The prototype is intended to assist project managers in planning a safe and efficient site. The system allows users to create their own site layout environment, manually select and place objects representing facility and equipment, and create walk-through to view the virtual facility.

5.6   VR as an Analysis Tool

Dawood and Marasini [50, 51] developed a visualization and simulation model to address the problem which is experienced by the pre-cast concrete products industry. Problems include space congestion and long vehicle waiting times on stockyard for both the storage and retrieval of concrete products (Figure 15). The visualization model was developed using ILOG Views and the simulation model used a general-purpose simulation language, ARENA/SIMAN.

5.7   VR for Architectural Walkthrough and Preservation

Various projects and research efforts have utilized VR for architectural preservation and walkthrough purposes including: the VRND project [52] (developed by Digitalo Inc. and funded by UNESCO to electronically preserve the cathedral's historical and architectural values) and the Virtual Graz project [53] developed by Bongfish (see Figure 16). In 2000, Shiratuddin *et al* [31] utilized 3D Game Engine technology to develop a model of a mosque (Figure 17).

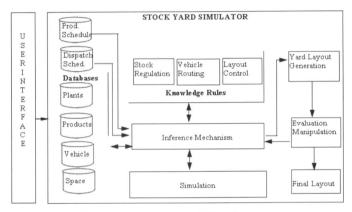

Figure 15: Stock Yard Simulator

Use of VR for visualization in the industry is demonstrated by many companies including WS Atkins, UK [54]. Various applications were developed for design review, external/internal inspection, below ground inspection, assessment of land take and space usage, sight lines, and construction sequencing. VE applications are developed from 3D models imported into native VRML.

Figure 16: Sample images from the Virtual   Figure 17: The interior of the
              Graz project   Virtual Mosque

6 VR in Construction – Current Research Work at Virginia Tech

Through a collaborative effort between the Department of Building construction and the department of Computer Science at Virginia tech, several research efforts have been initiated to investigate the use of VR technologies in construction. This section provides an overview of some of the main VR research activities currently underway.

6.1 Evaluating the Effectiveness of VE Displays

This research project evaluates and compares the applicability, usability and effectiveness of various virtual environment (VE) displays versus other 2D and 3D representations for design review and pre-construction planning. This will allow for assessing the benefits that may be gained by adopting VE technology – benefits such as making design changes/modifications and planning decisions faster, easier, more efficiently, and more accurately. The research intends to create a mapping between display types and design/planning tasks, identifying one or more effective display modalities for each task.

Four VE displays are used in the research study; Fakespace CAVE™, HMDs (Virtual Research V8s), a tabletop display (Fakespace Immersive Workbench), a personal half-dome display (Elumens visionStation). Several industry partners, including design firms, contractors and owners, have begun to take part in this study. Each subject will view and navigate 3D models of buildings and assemblies in various stages of design and construction, using multiple displays and representations (e.g. egocentric vs. exocentric views). The subject will be asked to make decisions regarding changes to the design, the constructability of some aspect of the structure, the placement or resources, and the like, using the 3D display. Metrics will include time to reach a decision, the quality of the decision, and the subject's opinion about the suitability of the display/representation for making the decision.

6.2 The Virtual Construction Environment (VCE)

Pre-construction planning contributes significantly to the successful development and execution of construction projects. To allow for a more efficient approach to perform macro planning, a framework for a Virtual Construction Environment (VCE) for project planning has been proposed and presented [55]. The framework suggests the use of virtual reality (VR) coupled with 3D CAD and object oriented technologies to develop an interactive collaborative environment for thinking and visualizing projects in a near reality sense, prior to the actual start of construction (See Figure 18).

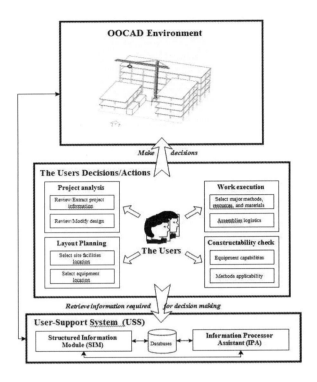

Figure 18: Proposed framework of VCE

6.3   Using VEs for Real-time Design Review and Modification

Immersive VEs can improve the design review process by allowing team members to walkthrough a realistic, life-sized model of the facility. We claim, however, that VEs can provide even more benefits to architects if team members are allowed to *interact with* and *modify* the design of the facility in *real time*, while they are *immersed* within it. This research project will implement a VE system allowing users to perform conceptual design, design review, and modification, including three interaction modules: Move, scale, and remove elements module, display/hide elements module, and massing studies module.

6.4   A Collaborative VE Utilizing 3D Game Technology

This research project [56] explores the use of game engines for implementation of a collaborative virtual environment with remote connectivity capabilities. The research utilizes the Unreal Tournament 3D Game Engine (3DGE) from Epic Games [33] to develop the framework of a collaborative environment that will allow geographically dispersed design members to virtually meet for design review (shown in Figure 19). The Unreal Tournament engine comes with a built-in multi-user support that allows for up to 16 multiple users (i.e. clients) to connect to the VE server and communicate via a basic text chat tool. The proposed work is being implemented using the Unreal Script programming language.

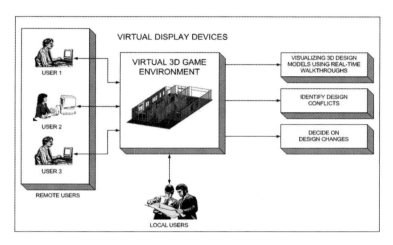

Figure 19: Remote Collaborative Virtual Environments

Research tasks include:

1. Develop additional communication mechanisms among the users connected. This task will develop advanced text messaging and voice communication modules.
2. Investigate capabilities to allow users to modify/manipulate 3D graphical elements in real time, while navigating the game engine VE.
3. Design a database structure integrated in the game engine environment to capture user design comments and store design information for design components and assemblies.

6.5   3D Visualization Using the Pocket PC

This research project [36] explores the use of the Pocket PCs for visualization of 3D models using several VR applications that support the Windows CE platform. Figure 20 depicts the proposed framework under development for investigating the current benefits and limitations of the Pocket PC and their potential use as a stand alone visualization platform.

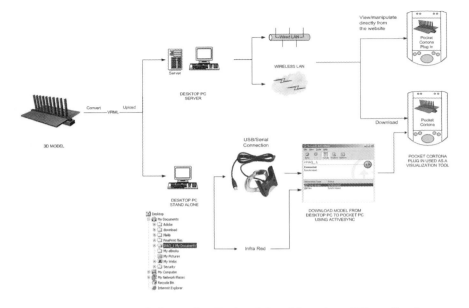

Figure 20: Framework for testing Pocket PCs with various VR applications

The development of the proposed framework currently involves the following two major tasks:

1. Identify VE applications that support the Pocket PCs and the Windows CE platform. We are currently exploring two applications; Pocket Cortona (a VRML viewer), and Pocket Quake (a compact version of the Quake game engine). Pocket Cortona, developed by Parallel Graphics [57], allows for viewing VRML files and support various navigation functions (Figure 21). Pocket Quake is a beta released 3D game environment for viewing 3D models on the Pocket PC [38].

2. Test the communication of the VE applications running on the Pocket PC with the desktop platform using a wired LAN, and wireless connection (WLAN) to investigate the different options for downloading and uploading 3D models between the two platforms, and identifying the benefits and limitations.

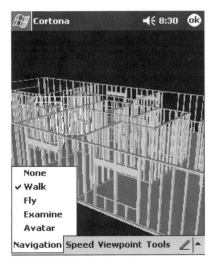

Figure 21: Interface of Pocket Cortana

6.6   An Immersive VE Training System for Disaster Relief and Assessment

Various organizations across the US are continuously creating a database of engineers that can be called on when a disaster occurs. However, many of these engineers do not have the skills necessary to assist in these situations. For example, after the World Trade Center collapse, hundreds of volunteers were needed for structural assessment of the buildings in the surrounding area. This project will develop a prototype immersive VEs to train and prepare rescue and assessment volunteers for relief and assessment efforts in collapsed buildings as a result of a large-scale disaster such as a terrorist attack, earthquake, or fire.

The short term (1-year project) of this project is to develop a detailed 3D graphical model of a collapsed facility (e.g. building) and allow users to walk through it in an immersive VE such as the CAVE. The users will be able to interact in real-time with the environment to simulate a training scenario. The level of interaction implemented will be limited. The long-term vision of this project involves the design of advanced interaction techniques. The system will also be designed to graphically predict and simulate the failure of the facility for different types of disaster scenarios.

7   Conclusion

A computer graphics problem may involve the drawing of three-dimensional (3D) objects and can be achieved either by writing a computer program using a programming language (e.g. C^{++}), or using any CAD packages (e.g. AutoCAD™). A computer animation problem introduces the dimension of time, and the goal is to make objects move in the 3D graphical environment. Implementation is achieved by writing high-level programs, but is generally helped by computer graphics libraries. An integrated animation system can make the development easier by allowing objects to be modeled and animated in one environment. A virtual reality problem is even more complex as it involves creating a 3D environment that is animated, interactive and run in real time. Virtual reality is not just concerned with moving about a virtual environment to gain a different view of a model, VR is about true interaction that can occur with all aspects of the virtual domain [2]. This might involve the touching of objects, moving objects to new positions, modifying object sizes, changing light levels, and so on.

This paper has reviewed some of the work currently being implemented to use VR in construction. There is a lot of other ongoing work that has not been discussed in this paper due to time and space limitations. However, there is much more work to be done in this field to explore its tremendous applications that can greatly benefit the construction industry as a whole.

References

[1] J. Vince, "Essential Virtual Reality Fast", Springer-Verlag Ltd., 1998.

[2] J. Vince, "Virtual Reality Systems", Addison-Wesley, 1995.

[3] J. Isdale, "What is Virtual Reality", 1998.
Web-site: http://www.cse.dmu.ac.uk/~cph/VR/whatisvr.html

[4] S. Aukstakalnis, and D. Blatner, "Silicon Mirage: The Art and Science of Virtual Reality", Peach Pit Press, 1992.

[5] W. Barfield and T. Furness, "Virtual Environments and Advanced Interface Design", Oxford University Press.

[6] D.K. Boman, "International Survey: Virtual Environment Research", Computer, 28(6), 57-65, 1995.

[7] M.R. Halfawy, F.C. Hadipriono, J.W. Duane and R.E. Larew, "Visualization of Spatial and Geometric Databases for Construction Projects" Proceedings of the Conference on Computing in Civil Engineering, ASCE, 920-926. 1996.

[8] W.R. Sherman, and A. B. Craig, "Literacy in Virtual Reality: a new medium," Computer Graphics, 29(4), CAM Press, November 1995.

[9] M. Slater, M. Usoh, A. Steed, "Depth of Presence in Virtual Environments" Presence: Teleoperators and Virtual Environments, 3(2): 130-144, 1994.

[10] L. Poupyrev, et al., "The Go-Go Interaction Technique: Non-linear Mapping for   Direct Manipulation in VR", Proceedings of the ACM Symposium on User   Interface Software and Technology, 79-80, 1996.

[11] M. McKenna, "Interactive Viewpoint Control and Three-Dimensional Operations", Proceedings of the ACM Symposium on Interactive 3D Graphics, 53-56, 1992.

[12] C. Buehler, "Unstructured Lumigraph Rendering", Proceedings of ACM SIGGRAPH: ACM Press, 425-432, 2001.

[13] P. Debevec, C. Taylor, and J. Malik, "Modeling and Rendering Architecture from   Photographs: a Hybrid Geometry- and Image-Based Approach", Proceedings of ACM SIGGRAPH: ACM Press, 11-20, 1996.

[14] K. Pimental and K.Texeira, "Virtual Reality: Through the New Looking Glass", Windcrest, ISBN: 0070501688, 2nd edition, November 1994.

[15] T. Hoellerer, et al., "Exploring MARS: Developing Indoor and Outdoor User Interfaces to a Mobile Augmented Reality System", Computers and Graphics, 23(6): p. 779-785, 1996.

[16] Steven Feiner, Tobias Höllerer, Elias Gagas, Drexel Hallaway, Tachio Terauchi, Sinem Güven, and Blair MacIntyre, "MARS - Mobile Augmented Reality Systems", Columbia University Computer Graphics and User Interfaces Lab, 1999.
Web-site: http://www.cs.columbia.edu/graphics/projects/mars/mars.html

[17] C. Cruz-Neira, D. Sandin, and T. DeFanti, "Surround-Screen Projection-Based Virtual Reality: The Design and Implementation of the CAVE", Proceedings of ACM SIGGRAPH, 1993.

[18] J. Webster, "The Hemispherium(TM) Experience: Fasten Your Seat Belts", Proceedings of TILE '99, London, UK, 1999.

[19] I. Sutherland, "A Head-mounted Three Dimensional Display", Proceedings of

the Fall Joint Computer Conference, 757-764, 1968.

[20] F. Brooks, "What's Real About Virtual Reality?" IEEE Computer Graphics & Applications, 19(6): p. 16-27, 1999.

[21] L. Hodges *et al.*, "Virtual Environments for Treating the Fear of Heights", IEEE Computer, 28(7), 27-34, 1995.

[22] J. Noyes, "Chord Keyboards", Applied Ergonomics, 55-59, 1983.

[23] D. Bowman, *et al.*, "Using Pinch Gloves for both Natural and Abstract Interaction Techniques in Virtual Environments", Proceedings of HCI International, New Orleans, Louisiana, 2001.

[24] Sense8 Virtual Reality 3D Software. Web-site: http://www.sense8.com

[25] Multigen-Paradigm Incorporated. Web-site: http://www.multigen.com

[26] Unrealty: Real-time 3D Virtual Environments. Web-site: http://www.unreality.net

[27] The Virtual Reality Modeling Language Specification, Version 2.0, August 4, 1996. Web-site: http://web3d.org/VRML2.0/FINAL/

[28] VRML FAQ, 1996. Web-site: http://www.web3d.org/VRML_FAQ.html#whatIsVRML

[29] Genesis 3D Open Source Engine. Web-site: http://www.genesis3d.com

[30] 3DState. Web-site: http://www.3dstate.com

[31] M.F. Shiratuddin, A.R. Yaakub, and A.S. Che Mohamed Arif, "Utilising First Person Shooter 3D Game Engine in Developing Real World Walkthrough Virtual Reality Application: A Research Finding", CONVR 2000, University of Teesside, Middlesbrough, UK, 105-114, 2000.

[32] M.F. Shiratuddin and W. Thabet, "Making the Transition toward an Alternative VR", AVR II & CONVR 2001, 4th-5th October 2001, CHALMERS University of Technology Goteborg, Sweden, 2001.

[33] EpicGames. Web-site: http://www.epicgames.com

[34] Id software. Web-site: http://www.idsoftware.com

[35] M.F. Shiratuddin and W. Thabet, "Virtual Office Walkthrough Using a 3D ame Engine", Special Issue on Designing Virtual Worlds, International Journal of Design Computing, Vol. 4, 2002. Web-site: http://www.arch.su.edu.au/kcdc/journal/vol4/index.html

[36] M.F. Shiratuddin, J.L. Perdomo and W. Thabet, "3D Visualization Using the Pocket PC", European Conference of Product and Process Modeling (ECPPM), Portorož, Slovenia, Sept. 9-11, 2002.

[37] Planet Quake. Web-site: http://www.pocketquake.com

[38] Pocket Quake. Web-site: http://quake.pocketmatrix.com

[39] J. Whyte, N. Bouchlaghem, A. Thorpe, and R. Mccaffer, "A Survey of CAD and Virtual Reality Within the House Building Industry," Journal of Engineering Construction and Architectural Management, 6(4), 371-379, 1999.

[40] J. Whyte and D. Bouchlaghem, "VR applications and Customer Focus in Japanese house building", CONVR2000, Middlesbrough, UK, 81-88, 2000.

[41] H., Neville, Workplace accidents Industrial Management, 40(1), 1998.

[42] A. S. Barsoum, F.C. Hadipriono, and R. E. larew, "Avoiding falls from Scaffolsing in Virtual World," Proceedings of the 3rd Congress in Computing

in Civil Engineering, ASCE, 906-9 12, 1996.

[43] D. R. Soedarmono, F. C. Hadipriono, and R. E. Larew, "Using Virtual Reality to Avoid Construction Falls," Proceedings of the 3rd Congress in Computing in Civil Engineering, ASCE, 899-905, 1996.

[44] A. Retik, "Planning and monitoring of construction projects using virtual reality", University of Strathclyde, UK, Project Management 3/97, Research paper, 1997.

[45] N. Dawood, B. Hobbs, A. Akinsola, and Z. Mallasi, "The Virtual Construction Site (VIRCON) - A Decision Support System for Construction Planning," Proceedings of the Conference on Construction Applications of Virtual Reality (CONVR 2000), University of Teesside, Middlesbrough, UK, 3-10, 2000.

[46] J. Mahachi, L. Chege, and R. Gajjar, " Integrated Construction Scheduling, Cost Forecasting and Site Layout Modeling in Virtual Reality," Proceedings of the Conference on Construction Applications of Virtual Reality (CONVR 2000), University of Teesside, Middlesbrough, UK, 74-79, 2000.

[47] W. Kim, H. Lim, O. Kim, Y.K Choi, H. Lee. "Visualized Construction Process on Virtual Reality - Information Visualization," Proceedings of the Fifth International Conference, IEEE, 2001.

[48] A. Webster, S. Feiner, B. Macintyre, W. Massie and T. Krueger, "Augmented Reality in architectural Construction, Inspection and Renovation" Proceedings of the 3rd Congress in Computing in Civil Engineering, ASCE, 913-919, 1996.

[49] A.H. Boussabaine, B. Grew, S. Cowland, and P. Slater, "A Virtual Reality Model for Site Layout," Proceedings of the Conference on AICIVIL-COMP99 and CIVIL-COMP99, 77-82, 1999.

[50] N. Dawood, and R Marasini, "Visualization of a Stockyard Layout Simulator: A Case Study in Pre-Cast Concrete Products Industry," Proceedings of the Conference on Construction Applications of Virtual Reality (CONVR 2000), University of Teesside, Middlesbrough, UK, 98-105, 2000.

[51] N. Dawood and A. Molson, "An Integrated Approach to Cost Forecasting and Construction Planning for the Construction Industry", The Fourth Congress on Computing in Civil Engineering, Philadelphia, 535-542, 1997.

[52] VRND - Notre-Dame Cathedral. Web-site: http://www.vrndproject.com

[53] BONGFISH - VR GRAZ. Web-site: http://www.bongfish.com/e2k/index.html

[54] S. Kerr, "Application and use of Virtual Reality in WS Atkins," Proceedings of the Conference on Construction Applications of Virtual Reality (CONVR 2000), University of Teesside, Middlesbrough, UK, 3-10, 2000.

[55] A. Waly, and W. Thabet, "A Virtual Construction Environment (VCE) for Project Planning," Proceedings of the Conference on Construction Applications of Virtual Reality (CONVR 2000), University of Teesside, Middlesbrough, UK, 43-53, 2000.

[56] M.F. Shiratuddin & W.Thabet, "Remote Collaborative Virtual Walkthroughs Utilizing 3D Game Technology", Conference of Product and Process Modeling (ECPPM): eWork and eBusiness in AEC, Portorož, Slovenia, Sept. 9-11, 2002.

[57] Parallel Graphics. Web-site: http://www.parallelgraphics.com

©2002, Saxe-Coburg Publications, Stirling, Scotland
Engineering Computational Technology
B.H.V. Topping and Z. Bittnar, (Editors)
Saxe-Coburg Publications, Stirling, Scotland, 53-74.

Chapter 3

Knowledge Based Systems in Structural Steelwork Design: A Review

T.J. McCarthy
Manchester Centre for Civil and Construction Engineering
UMIST, Manchester, United Kingdom

Abstract

Computer based design of structural steelwork has been around since the 1970's. Knowledge based systems have been developed in structural engineering applications since the 1980's. This paper takes a retrospective view of the developments in KBS research in structural steelwork design and in doing so exemplifies the broader shifts in engineering application software development. The paper examines how engineers have moved from developing software to do calculations rapidly and accurately to adding a higher level of decision support. The nature of structural design has evolved beyond simply providing structures that are strong enough to ones that meet other requirements such as buildability and cost reduction. The paper concludes with the author's personal view on the future developments in this field.

Keywords: knowledge based systems, expert systems, KBS, KBES, ES, design, structural steelwork, review.

1 Introduction

Structural steelwork accounts for approximately 50% of the market for building frames in the UK. Across Europe the market share for steel frames versus concrete frames is slightly less although there are sharp regional differences. A key factor in structural steelwork maintaining its share of the market and its dominance in certain sectors is the availability of high quality guidance in codes of practice, design guides, software and information and advisory services.

Organisations such as European Coal and Steel Community, the Steel Construction Institute [1] in UK, Australian Institute of Steel Construction [2], the American Institute of Steel Construction [3] and many other national trade organisations have always seen their role to promote steel to design practitioners.

The best way to do this promotion has been to develop knowledge and help structural designers to become better informed about steelwork. The range of publications produced by these organisations is vast and covers most aspects of structural steelwork in buildings, bridges and other structures. These paper based "knowledge bases", such as the excellent "Steel Designers Manual" [4] contain the distilled expertise of the best practitioners in the industry.

As the quality and quantity of these publications has improved so too has computer hardware and software technology. It is now easier than ever to create electronic books with hypertext explanations and user friendly navigation features. Researchers have worked for decades to find the most appropriate computational mechanism to deliver this expertise to users in a timely and efficient manner. This paper looks at the evolution of the use of Expert Systems, also known as Knowledge Based Systems, in this search for an efficient paradigm.

The paper argues that, although the combined effort in developing numerous expert system prototypes has not produced the so called "killer application" that solves all our steelwork problems, it has resulted in a necessary restructuring of the way we think about design. It has paved the way for today's buzz words: knowledge economy, knowledge management, organisational learning, etc.

One reason why the "killer app" has not been produced to date is that our collective understanding of information modelling has been inadequate. On top of this, the software development environments that were available were inadequate for generic solutions. They were good for individual prototypes, as is demonstrated in the many examples cited below. However, it was difficult, if not impossible, to create extensible solutions which could build effectively on previous solutions. In the researchers' desire to remain at the leading edge of technology we kept having to re-invent things. The shelf life of computer software was just too short.

Work begun by Hirst and Watson [5] in the late 1980's has resulted in the emergence of robust and extensible data modelling standards such as ISO10303 also known as STEP [6]. With the increasing number of STEP Application Protocols we now have a framework in which to describe information. The advent of HTML and XML makes it easier for us to create electronic resources that have a longer shelf life (from an information technology perspective). It is also easier to build models collaboratively and to extend information models.

With the congruence of this new development environment and our improved understanding of the importance of data and information modelling this paper concludes that the first decade of the 21st century will be the age of the Expert System. By working collaboratively in a formal framework, the efforts of researchers will be combinatorial and extensible. By better managing and modelling our information we can make use of the advances in computer science by truly separating the knowledge from the computational mechanism.

2 Development of Expert Systems

2.1 General overview

2.1.1 Codes of Practice – the First Expert Systems?

The steel industry has been at the forefront in producing books and design guides, national and international standards to assist structural designers in selecting member sizes and formulating safe structures. In Britain, the main design standard used for steelwork from 1932 to 1969 (and a number of years after) has been BS449 [7].

BS449 was a permissible stress code of practice that contained, in one short document, all the design calculations necessary for creating safe steel structures for most building applications. It can be argued that BS449 was the first steel design "expert system" to be widely used in the United Kingdom. It bears most of the hallmarks of a conventional expert system. The knowledge was elicited from the experts of the day, rules were formulated and the knowledge representation was developed so as to be user friendly. The one component of an expert system that BS449 missed was the explanation facility.

Although, BS449 was an easy to use and popular code of practice, many of its rules were opaque and lacked explanation. Indeed one criticism that was levelled at the nature of steelwork design was: *Design in steel used to be regarded as a 'black art' where one only reached a level of competence after 20 years of hard-won experience*[8]. There was a desire for representation of the deeper knowledge so that engineers could not only produce safe structures but also understand why they were safe.

It can be debated, as was done by Rosenman and Gero [9], whether the codes of practice, books and design guides really represent crude forms of knowledge based systems. True, the system, printed words on paper is a bit old fashioned. However, the knowledge base aspect cannot be denied. The inference engine resided in the skull of the designer. Latterly, with the development of Limit States design codes, a deeper knowledge and better explanation has been available. Indeed, newer standards such as BS5950[10] and Eurocode 3[11] have been written in such a way as to facilitate the development of conventional and knowledge based software. In section 2.2.2, below, it will be seen that many developers concur with these views.

2.1.2 The Early Years of Computer Based Expert Systems

Expert systems are a development from conventional computer science research in the field of artificial intelligence. It is not the intention here to give a history of AI since this has been done admirably by many authors, in particular the book by Russell and Norvig [12].

Some of the earliest work in the field of knowledge based systems in structural engineering was at NASA where Lazlo Berke was involved with producing systems for aerospace structures [13, 14]. NASA produced an integrated design system linking together the chief engineering activities of preliminary design, finite element analysis, structural optimization and Cad graphics. NASA also produce the CLIPS [15] expert system shell which is still an important development environment. This is even more so with the advent of the internet version of CLIPS, JESS (Java Expert System Shell) [16].

The seminal research in the use of artificial intelligence for civil engineering and building structures was carried out a Carnegie Mellon University by Stephen Fenves and his doctoral students. Maher's program, HI-RISE [17, 18] is amongst the most cited of the early KBS building design systems. Other early researchers in the application of AI to construction were Grierson at Waterloo University [19, 20] and Gero at University of Sydney [21]. Adeli's book in 1988 [22] gives a good snapshot of developments at that time. In the United Kingdom, the pioneers of applying artificial intelligence to design in construction were Allwood [23], Topping who instituted the AI CivilComp series of conferences [24], Miles and Moore at Cardiff University [25].

In the late 1980's, with the advent of commercially available expert system shells such as Kappa and Kappa-PC [26] and the LISP based Goldworks 2 and 3 [27], a number of other researchers joined in. This is evidenced by the rapid expansion of such conferences as AI CivilComp and ASCE ICCCBE. Many of the "new" arrivals began applying AI to specific and well defined domains such as structural steelwork rather than the larger questions of how is design accomplished. While the early work developed the most appropriate software architectures and communications schema, the later work was concerned more with application development and using AI to learn more about the domain.

2.1.3 Expert system architecture

The conventional view of what an expert system, or knowledge based system should comprise was given by Allwood [23]:
(a) some method or tool for knowledge elicitation
(b) knowledge representation scheme (the knowledge base)
(c) an inference engine
(d) an explanation facility
(e) user interfaces.

Although most researchers in the field of ES aim to include all of the above features, there is little evidence that the automatic explanation facilities have had much success. In this author's experience of a number of development shells, the explanation facility has been unwieldy. It may be appropriate for use by the developer of the expert system as a debugging aid and for validation, however the presentation of explanations to the end user is best controlled via the user interface (which in turn may use the automatic explanation facility).

2.1.3.1   Production rules, assertions and facts

The bulk of expert systems that have been developed rely of the use of rules of the form IF (Condition) Then (Do something or Assert something). By decomposing the domain into a series of stand alone rules developers can create quite successful expert systems.

Things that are known to be true can be asserted as facts. The total set of facts is called the context. The expert system will then combine the facts with its rules to infer new facts and so on until the solution is found.

The rules are not necessarily stored in the sequence in which they will operate. The rule firing sequence is controlled by the inference engine. In a typical operation, the inference engine compares the antecedent or condition part of all the rules with the current context or known facts. Those rules whose condition is satisfied are placed in a "ready to fire" state. Then the inference engine selects one rule to fire. This may be done using a predetermined priority scheme where some rules have a higher priority than others or may be done at random. When the rule has fired it may assert a new fact which changes the context and the procedure starts again. Some rules may be added to the "ready to fire" selection and some may be removed because of the new context.

This is a simple description of "forward chaining". In a well constructed rule one should not require the over use of priorities – since by setting priorities one is implicitly creating a new rule. It is better to formulate meta-rules that explicitly state the knowledge that is used to control a given sequence.

Some expert systems operate by "backward chaining". A typical example would be a diagnostic type of ES. Here the goal is known and we want to find out the condition that caused it. Some problems require mixed mode chaining i.e. a combination of a forward search and leading to conjectures which are then tested by a backward search. Structural design sometimes operates like this: based on some data, the engineer makes an educated guess at the solution and then uses analysis techniques to test the viability of the guess.   An in depth description of production rules can be found in [23] and [25].

2.1.3.2   Frame based and object oriented systems

While rules are an effective representation for heuristics, they are not very good for describing objects. In structural design we use rules, equations etc in order to create structural objects.   Furthermore, the objects we wish to create are usually interrelated. For example, a column supports a beam and a beam supports a floor. The beam is connected to the column by a joint. The floor rests on the beam. In a building we will have families of elements (e.g. the family of columns) all of which have similar behaviours and functions.

In object oriented representations, a hierarchy is set up to describe the families of objects. The expert system, HYCON, was developed to design hybrid

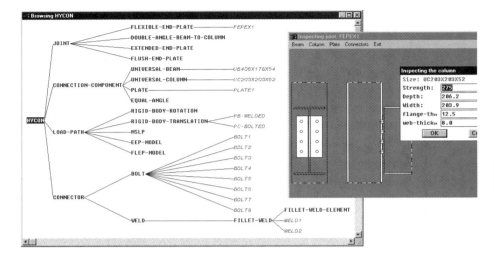

Figure 1 Hycon Frame based system.

welded/bolted steelwork connections [28]. An extract of the hierarchy of objects is shown in Figure 1 which shows four main objects, Joint, Connection-Component, Load-path and Connector. The Joint object has four children, Flexible-End-Plate etc. In the frame based system all the common attributes are defined in the higher level object.   The things that make the Flexible-End-Plate different from an Extended-End-Plate appear in the lower level object. The objects at the far right of the tree in Figure 1 are the instances of the objects.

Frames provide an efficient way of organising objects.   The children automatically inherit the properties from their parent. In some systems, objects may have multiple parentage. In Hycon, all joints consist of a beam and column, either angles or end plates and connectors (welds and bolts). The objects have rule bases associated with them and also perform functions using "methods". A method is simply a piece of computer code that performs a calculation or sends a message to another object.

In the Hycon example, the frames are also used to control the user interface. If we send a message to a flexible-end-plate joint to draw itself on the screen, the object passes on the message to each of its component parts to draw itself at a particular location. Exactly the same method is used to draw any connection type.

With object oriented approaches, all knowledge about the object is encapsulated and stored in the object. Within a system like HYCON, the objects can be re-used at will. However, these objects, in their current form, are not available to other systems. If a standardised knowledge representation scheme is employed then the objects can be re-used and shared by any application that understands the scheme. This is the goal of information and product data modelling.

2.1.4 Knowledge Acquisition and Elicitation

Once a domain has been selected for the development of an expert system, the first important task is to acquire the appropriate knowledge. Knowledge may considered as tacit knowledge or explicit [29].

Explicit knowledge can be acquired from various sources such as printed documents, case studies and so forth. The difficult job in knowledge elicitation is to try to capture the tacit knowledge which mostly resides in the memories of human individuals. Terms such as "experiential knowledge" or heuristics have been applied to this form of knowledge. In Chapter 5 of their book, Miles and Moore cover a range of interviewing techniques which are appropriate when eliciting knowledge from experts [25]. At the early stage of development of an expert system, one may opt for an unstructured interview with or without props. This helps to decide the scope of the system. As the system is developed and more specific information is required, semi-structured, structured interviews and/or questionnaires can be employed. The goal of the elicitation is to take tacit knowledge and to formulate it into explicit knowledge.

Other forms of knowledge acquisition include the analysis of case studies, parametric studies or data mining. This form of acquisition seeks to derive new heuristics from improved analysis of available material. MacKenzie and Gero [30] used results from optimisation studies to derive rules for the design of concrete slabs.

Knowledge is also classified as shallow knowledge and deep knowledge. Shallow knowledge equates to the rules we know to be valid but which are not based on a fundamental understanding. These are sometimes termed "rules of thumb" and are the mainstay of most expert systems. Deep knowledge refers to relationships which can be derived from first principles. Some consider this to be the mainstay of algorithmic programming. Shallow knowledge enables us to take short cuts in the solution process and to solve otherwise intractable problems. Deep knowledge can then be used to verify that the solution is valid.

Knowledge can be considered as "information with context". The context gives the fact some meaning. Information in turn is data with a structure. A key factor in the success of any knowledge based system will therefore be dependent on the ability to structure the data and represent the context. That is, we must not only acquire the knowledge for the system but we must also find the most useful way to store it. Many of the early expert systems missed the importance of formal information structures and so when their knowledge was acquired it became locked within the system.

2.1.5 Validation and Evaluation of Expert Systems

An essential part in the development any piece of software is the validation/verification and evaluation phase. This is necessary to check that the rules and knowledge included in the system are correct and that they produce valid solutions when presented with a problem to solve. The key test of an expert system is whether is produces a solution that an expert agrees is valid. The ideal situation is when the system produces an answer that the expert agrees is the best solution.

To date, no formal validation techniques have been established for use with expert systems. In other areas of software development, eg databases and e-commerce, rigorous standards exist for the formal validation. IEC 61508 is a proposed validation standard for safety critical systems, including software [31]. Broadly, the validation process should check that the information has been coded correctly. It should then verify that under the specified usage conditions this coded information produces the correct results.

The evaluation of expert system software looks at whether it fulfils the objective of the project. While validation or verification merely checks accuracy, evaluation examines usefulness. The real success of any software, in practical terms, is an amalgamation of technical excellence in the coding and a well designed interface. Yusuf et al [32] used a study of the tasks undertaken by structural steel designers to design a user interface that facilitated those tasks. By presenting users with a tool that makes their job easier, one is more likely to get the software accepted. The users do not need to know or see any evidence of the excellent computer code that drives the application.

2.2 Knowledge Based Systems in Steelwork

2.2.1 Early developments

Structural Engineers began active work in research related to the application of expert systems to steel design in the mid-1980's. After the seminal work at Carnegie Mellon University in 1984 [17] which covered conceptual structural design of high rise buildings, Grierson and Cameron published their expert system which designed multi-storey steel frames to achieve minimum weight [19, 20].

Grierson's work is interesting since it integrated an existing traditional FORTRAN design program with a rule based expert system. The design involved a preliminary stage, a solution stage and a critique stage. The preliminary stage used a rule base to find a near correct solution and then the solution stage uses formal optimisation to find the minimum weight solution. In the critique phase a rule base is employed to check that designer/fabricator preferences are met, e.g. grouping members for design or looking for potential lateral stability problems. If the critique stage suggested improvements, then they would be applied and the solution stage re-started.

The conference Applications of Artificial Intelligence in Engineering Problems organised by Sriram and Adey in 1986 [33] was followed in 1987 by the first in the long running AI Civil Comp Conferences [24]. This year also saw the publication of an important special issue of ASCE Journal of Computing in Civil Engineering devoted to Expert Systems in Civil Engineering [34]. Allen reported on 23 ES development as "work in progress" in 1987. Included in his review were systems for welding, seismic risk, damage assessment and bridge design. The main commercial thrust was in the aerospace industry, while most of the building related systems were being developed in universities. Since the mid-nineteen eighties there has been a steady flow of research papers covering ES and KBS in structural design. The reader is referred to Topping and Kumar's extensive 1989 bibliography for the pre-1990 picture [35].

2.2.2 Standards Representation

Earlier in this paper it was mentioned that codes of practice and design standards represent an accumulation of expertise and knowledge about how to make structures strong and safe. A number of researchers have grappled with the problem of how to take these standards and represent them in a form usable by computer software.

Early work on representing codes of practice was done in Australia where Rosenman and Gero created a PROLOG system based on the Australian Model Uniform Building Code[9, 36]. Later work in representing standards has been done by Fenves and Garrett at Carnegie Mellon University and by Kumar at Heriot Watt and Strathclyde Universities. The earliest CMU work related to decision tables and predicate rules. Garrett and Hakim [37] developed the object oriented approach to the AISC LRFD standard using CLOS (Common LISP Object System). An improved approach was presented by Garrett in 1994 incorporating a so called "Context oriented model" [38]. Yabuki and Law developed a "hyperdocument" model for design standards documentation using hypercard in 1993 before the advent of HTML [39]. A second hypertext representation was given by Neilson et al in 1995 was aimed at providing a web enabled information base to supplement an electronic version of the steel design codes in the UK [40].
Kumar and Topping [41] used a PROLOG based representation for the British Standard BS5950. This work represented the steel design standard as a set of facts which are then processed by a series of generic rules.

One of the abiding problems with knowledge based representations of design codes is that the documents themselves are an accumulation of technical data, expertise and vested interests. No one representation scheme will suit a complex design standard. DeGelder and Steenhuis [42] have considered the problem of representing complicated formulae related to elastic-plastic behaviour of beam-columns. The particular problem relates the to the draft EC3 Design Code of having possible rule conflicts and incompleteness. The solution proposed by DeGelder and Steenhuis is to create a hierarchy of decision tables. Whereas the design code rules are difficult to apply in their raw form, the authors claim that by being able to consult the knowledge that resides in the decision tables, the designer will get a complete and consistent result. Furthermore they suggest that these decision tables in a hierarchical tree have a role to play when the code rules are being drafted so as to avoid the current difficulties with Eurocode 3.

Much work is required in the field of knowledge and information management to (a) facilitate the drafting and maintenance of structural design codes and (b) to make standard machine readable.

2.2.3 Steel Bridge Design

Adeli and Balasubramanyam [43] presented a heuristic approach to obtain the maximum compression and tensile forces in bridge trusses subject to rolling loads while Adeli and Mak [44] used coupled expert systems for the optimum design of plate girder bridges.

Conceptual design of steel bridges was tackled by Miles and Moore in 1989 and in subsequent years [45-47]. This work at Cardiff University was extended by Philbey et. al in 1993 [48] and Clayton et al in 1996 [49]. This team has not only developed a number of tools for decision support in the early design stages for road and river crossings but have also covered methods of evaluating and maintaining expert systems. With reference to section 2.2.2 above, the maintenance of software which depends on codes of practice is a particularly time consuming problem. Hence, the desire for machine readable codes of practice.

Moving from the early design phase of bridges, work by Roddis et al was concerned with eliminating fabrication errors in steel bridge construction. [50, 51]. An expert system, BFX, was developed by compiling a case base of fabrication errors directly from the fabrication shops, the Kansas State inspectors field notes and KDOT documents. In 66% of test cases the expert system produced the correct solution. No solutions were generated for the remaining 33% which were then shown to be beyond the scope of the system.

Interestingly, both the team at Cardiff University and University of Kansas have more recently moved from straightforward expert systems approaches to case based reasoning (CBR). Roddis and Bocox [52] have extended the BFX system using the CBR tool MEM-1. The CBR approach has improved the success rate from 66% to 82% while a combined ES/CBR approach gave 91%. Furthermore, the case base approach is seen to be much easier to maintain and expand than the traditional expert system. Lehane and Moore [60] at Cardiff applied CBR to the design of motorway bridges. While the Kansas system was developed in Lisp and MEM-1 the Cardiff system was prototyped using the CBR shell, ReMind and then transferred to a MS Visual C++ system developed in house.

2.2.4 Design of structural components and buildings

Work at UMIST has been concerned with the design and costing of steelwork connections. The first attempt was to develop a simple system in AutoLISP for the design of column baseplates [53]. This had the useful by-product of being easily able to produce the CAD drawings as well as the structural design. It was clear that for a more substantial design problem a more elaborate knowledge base development tool was required. Goldworks 3 and Golden Common LISP were used to develop Hycon, a design system for hybrid bolted/welded connections [28].

HYCON which is show in Figure 1 produced retrofit strengthening schemes for bolted connections. By adding welds to certain parts of the beam endplate, increases in load capacity could be achieved. This strengthening depended on the flexibility of the endplate and the location of the welds. The expert knowledge came from a 10 year research programme on strengthening at UMIST and combined many full scale tests and 3D finite element analyses. HYCON has been used as an in house tool for designing strengthening schemes in place of a trial and error approach using FE models. The KBS produces answers in minutes which had previously taken days and weeks to achieve.

Also concerned with steelwork connections, KBCCM was a knowledge based system for the costing of beam to column connections [54]. This program, developed in Goldworks 3 was an object oriented model of a steelwork fabrication shop. Each machine was modelled and instances created in a frame based system. The connection components also existed as instances of frames. An activity model of the fabrication operations controlled the system. A rule base determined which equipment is available and capable of completing each task. Then a second rule set chooses the best equipment based on time taken and the shop schedule. The cost is calculated in terms of workshop hours. A Java fabrication simulator was subsequently added based on the scheduling rules [55].

EXSEL, a KBES for designing structural steelwork elements was presented in 1993 [56]. This is a combination of production rules, C functions and a relational database with a backward chaining algorithm. It is a good example of how different software components can be linked to build a practical design system.

Moving from individual components to steel building design, Pasley and Roddis presented a knowledge based system called Steelteam in 1996 [57]. This system does two things. Firstly it provides a communication environment for co-ordinating design and drafting and secondly it acts as an intelligent advisor to the team members. The aim is to avoid downstream problems by addressing the concerns at an early stage.

Early work on a KBES for steel roof trusses was done by Adeli and Al-Rijleh in 1987 [58]. In 1996 Adeli and Kao [59] presented a complex system of knowledge bases and structural analysis tools for the design of very large steel structures. This work builds on the previous work by Paek and Adeli (STEELEX) [60] and employs a hierarchy of knowledge bases for integrated design of large structures containing 100's or 1000's of members. Adeli and Kao present a blackboard approach to control the communication between the various knowledge bases. The implementation language is C++. They obtained designs for a 35 storey building according to American allowable stress design codes and LRFD codes. The LRFD produced a 7% weight saving over the allowable stress design.

TALLEX is another expert system for the preliminary design of tall buildings [61]. This uses a rule base of some 225 rules to calculate weightings for factors which are used to determine a "vitrual number of floors" and the acceptance value for each of 16 available structural schemes. The structural schemes vary from flat slab and R/C columns (suitable for up to 10 floors) to Cellular construction using tubular steel (110 to 120 storeys). For a given building, TALLEX displays a "rate of acceptance" for each of the 16 structural systems.

From the tallest buildings, we now move to single storey frames. ADAIL, a system for selecting the most appropriate structural steelwork solution for single storey industrial units is presented by Tizani et al [62]. Similarly to TALLEX, this system aims to select the best system from a set of possible solutions. There are 6 possible structural forms ranging from steel portal frames to lattice girders. The roof

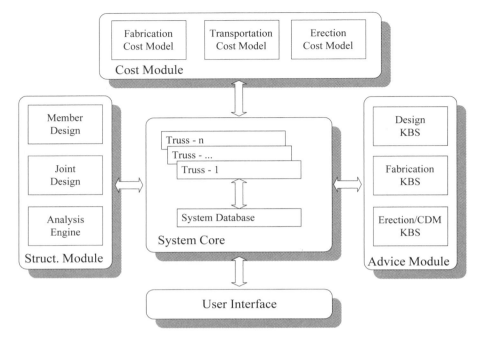

Figure 2 IDS System Architecture.

can be either flat or pitched. The expert systems caters for single bay and multi-bay buildings and employs a costing model based on a parametric study by Horridge and Morris [63]. ADAIL was developed in the PROLOG based shell, Arity/Expert and employs a hybrid rule and frame based approach.

2.2.5 Construction-led design – incorporating fabrication and construction

Adeli and Kao's definition of integrated design consisted of conceptual design, premilinary design, frame analysis and member selection to meet code requirements [59].   In this section a number of papers are summarised which employ expert systems to go one or two steps further, namely, the integration of the fabrication and construction requirements within the design cycle.

Zayyat, Smith and Nethercot [64] presented a paper in 1989 concerning the problem of designing heavy steelwork box columns. Their knowledge based system was one of the earliest examples of a program for designing structural steelwork with production and cost as well as strength a major influence on the decisions. The knowledge base was derived primarily from parametric studies and implemented as mathematical functions and If-Then rules. This theme of production-led knowledge based design systems has been carried forward by Nethercot, Smith, McCarthy, Tizani and Davies in a series of papers from 1993 to 2000 [32, 65-68].

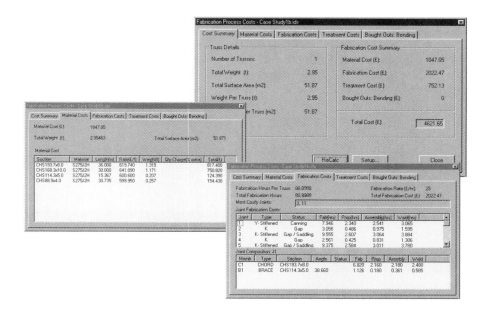

Figure 3 IDS Fabrication Cost.

The work of this group centred on the development of an integrated design tool for tubular steelwork trusses in roof applications. The IDS (Integrated Design Software) program began as a traditional expert system developed in the Goldworks 3 shell. While the shell facilitated the concept of the integrated system and allowed the development of an advanced prototype, the limitations of the shell (and the cost of licensing) inhibited its use as a really practical tool. In the later development of IDS, the expert system shell was replaced with a more conventional C++ object oriented approach with a much simplified inference mechanism.

Figure 2 shows a schematic representation of IDS. While the structural design module is a fairly traditional approach with matrix analysis, code checking etc, it had novelty in the user interface and Advice Modules. In addition, IDS produces a very detailed fabrication cost (see Figure 3), highlighting the high cost items in the structure. For the first time, structural designers had the facility to appraise their design for not only code compliance but also for cost, transportation and site safety issues. The site safety issue is now increasingly important in Europe as structural designers are responsible for the site safety consequences of their decisions under EC Directive 92/57/EEC. In the United Kingdom, this directive has been implemented in the form of the Construction Design and Management (CDM) regulations[69].

The IDS work has shown that for the case of tubular trusses, the minimum weight design does not generally give the lowest cost. Tubular joint design usually governs the economy of the structure since it is often cheaper to use thicker tubes if this helps to avoid the need for local stiffening of the joints.

2.2.6 KBS in Steel manufacturing and fabrication

There have been many expert systems developed for the design of steel alloys, castings and for steel production. The following is a brief and incomplete review which serves to introduce some of the topics covered in the literature.

Hung and Sumichrast [70] have produced an expert system to assist with the material ordering and cutting problem for efficiently combining steel plates into elements such as plate girders. They have developed the expert system from knowledge elicitation using multiple experts. Their system was tested on a number of historical cases from the Lien-Kang Heavy Industrial Company (LK). In all cases the cutting schedule proposed by the expert system was cheaper than the historical solution. The system has now been incorporated in LK's decision support system.
Hot strip mill production is the subject of a system developed by Sanfilippo et al. in 1996 [71]. Their system combines a diagnostic component that monitors the electromechanical equipment and a second system to control productivity and quality. Intelligent monitoring of a steel rolling mill is also the subject of Zhong et al. with the aim of predicting faults in the production [72].

Fabrication or more specifically welding is the subject of Goel et al's expert system presented in 1993 [73]. Their expert system was aimed at planning operations and training welders in shielded metal arc welding. They gave special emphasis to the explanation facility implemented in PROLOG so that it can show the operators the reasoning behind the conclusion.

Cold bending of structural steelwork is becoming more popular. Current work at UMIST concerns the development of a KBS advisor for structural engineers based on the new Steel Construction Institute Guide on the Design of Curved Steel [74]. This system advises engineers on the amount of curvature that can be achieved depending on section size and grade of steel. In addition it will advise on how to analyse the curved members according to the methodology developed by SCI.

A recent KBS relating to the cold bending of tubes was presented by Jin et al [75]. This system is aimed at those designing the fabrication process for bending relatively small pipes and tubes. Some of the bending techniques are applicable to small structural elements. Their KBS, developed in Level 5 Object ES shell has significantly reduced the number of defects (e.g. excessive thinning of the walls or wrinkling of the compression side) in the finished products.

2.2.7 Life cycle of structures

Having designed and constructed steel structures it is appropriate to finish the review with two expert systems that help us to detect damage in a monitored structure and then assist with the demolition task.

Expert systems have been used for diagnosis problems in many domains. Kim and Chen presented an object oriented system for health monitoring of instrumented steel structures to detect damage [76]. Knowledge about structural monitoring and diagnosis is formalised and represented in a logic based object oriented modelling environment. Kim and Chen have used the system successfully in a laboratory environment to detect damage in a successively damaged steel structure.

Fox and Butler dealt with the tricky problem of estimating the mass of a structure in situ [77]. This is needed if the structure is to be demolished. They captured the reasoning from an experienced demolition estimator and employ a backward chaining rule based structure. When given detailed data on the steel roof truss, the system can achieve answers within 6% of the correct value. When only sketchy details are available the accuracy is to within 24% of the correct mass.

2.2.8 Product and Information models

While many structural engineering researchers were exploring the novel computational tools that AI was providing, few of us were fully aware of what we were doing with information. We were structuring data and knowledge to fit into expert systems and by and large a lot of duplication resulted. That is not to say that the effort was wasted – rather it was a learning curve that we had to go through.

An exception to this is the work of Alistair Watson at Leeds University. Watson's group has been at the forefront of developments to create efficient and re-usable product data models. He had realised that we needed our systems to be able to talk to each other and that this required a common language for our data. In 1989 he presented a Prolog based database management system founded on a formal and robust data models [78]. In 1993 he presented the emerging ISO 10303 STEP [6] standard and the Application Protocols (AP) that were to deliver the data model to the structural steelwork community [79].

There has been a sluggish but steady progress on developing the steelwork AP. This has been led by Watson and the European Steel industry but the rest of the world eventually joined in.

In recent years, XML has made this product data and information modelling approach more readily available to all. Software tools such as XMLSpy [80] allow us to create usable data models. While information modelling in EXPRESS (the STEP information modelling language) has been cumbersome in the past a number of data management tools are emerging to shield the application developer from the tedium. Furthermore, graphical tools which help in the creation and implementation of abstract information models are now available. Standard applications such as MS Visio include EXPRESS-G representation. AI0 from KBSI allows users to quickly generate activity models in IDEF0[81].

The great advantage of using these formal models is that software tools are available that can interpret them. Provided the developer adheres to the principles in the AP and creates well formed XML files, the rest can be done "automatically". Indeed, the final goal is that the developer will use a software tool that adheres to the AP and produces the model for him. Kahn et al have demonstrated a generic framework for transformation of EXPRESS models to other formats [82]. This technology is key for the re-use of information from, say, design to construction and beyond.

3   The Future Development of Expert Systems

The last two decades have seen the evolution of the expert system from an experimental research tool to an integral part of many design systems. Computer software has moved from the era of stand-alone strange languages to integrated web-enabled architectures. Whereas in the early days the complete system would be developed within one language, say LISP, today it is possible to link together many software components written in different languages. We no longer have to force PROLOG to be a graphics interface or to be a user-friendly database. Now we can let PROLOG do what PROLOG is good at and use Visual Basic or some other tool to generate the graphics. Information that should reside in a database can do so.

The creation of information and data modelling standards and the greater acceptance of open standards such as XML will allow us to move towards systems that communicate much better with each other. While information modelling itself is rather tedious, it is an enabling technology. Just like building a modern highway system is time consuming and repetitive but when it is finished (or even partly so) it enables us to zoom around with greater speed and efficiency. With well developed information models and application protocols, our software will be able to interconnect with greater resilience and accuracy. We will be able to build software that is extensible and that communicates with other software more easily.

The eValid project currently running at UMIST aims to demonstrate how activity and information models can be developed to facilitate the information flow from detailed design to construction and on to commissioning and validation of complex facilities[83]. The goal is to provide information within formal structures so that it can be transformed and used by intelligent systems. The application domain for eValid is Validation of Pharmaceutical facilities but the approach is applicable to verification of designs and commissioning of any type of structure. When information is available in large and usable quantities, it is then that Expert Systems will come to the fore.

4   Conclusion

Knowledge based expert systems have been developed in the field of structural steelwork design for about 16 to 18 years. There was a rapid expansion in the study of applications of artificial intelligence in the later 1980's and 1990's. New conferences and journals appeared and expert systems proliferated. And still, despite the activity and the ingenuity, the great break though never came – or if it did it crept up on the AI community.

After the initial exuberance surrounding AI, there was a reality check. Expert systems were re-branded as Knowledge based systems. We have now rid ourselves of our hang-ups and can concentrate on making software that solves problems and enables intelligent people to do good steelwork design. Even though we did not produce the "killer application" in expert systems, we have learned a lot about knowledge bases and about knowledge itself.

Expert systems are more likely to be called decision support systems and are more likely to be embedded in more conventional software packages. It is acknowledged that the Expert system cannot function alone but must be part of an overall package the structural engineers' needs. These needs include, good analysis tools, good advice and good information.

The advent of good information models and an improved understanding of how to make them will facilitate the creation of integrated knowledge based systems. These will link with case based systems and engineering databases across the INTERNET to deliver the required information in a timely manner to enable the structural designer to make good decisions.

With more and more knowledge readily available and in highly usable forms, the age of the knowledge based system is only just beginning.

References

[1]   SCI, "http://www.steel-sci.com/," The Steel Construction Institute, 2002.

[2]   AISC, *Economical Structural Steelwork*: Australian Institute of Steel Construction, 1990.

[3]   American Institute of Steel Construction, *Specification for the Design Fabrication and Erection of Structural Steel for Buildings*. New York: American Institute of Steel Construction, 1978.

[4]   G. W. Owens and P. R. Knowles, "Steel Designers Manual," 5th ed. Oxford: The Steel Construction Institute and Blackwell Science, 1994.

[5]   P. B. Hirst and A. S. Watson, "Functional Modelling of the Constructional Steelwork Industry," presented at Proc 4th Int Conf on Civil and Structural Engineering, 1989.

[6]   ISO10303-11, "Industrial automation systems and integration: Product data representation and exchange: Part11: Description methods: the Express language reference manual," ISO, Switzerland 1994.

[7]   British Standards Institution, "BS449 Code of practice for the use of structural steelwork in building," London 1969.

[8]   ESDEP WG 1A, "Lecture 1A," The ESDEP Society, http://www.esdep.org 1992.

[9]   M. A. Rosenman and J. S. Gero, "Design codes as Expert Systems," *Computer Aided Design*, vol. 17, pp. 399-409, 1985.

[10]   British Standards Institution, "BS5950 Pt 1 2000, Code of Practice for Design in simple and continuous construction: hot rolled sections," BSI, London 2001.

[11]   British Standards Institution, "DD ENV 1993-1-1: 1992 Eurocode 3: Design of Steel Structures Pt 1.1 General rules and rules for buildings," BSI, London 1992.

[12]   S. Russell and P. Norvig, *Artificial Intelligence - a modern approach*: Prentice Hall, 1995.

[13]   R. J. Melosh, L. Berke, and P. V. Marcal, "Knowledge-Based Consulting for Selecting Structural Analysis," presented at Symposium on Research in Computerized Structural Analysis and Systhesis, 1978.

[14]   H. Kamil, A. K. Vaish, and L. Berke, "An Expert System for Integrated Design of Aerospace Structures," in *Artificial Intelligence in Design*, J. S. Gero, Ed.: Springer Verlag, 1989, pp. 41-55.

[15]   CLIPS, "User Guide," Artificial Intelligence Section, Johnson Space Center 1988.

[16]   JESS 5.2, "The Java Expert System Shell," Sandia National Laboratories, (http://herzberg.ca.sandia.gov/jess), Livermore 2001.

[17]   M. L. Maher, "HI-RISE: A knowledge-based expert system for the preliminary Structural Design of High Rise Buildings," PhD Thesis *Department of Civil Engineering*. Pittsburgh: Carnegie Mellon, 1984.

[18]   M. L. Maher, "Expert systems for structural design," *J Computing in Civil Engineering*, vol. 1, pp. 276-283, 1987.

[19]   D. E. Grierson and G. E. Cameron, "A Knowledge Based Expert system for Computer Automated Structural Design," in *The Application of Artificial Intelligence Techniques to Civil and Structural Engineering*, B. H. V. Topping, Ed.: Civil-Comp Press, 1987, pp. 93-97.

[20]   G. E. Cameron and D. Grierson, "Developing an Expert System for Structural Steel Design: Issues and Items," in *Artificial Intelligence in Design*, J. S. Gero, Ed.: Springer Verlag, 1989, pp. 15-39.

[21]   J. S. Gero and M. A. Rosenman, "A conceptual framework for knowledge based design research at Sydney University's Design Computing Unit," in *Artificial Intelligence in Design*, J. S. Gero, Ed.: Springer Verlag, 1989, pp. 363-382.

[22]   H. Adeli, "Expert systems in Construction and Structural Engineering," Chapman Hall, 1988.

[23]   R. J. Allwood, *Techniques and Applications of Expert Systems in the Construction Industry*: Ellis Horwood, 1988.

[24]   B. H. V. Topping, "The Application of Artificial Intelligence Techniques to Civil and Structural Engineering," Civil-Comp Press, 1987.

[25]   J. Miles and C. Moore, *Practical Knowledge-Based Systems in Conceptual Design*: Springer Verlag, 1994.

[26]   Kappa-PC, "User Guide." California: Intellicorp inc, 1993.

[27]   Goldworks 3, "User Guide." Cambridge, Mass: Goldhill inc., 1991.

[28]   Z. Nouas and T. J. McCarthy, "Knowledge-based system interface design: an object oriented approach," in *Information Technology for Civil & Structural Engineers*, B. H. V. Topping, Ed.: Civil-Comp Press, 1993, pp. 159-164.

[29]   M. B. Patel, T. J. McCarthy, and P. W. G. Morris, "The role of IT in capturing and managing knowledge for organisational learning on construction projects," presented at Construction Information Technology, Reykjavik, 2000.

[30]   C. A. MacKenzie and J. S. Gero, "Learning Design Rules from Decision and Performances," *Int J Artificial Intell*, vol. 2, pp. 2-11, 1987.

[31]   IEC, *IEC 61508:1998 Functional Safety of Electrical Electronic and Programmable Electronic Safety Related Systems*: BSI, 1998.

[32]   K. O. Yusuf, W. Tizani, and T. J. McCarthy, "A User Interface for Engineering Decision Support in the Fabrication-Led Design of Tubular Trusses," in *Developments in Computer Aided Design and Modelling for Structural Engineering*, B. H. V. Topping, Ed.: Civil-Comp Press, 1995, pp. 229-235.

[33]   D. Sriram and R.Adey, "Applications of Artificial Intelligence in Engineering Problems, Proc 1st Intnl Conference." Southampton, UK: Computational Mechanics Publ, 1986.

[34]   R. N. Palmer, "Expert Systems in Civil Engineering," *J Computing in Civil Engineering*, vol. 1, 1987.

[35]   B. H. V. Topping and B. Kumar, "The Application of Artificial Intelligence to Civil and Structural Engineering - A Bibliography," in *Artificial Intelligence Techniques and Applications for Civil and Structural Engineering*, B. H. V. Topping, Ed.: Civil-Comp Press, 1989, pp. 285-303.

[36]   M. A. Rosenmann, J. S. Gero, and R. Oxman, "An Expert System for Design Codes and Design Rules," presented at Applications of Artificial Intelligence to Engineering, 1986.

[37]   J. H. Garrett and M. M. Hakim, "Object Oriented Model of Engineering Design Standards," *J. Computing in Civil Engineering*, vol. 6, 1992.

[38]   J. H. Garrett, "The context oriented model: An Improved Approach to Representing and Processing Design Standards," *J Computing in Civil Engineering*, vol. 8, pp. 145-152, 1994.

[39]   N. Yabuki and K. H. Law, "Hyperdocument Model for Design Standards Documentation," *J Computing in Civil Engineering*, vol. 7, pp. 218-237, 1993.

[40]   A. Neilson, B. Kumar, and I. A. McLeod, "Information base for Design Checking," in *Computer Aided Design and Modelling for Structural Engineering*, B. H. V. Topping, Ed.: Civil-Comp Press, 1995, pp. 65-70.

[41]   B. Kumar and B. H. V. Topping, "A prolog based representation of standards for structural design," in *Artificial Intelligence Techniques and Applications for Civil and Structural Engineers*, B. H. V. Topping, Ed.: Civil Comp Press, 1989, pp. 165-169.

[42]   J. de Gelder and M. Steenhuis, "A Knowledge-based system approach for code checking of cross sections in steel structures according to Eurocode 3,"

in *Information processing in Civil and Structural in Civil and Structural Engineering*, B. Kumar, Ed.: Inverleith Spottiswoode, 1996, pp. 253-261.

[43]   H. Adeli and K. V. Balasubramanyam, "A heuristic approach for interactive analysis of bridge trusses under moving loads," *Microcomputers in Civil Engineering*, vol. 2, pp. 1-18, 1987.

[44]   H. Adeli and K. Y. Mak, "Application of a Coupled System for Optimum Design of Plate Girder Bridges," *Engineering Applications of Artificial Intelligence*, vol. 2, 1988.

[45]   J. C. Miles and C. J. Moore, "Expert system for the conceptual design of Bridges," in *Artificial Intelligence Techniques and Applications for Civil and Structural Engineering*, B. H. V. Topping, Ed.: Civil Comp Press, 1989, pp. 171-176.

[46]   J. C. Miles and C. J. Moore, "Conceptual Design: Pushing back the boundaries using Knowledge Based Systems," in *AI and Structural Engineering*, B. H. V. Topping, Ed.: Civil Comp Press, 1991, pp. 73-78.

[47]   M. S. Lehane and C. J. Moore, "Applying Case-based reasoning in Bridge Design," in *Information Processing in Civil and Structural Engineering*, B. Kumar, Ed.: Inverleith Spottiswoode, 1996, pp. 1-5.

[48]   B. T. Philbey, C. Miles, and J. C. Miles, "Empirical Evaluation of Engineers Requirements for the User Interface of a Conceptual Bridge Design Expert System," in *Knowledge Based Systems for Civil and Structural Engineering*, B. H. V. Topping, Ed.: Civil Comp Press, 1993, pp. 297-302.

[49]   N. Clayton, J. C. Miles, and C. J. Moore, "A Decision Support system for the conceptual design of River Bridges," in *Information Processing in Civil and Structural Engineering Design*, B. Kumar, Ed.: Inverleith Spottiswoode, 1996, pp. 97-101.

[50]   W. M. K. Roddis, H. Melhem et al., "BFX: Operational expert system for bridge fabrication," *Transp Res Record*, pp. 62-68, 1995.

[51]   H. Melhem, W. M. K. Roddis et al., "Knowledge acquisition and engineering for steel bridge fabrication," *J Computing in Civil Engineering*, vol. 10, pp. 248-256, 1996.

[52]   W. M. K. Roddis and J. Bocox, "Case-based approach for steel bridge fabrication errors," *J Computing in Civil Engineering*, vol. 11, pp. 84-91, 1997.

[53]   T. J. McCarthy and Z. Nouas, "An experimental expert system for the design of column base plates," in *Artificial Techniques and Applications for Civil and Structural Engineers*, B. H. V. Topping, Ed.: Civil Comp Press, 1989, pp. 217-220.

[54]   T. J. McCarthy, Z. Nouas, and A. Muhammed, "Knowledge based system for steel fabrication," *Structural Engineering Review*, vol. 3, pp. 255-264, 1991.

[55]   T. J. McCarthy and C. H. Ong, "Simulation of Steel Fabrication using JAVA," in *Innovation in Civil & Construction Engineering*, M. B. Leeming and B. H. V. Topping, Eds.: Civil Comp Press, 1997, pp. 281-286.

[56]   D. K. Ghosh and V. Kalyanaraman, "KBES for Design of Steel Structural Elements," *J Computing in Civil Engineering*, vol. 7, pp. 23-35, 1993.

[57]   G. P. Pasley and W. M. K. Roddis, "Decision Support Environment for Structural Steel," presented at 12th Conf on Analysis and Computation, Chigago, 1996.

[58]   H. Adeli and M. M. Al-Rijleh, "A Knowledge-based expert system for design of roof trusses," *Microcomputers in Civil Engineering*, vol. 2, pp. 179-195, 1987.

[59]   H. Adeli and W. M. Kao, "Object-oriented blackboard models for integrated design of steel structures," *Computers and Structures*, vol. 61, pp. 545-561, 1996.

[60]   Y. Paek and H. Adeli, "STEELEX: a coupled expert system for integrated design of steel structures," *J Engineering Appl Artificial Intell*, vol. 1, pp. 170-180, 1988.

[61]   A. R. Sabouni and O. M. Al-Mourad, "Quantitative knowledge-based approach for preliminary design of tall buildings," *Artificial Intell in Eng*, vol. 11, pp. 143-154, 1997.

[62]   W. Tizani, G. Davies, and A. S. Whitehead, "A Knowledge based expert system to advise on the selection of cost effective steel frames for single storey industrial buildings," in *Developments in Artificial Intelligence for Civil and Structural Engineering*, B. H. V. Topping, Ed.: Civil Comp Press, 1995, pp. 211-217.

[63]   J. F. Horridge and L. J. Morris, "Comparative costs of single storey steel framed structures," *The Structural Engineer*, vol. 64A, pp. 177-181, 1986.

[64]   M. M. M. Zayyat, N. J. Smith, and D. A. Nethercot, "Production related design of heavy steelwork box columns," presented at Proc 4th Int Conf on Civil & Structural Engineering Computing, London, 1989.

[65]   W. Tizani, G. Davies, and T. J. McCarthy, "A construction-led design process for tubular trusses," *Design Studies*, vol. 3, pp. 248-259, 1994.

[66]   W. Tizani, G. Davies, T. J. McCarthy, D. A. Nethercot, and N. J. Smith, "A knowledge-based approach to construction-led structural design," in *Informing Technologies for Construction, Civil Engineering and Transport*, J. A. Powell and R. Day, Eds. Uxbridge, UK: Brunel University and SERC, 1993.

[67]   W. Tizani, D. A. Nethercot, G. Davies, N. J. Smith, and T. J. McCarthy, "Object-oriented fabrication cost model for the economic appraisal of tubular truss design," *Advances in Engineering Software*, vol. 27, pp. 11-20, 1996.

[68]   K. O. Yusuf, T. J. McCarthy, N. J. Smith, and W. Tizani, "A decision support system for construction led design," in *Computing in Civil Engineering*, T. Adams, Ed.: ASCE, 1997, pp. 183-198.

[69]   HMSO, *The Construction (Design and Management) Regulations 1994*. London: Her Majesty's Stationery Office, 1994.

[70]   C. Y. Hung and R. T. Sumichrast, "Multi-expert system for material cutting plan generation," *Expert Sys Appl*, vol. 19, pp. 19-29, 2000.

[71]   F. Sanfilippo, F. Martini, and M. Oriati, "Advanced integrated plant-product management system for rolling mills," *Iron and Steel Engineer*, vol. 73, pp. 17-21, 1996.

[72]   J. Zhong, Z. Wang, Q. Wang, and X. Tu, "Intelligent monitoring and fault detection and diagnosis system in steel rolling mill," presented at IEEE Intnl Conf on Industrial Technology, 1996.

[73]   V. Goel, T. W. Liao, and K. S. Lee, "Shielded metal arc welding expert system," *Computers in Industry*, vol. 21, pp. 121-129, 1993.

[74]   C. King, *Design of Curved Steel*: The Steel Construction Institute, 2001.

[75]   Z. Jin, S. Lou, and X. D. Fang, "KBS-aided design of tube bending processes," *Engineering Applications Artificial Intelligence*, vol. 14, pp. 599-606, 2001.

[76]   S. Kim and S. S. Chen, "Hybrid data model for structural health monitoring," presented at 12th Conf on analysis and Computation, Chigago, 1996.

[77]   D. S. Fox and J. E. Butler, "A knowledge-based system approach to the estimation of structural steelwork mass," in *Artificial Intelligence and Object Oriented Approaches for Structural Engineering*, B. H. V. Topping and M. Papadrakakis, Eds.: Civil Comp Press, 1994, pp. 9-13.

[78]   A. S. Watson and S. H. Chan, "A Prolog based object oriented engineering DBMS," in *Artificial Intelligence Techniques and Applications for Civil and Structural Engineers*, B. H. V. Topping, Ed.: Civil Comp Press, 1989, pp. 37-48.

[79]   A. S. Watson and A. Boyle, "Product Models and Application Protocols," in *Information Technology for Civil and Structural Engineers*, B. H. V. Topping and I. Khan, Eds.: Civil Comp Press, 1993, pp. 121-129.

[80]   XML Spy: http://www.xmlspy.com/, 2001.

[81]   KBSI, "AI0Win," www.kbsi.com, 2002.

[82]   H. Kahn, N. Filer, A. Williams, and N. Whittaker, "A Generic Framework for Transforming EXPRESS Information Models," *Computer Aided Design*, vol. 33, pp. 501-510, 2001.

[83]   S. V. Lord, H. Aleem, and T. J. McCarthy, "Process Modelling of Validation and Product Lifecycles," to be presented at 8th European Conference on Product and Process Modelling, Portoroz, Slovenia, 2002.

Chapter 4

©2002, Saxe-Coburg Publications, Stirling, Scotland
Engineering Computational Technology
B.H.V. Topping and Z. Bittnar, (Editors)
Saxe-Coburg Publications, Stirling, Scotland, 75-104.

Integrating CAE Concepts with CAD Geometry

C.G. Armstrong†, D.J. Monaghan†, M.A. Price†
H. Ou† and J. Lamont‡
† School of Mechanical & Manufacturing Engineering
  Queen's University, Belfast, Northern Ireland
‡ Transcendata Europe plc, Cambridge, England

Abstract

The integration of CAE simulation models into the design process is a key factor to enabling enhanced product development. Preliminary design investigations are routinely carried out on simple conceptual aerospace, automotive and construction models that have little or no linkage to the subsequent detailed designs. However, most of the information required to generate an initial 3D model is implicit in the property attributes of initial beam/shell models. As detailed designs evolve, equivalent properties in the global model need to be updated to account for cumulative and detailed modifications. Due to the different requirements of the various classes of simulation, procedures for abstracting simpler analysis models from detailed geometry are needed. An assessment of the modelling errors arising from any given idealisation is a necessity at all stages of the design process. There has been substantial progress in these areas, but many gaps in functionality remain. An attempt to provide a coherent description of the future requirements for integrating simulation into conceptual design processes is presented.

Keywords: CAD, CAE, feature suppression, dimensional reduction, dimensional addition, idealisation.

1 Introduction

To remain competitive, high-quality/high-value manufacturers seek to improve product excellence, while reducing costs and lead times, by producing 'right first time' products. Reducing cost and time-to-market while increasing product performance is fundamental to increasing market share and competitive business performance. One of the major challenges towards achieving this objective is to integrate the complete process, from the initial concept stage, through design and verification, initial manufacture, testing, and full-scale production.

The rapid evaluation of designs at the early conceptual stage is fundamental to the eventual success of any product but this is difficult, especially for complex products such as are typical in the automotive or aerospace industry. The design process most commonly used today constrains the development of a product by restricting the number of possibilities which can be tried at the concept stage before undertaking detailed design and analysis.

Although the sophistication of design tools for CAE has increased dramatically, the actual procedure and sequence has not changed significantly over the years. Initial concepts are studied at a high level, from which candidate designs are developed further by the addition of detail. Detail is added and modified to a chosen concept according to the results of performance and cost analyses until a satisfactory result is obtained.

In fields such as aerospace, the complexity of design geometry and physics often means that simulation is difficult and time consuming and as a result highly idealised or simplified models are used to assess component performance. There is a serious risk that cumulative detailed modifications eventually invalidate the global concept, since it is difficult to feed back the effects of detailed component simulations into global models of system behaviour. Thus, detailed component modifications may eventually change the load paths through a structure and invalidate the component loadings used in detailed design. Furthermore since the analyses of different phenomena require different models (e.g. structures vs fluids), the simulation models tend to diverge as the design progresses and more details are added. Although improvements in computer power may reduce analysis time by a factor of tens, *appropriate* modelling at each design stage has the potential to reduce it by a factor of tens of thousands.

Integration of design and evaluation at multiple levels of detail for different phenomena requires an attack on the problem from several angles. Firstly to develop techniques at the conceptual design stage so that idealised analysis models are used to drive the detailed design. Secondly, to provide tools in detail design analysis so that the results from detailed models can be used to update the global idealisation, and thirdly, procedures to assess the modelling errors in a given simulation are needed. A generic framework is shown in Figure 1.

The continuing relevance of simple but appropriate models of behaviour has been reinforced recently by Drela [1], who won an AIAA award for his integrated model for aerodynamic, structural and control simulation of a flexible aircraft in extreme situations. The technology incorporated in this model is 50 years old, (the wings, fuselage and tail are modelled with simple beams, the aerodynamics by lifting lines with trailing vorticity), but the final numerical model is small enough to allow interactive computation and the overall approach allows quick generation of a robust preliminary design which can serve as the basis for subsequent detail design. The next logical step in the exploitation of such models is to use them to drive the generation of the detailed designs, and to use detail design results (when they become available) to correct or update the properties of the global model.

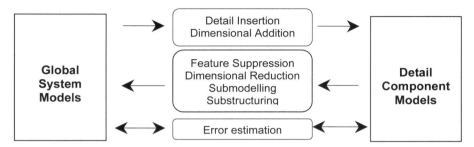

Figure 1: Framework for design evaluation at multiple levels of detail

In order to provide the necessary linkage between these models, tools are needed to quantify and assess errors inherent in the models and to provide the ability to modify idealisations to reduce the errors. In the SAFESA system, Morris and Vignjevic [2] make a distinction between the real world, the structural world and the idealised world. Moving from one world to a more idealised one introduces modelling errors. Morris and Vignjevic also observe that developing confidence in virtual testing by FE requires the creation of some form of objective error control methodology. The major issues are errors and uncertainties – errors need to be controlled whilst providing bounds on uncertainties.

The integration of design and simulation places the following requirements on CAE systems:

- Fundamental geometric modelling tools to support a flexible, extensible modelling environment in which simplified conceptual models can co-exist with design detail
- Support for design by simulation feature, so that for example simple stiffened shell models can be used to optimise equivalent 3D solids containing thin sheets, connections and long slender parts
- Tools for feature recognition, so appropriate simulation models can be automatically identified for parts not designed with simulation in mind
- Tools for simulation model derivation, where appropriate analysis models can be derived from a more complex design geometry
- Methods for the assessment of modelling error, with which the appropriateness and accuracy of a given simulation can be assessed.
- Open frameworks whereby collaborative, distributed design can be accomplished with heterogeneous, best-in-class modelling and simulation systems

Each of these requirements are discussed in turn in the following sections.

2   Fundamental Geometric Modelling Requirements

At the highest level, the interaction between analysis and geometric modelling has the following requirements:
- To represent the underlying geometry upon which an analysis mesh should be constructed;
- To identify regions of the model for the specification of analysis attributes;
- Mesh generation to produce the analysis model;
- Geometric modifications to improve the design based on analysis results.

Therefore, the following modelling operations, which facilitate suitable design procedures, are identified:

2.1   Multi-dimensional modelling

Traditional geometric modelling systems are concerned with objects of a single dimensionality embedded in two- or three-dimensional space: for instance, drafting systems (one-dimensional arcs in two-dimensional space) and three-dimensional design systems (two-dimensional surfaces or three-dimensional solids in three-dimensional space). Such systems do not meet the needs of applications that require the ability to model geometric entities of different dimensions simultaneously.

In simulation-driven design, surfaces might represent boundaries of solids or, as a sheet body together with a thickness attribute, solid shells. Thin-sheet representations are widely used in the shipbuilding, automotive and aerospace industries and in these applications there is no requirement that a given edge should be shared by only two faces, as there is in legacy solid modelling systems.

Curves can be used to represent machine-tool cutter paths, the edges of sheet and solid objects or, together with a cross-section, sheet or solid objects. Trusses and frameworks can be regarded as 1D wire bodies with appropriate connectivity and cross-sectional attributes, upon which elements are generated.

2.2   Cellular modelling

A primary requirement in the specification of an analysis is to be able to partition or subdivide the geometric model. Point, edge, face, volume partitions are needed for the assignment of simulation attributes such as physical and mechanical properties, initial conditions, boundary conditions, element types and allowable errors. They are always associated with a region in space (a cell or set of cells of the object), which can be separately identified.

These subregions of the object will not generally correspond with the conventional boundary representation, B-Rep, decomposition of the object, so a more general scheme for creating additional cellular structure is required [3, 4]. Solids need to include internal membranes to represent regions of different material or even to represent solid parts and the surrounding space for a CFD analysis. The representation of a reinforced concrete structure might require a solid with internal lines representing steel reinforcing bars. Surfaces need to include internal points and linear scratches to represent point and line loads respectively.

Cellular modelling is a key enabling technology for the next generation CAE systems, and the basic infrastructure is already available in geometric modelling kernels such as ACIS [5] and Parasolid [6].

2.3 Partitioning and De-Partitioning

Normally, a geometric modeller will subdivide or partition boundaries using geometric continuity as the segmentation criterion – edges are created between faces at discontinuities of surface normal or curvature. In conventional B-Rep modelling this cellular structure of the boundary is recognized by the modeller, and operations such as Boolean set-combination, sweeping or tweaking modify the structure appropriately. However it is often necessary for applications to be able to modify the cell structure, using explicit partitioning and de-partitioning operations.

A convenient way of articulating such splitting is to use a second object as a 'pastry cutter', which imprints its own structure on to the first object. The resulting object has all cell boundaries from both operands in their region of overlap. Thus a circular region of a face to which a load is to be applied could be identified by partitioning it with a solid cylinder, or the individual plies in a composite laminate could be created by partitioning with an offset surface.

The complementary operation is de-partitioning, where cells are removed from an object by merging adjacent cells. If for example two adjacent patches define a single logical surface, then de-partitioning by removing the common edge unites the point sets of the patches. Note that no change to the geometric shape of the object is implied; the only change is that the subdivision of its boundary does not correspond exactly with that used in most conventional B-rep modellers. For example, de-partitioning all the surfaces of a cube will result in the topology of a sphere, but the underlying geometry will still be a cube.

This same functionality can be used for the suppression of small features by merging point sets. If the point set or the cell defining the hole in Figure 2 is merged with the block by de-partitioning with the faces and edges in common with the block, then the hole is effectively removed. For feature suppression, there is an implicit assumption that the target element size is sufficiently large that the lack of explicit geometry for the hole in the top surface is not an issue.

Modifying the subdivision of the object boundary without modifying the underlying shape has been called logical modelling or virtual topology [7, 8]. Combinations of partitioning and de-partitioning can be used to generate typical idealisations of geometry used for analysis. For example, a "collapse" function has been found useful in practice. This replaces a small surface with one equivalent node or a slender face with a line of element edges during the mesh generation process. These can also be regarded as dimensional reduction operations – a 2D face is meshed with a 0D point or a 1D edge mesh, Figure 2(b).

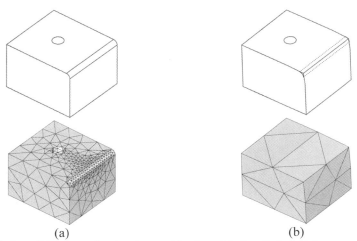

| (a) | (b) |

Figure 2: Detail Suppression by de-partitioning; (a) detailed model, target element size < feature size (b) simplified model, target element size >> feature size.

The choice of the location for the nodes on this reduced-dimensional mesh is somewhat arbitrary, since the suppression of the collapsed feature implies that it is below the scale of interest anyway – Bridgett [9] found that a simple averaging of points projected onto the collapsed feature surface and its boundary edges was adequate. In geometric modelling terms it can be represented as a partitioning of the surface to be suppressed along the line of 1D element edges, followed by de-partitioning or removal of the original edge marking the common boundary with the adjacent surfaces, Figure 2(b).

The concept of multi-dimensional cellular modelling was clarified in the Djinn project [3]. A cellular solid model can contain internal faces to partition the solid into different volumes, which could represent different materials or even regions that are to be meshed and analysed in different ways. Operations to partition and de-partition the model (i.e. remove the internal membranes) were also defined. Different analysis models can be represented by different subdivisions of the same model with different analysis attributes, using ideas similar to those illustrated in Figure 2.

2.4   Facetted models

Some objects can only be described by facetted representations. This may be because it is the only form available, e.g. as an STL file, but in many other situations it is the only form capable of capturing the complexity of the design data. For objects produced by manufacturing processes, such as thermoforming of plastics or forging of metals, the final shape and properties like sheet thickness or material state may only be available on a mesh, following a simulation of the manufacturing process.

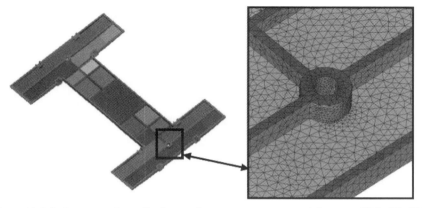

Figure 3: Mesh generation of a facetted geometry recovered from an STL file [10].

Figure 3 shows a watertight mesh representation of an imported STL file, in which individual surface facets are joined based on a join tolerance (60° in this example), resulting in a total of 146 surface meshes [10]. The underlying meshes have been subsequently decimated and refined to produce the simulation mesh of Figure 3, which has a radically different distribution of nodes and elements from the original STL mesh. A specified join angle is used to preserve sharp bends resolved in the surface meshes, i.e. to create edges in the analysis model topology. Any adjacent patches with similar surface normals will be effectively merged by this procedure.

For facetted models arising from manufacturing process simulations, there is an additional requirement that properties and results from the forming process simulation must be mapped onto downstream meshes for performance simulations and these may have a significantly different density. It has been repeatedly demonstrated in a number of disciplines that the material process history has a significant effect on the subsequent performance of a structure.

The regeneration of B-Rep models with analytical geometry from facetted representations, known as "beautification", is also a topic of current research and significant progress has been made [11].

2.5   Hierarchical models

Aerostructures comprised of composite laminates have many possible representations of interest. At the coarsest level, it may be considered as an anisotropic shell with equivalent properties. At a more detailed level, the distribution of sandwich core, skin plies and ply orientations may vary over the surface to provide a shell model with spatially-varying properties. Near free edges and local details such as hinges or actuator mountings, local 3D modelling of each ply may be required to capture the complexity of the design. On a micromechanics level, detailed modelling of fibres embedded in the matrix material may be necessary. Schemes for material modelling on multiple scales are now well established [12] and there is a need to extend this approach from the material modelling to the structural level, i.e. to provide mechanisms for representing structure at different levels of detail in a coherent manner, to allow global analysis models to provide boundary conditions to local details and to provide tools for updating the properties of simple schematic models using the results from detailed component designs.

STEP AP209 provides a conceptual framework within which an "analysis product structure" can associate a family of analyses with a given part in a Product Data Management system [13], but the derivation and control of the appropriate geometric idealisations is still very much a research issue.

3   Design by Simulation Feature

3.1   Feature-based modelling

Most current CAE systems utilise a feature-based approach to produce parametric models. Design features such as blends, holes or sweeps are constructed with parameters such as blend edge and radius, hole size and location, sweep profile, direction and length which are introduced to produce the desired shape. Design variants are generated by modifying parameter values subject to appropriate constraints, though the process may fail if an invalid shape is produced. Feature suppression allows variant models to be produced in which the design feature and its children have been omitted. Feature-based modelling greatly eases the integration with downstream applications such as CAM, since a standard treatment of a given feature is usually possible.

Feature suppression facilitates the temporary removal of features which are not relevant to a simulation if they are below the scale of interest. Thus algorithms for traversing a feature tree to find and suppress holes of less than a certain size are relatively straightforward to implement. Modification of feature parameters offers a simple method of allowing simulation results to optimise the CAD geometry.

There are a number of limitations to this approach. Firstly, the features defining the model construction may not coincide with those of interest: a hole may have been produced by subtracting a cylinder rather than as an explicit hole feature; a thin sheet may result from a deep pocket in a remote face. Current feature modelling systems have computational cost problems, suffer from dimensioning and modelling limitations due to their strong dependency on the chronological order of feature creation and to the use of unidirectional constraints, occasionally suffer from ill-defined semantics of modelling operations, and do not adequately maintain the semantics of features [14]. Whilst it will be shown later in this section that there are opportunities for modelling by 'simulation feature', there is a general need for algorithms to find geometric features of relevance to simulations.

A further problem with integrating feature modelling is that the differences amongst commercial systems in the description of feature information are even more pronounced than differences in model description, which makes the transfer of models between competing systems extremely difficult. One interesting current approach is through an abstract feature interface [15, 16], which allows an application to determine the features of a model, the geometric entities associated with these features, the feature suppression state and the feature parameters. No knowledge of the significance of a feature or its parameters is assumed, but new parameter values can be set and an update to the CAD model requested. Thus, feature parameter values are available as design variables in optimisation studies. These CAD queries and updates can even be conducted remotely on a computer which does not have the CAD system installed locally. Figure 4 illustrates a Unigraphics CAD model with sheet thickness as a design parameter which can be modified remotely from a simple HTML form. A limitation of the current approach is that the model and all its features must exist – there is currently no capability for creating new features. This would require a modeller-independent definition of feature type.

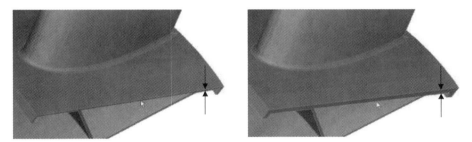

Figure 4: Remote update to CAD system parameter (sheet thickness) using CADscript [15], an implementation of OMG CAD Services [16].

A more radical approach, conducted under the auspices of the ISO Parametrics Group, is to transfer CAD models in procedural form, i.e. expressed in terms of the sequence of operations used to construct them [17]. Whilst this is an attractive option which may help capture design intent, there are formidable difficulties in developing a common representation which allows a reasonable coverage of the capabilities of commercial systems.

3.2   Detail suppression

The basis for the recognition of simulation features in elasticity problems is the Principle of St Venant, which implies that the details of how loads or restraints are applied only affects the stresses in the immediate locality. Equally, the presence or absence of a feature such as a notch or protrusion which is small compared to the amount of material it is embedded in will have a negligible effect on the solution a few feature sizes away. This implies that features such as blends, chamfers, holes, slots and protrusions can be suppressed in a coarse model of global behaviour, provided that they are relatively "small". Note that a small absolute feature size does not guarantee that the feature does not have a global effect.

If the mesh size is refined, either manually or automatically, then it is essential to re-instate the feature once the element size approaches the feature size, Figure 2. Mesh refinement in the vicinity of a sharp concave corner resulting from suppression of a blend will result in stresses which increase without limit with mesh refinement, as the elements try to capture the stress singularity implied by the sharp corner.

3.3   Dimensional addition

In many aerospace, automotive, shipbuilding and vehicle applications, preliminary design will be conducted on a stiffened shell model, where the thin solid sheet is represented as a surface with a thickness attribute and the stiffeners are represented as edges of the shell faces with cross-section attributes. This allows a model to be constructed with less detailed information than the equivalent solid, convenient exploration of design concepts and efficient optimisation with respect to shell thickness or stiffener cross-sectional shape.

The reduced dimensional beam and shell elements implicitly define the missing dimensions through attributes. The equivalent 3D solids, Figure 5, can be created by sweeping cross-sectional profiles along a 1D beam axis, or by sheet thickening, which can be regarded as sweeping a 1D generator over the surface of the shell. If the reduced dimensional elements were generated from a smooth wire or sheet body, then a body with smooth surfaces is obtained, rather than the facetted surface model, which would be generated if elements were swept individually. If the wire and sheet bodies and the subsequent dimensional addition were created in a feature-based modelling system, then feature suppression can be used to recover the reduced dimensional model directly. This does require that the geometric modelling system provides a coherent and consistent treatment of wire, sheet and solid bodies with respect to modelling operations such as Booleans and sweeps, as advocated by Djinn [3]. In most commercial systems, this functionality is currently limited.

3.4   Joint design

In a reduced-dimensional analysis, there is an implicit assumption that stresses which vary over lengths of the order of the beam cross-section or shell thickness are not required. Thus the design of connections, restraints or points of load application is a local problem for detail design – the most that can be obtained from the simplified model are beam forces and moments or shell stress resultants which can be compared with allowables derived from similar successful designs.

To facilitate the detailed design of joint or connections and appropriate updates to the higher level global behaviour two key elements are required: geometric tools to facilitate the creation of local geometric cells to represent the joint and simulation techniques for mixed dimensional modelling, where local 3D models can be embedded into global beam or shell models.

Figure 5: 1D model and equivalent 3D solid model formed by sweeping beam cross-section.

Figure 6: Solid model with extended sweeps.

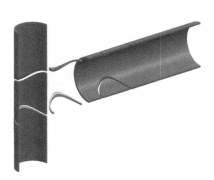

Figure 7: Solid cells produced by cellular union of extended members.

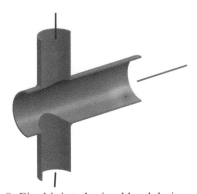

Figure 8: Final joint obtained by deleting unwanted cells and adding blends.

The creation of geometric cells in the proximity of joints in wire or sheet bodies is relatively straightforward. A 1D element can easily be partitioned at a distance of approximately one beam cross-section away from the joint. Sweeping the same cross-section along the 1D segments produces two separate solid cells, one in the vicinity of the joint, Figure 5, and one representing the rest of the beam remote from the joint. The one surrounding the joint can then be used to contain the detailed joint design, Figure 8. The equivalent subdivision of a shell model can be produced by a curve in the surface offset by a multiple of the sheet thickness, Figure 9.

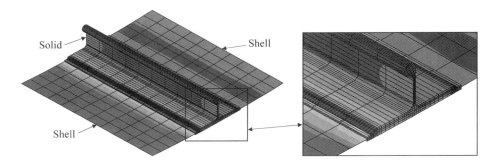

Figure 9: Von Mises stress on a solid-shell model of an aircraft stiffened panel. The solid-shell interfaces were defined as offsets from the line stiffener used in preliminary design.

3.5   Mixed Dimensional Coupling

The second necessary element to facilitate joint design is a capability for mixed dimensional modelling. Near beam ends or plate edges, boundary layers which are the size of the smallest dimension occur. In these areas the full problem must be analysed - in a modern context by a 3D FE analysis - and the reduced dimensional analysis is invalid. However, attempting to discretise the entire domain with 3D solid elements would lead to large, prohibitively expensive analyses.

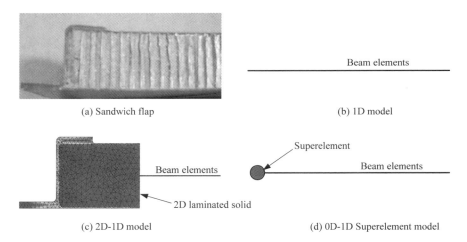

Figure 10: Detailed design of composite structure, followed by abstraction of detail in global model.

The use of beam elements implies that the distribution of stress arising from axial force, bending moment, torsional moment and shear forces over the beam cross-section is given by the conventional engineering theories. In a plate or shell element, equivalent assumptions are made about the distribution of stress through the thickness arising from in-plane and out-of-plane forces, bending and twisting moments.

In the procedure suggested by McCune [18], coupling between models of different dimension was achieved by assuming that the work done on the boundary of the reduced-dimensional model may be equated with the work done by the stresses on the 3D side of the same interface. Since the distribution of stress on the 3D side of the interface can be written in term of the forces and moments on the reduced dimensional side, the forces and moments can be eliminated, leaving multipoint constraint equations coupling the reduced dimensional displacements and rotations to the 3D displacements.

In Figure 10, this implies that a local 2D model of the edge of the composite construction of Figure 10(a) can be coupled to an adjacent sandwich beam element, since away from the edge an analytical solution is known for the distribution of bending and shear stress on the interface in terms of the beam forces and moments, Figure 10(c).

If the analysis model remains in the linear elastic range, the local 2D continuum can be substructured into an equivalent stiffness superelement, Figure 10(d). This approach has been utilised for coupling beams to 3D solids and shells [19], and coupling 3D shells to solids [20], though alternative approaches based on special transition elements are possible. The main benefits are that local, detailed models can be driven by global loading and the local models affect global behaviour.

3.6 Examples of design by simulation feature

3.6.1 Framework

A very simple illustration of this approach can be seen in a detail from a tubular framework illustrated in Figure 5 through Figure 8. For preliminary design a simple 1D framework model, Figure 5, is sufficient to determine candidate layouts and tubular section properties. Given the layout and section definitions, it is relatively straightforward to create a first iteration of the 3D model by appropriate sweeps of the cross-sections, Figure 5. During detail design of the joints in the framework, different layouts are possible. In the design of joints a facility to extrapolate the structures to be joined such that they completely overlap has been found useful, Figure 6. Figure 7 illustrates the series of solid cells formed by the Boolean cellular union of the members to be joined. (In fact the model shown here was created in a commercial CAD system as an assembly using a series of Boolean operations.) Many possible detail designs can be obtained by retaining different combinations of the possible cells, after which all the internal membranes can be removed to create a conventional manifold solid. Appropriate local operations such as blending to define design details could then be specified, Figure 8. Obviously there are opportunities to develop standard recipes for joint design.

A mixed dimensional model is then possible, incorporating a local 3D model of the connection into the 1D frame structure. For elastic analyses, this 3D model can be substructured at the nodes connected to the adjacent beam members to form a three-noded equivalent joint element. This means that the effect of the details of joint design can be incorporated into the global model of system behaviour. Once the local 3D joint model is analysed, joint stiffness data is fed back to the global model so that the global effect of any changes in joint design can be assessed.

In practice the model of the complete framework would be a cellular solid partitioned into different cells, representing long slender parts and chunky tube connections. The local joint models can be isolated by partitioning the 1D model a few cross-section sizes away from the joint. Depending on the analysis requirements, the tubular sections could be represented as 1D or 3D; the node sections could be represented as 3D, 0D nodes connecting the original beam elements, or 3-noded joint stiffness substructures [19].

Thus the initial 1D framework evolved into a detailed 3D design. The generation of new geometry (as details are added) is flagged as a change in the cellular structure of the master model and the simulation model is updated accordingly.

3.6.2   Gas turbine casing

Figure 11 shows a 1D skeleton with 1D cross-sections for a 30^0 segment of a gas turbine casing. Sweeping these 1D cross-sections produces a sheet body suitable for shell analysis, Figure 12. Local design of the connection between the swept bodies is again facilitated by extrapolation, cellular union and deletion of unwanted cells. In this case the outer segment of the aerofoil section strut is removed (though it can sometimes be used as a mounting for ancillary devices) and the section of the outer ring within the aerofoil section is removed to allow access to the engine core, Figure 14. There is a clear analogy with the design of the framework node above.

Figure 11: 1D conceptual layout of turbine casing segment.

Figure 12: Extended shell model before trimming.

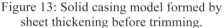

Figure 13: Solid casing model formed by sheet thickening before trimming.

Figure 14: Solid casing model after trimming.

A solid model can be produced either by sweeping a 1D line along the 1D cross-section to produce a 2D cross-section, or by thickening the shells, Figure 13 and Figure 14. At the trailing edge of the aerofoil, Figure 15(a), the design intent is not explicit in the shell model. Whether the sheets should be blended, have a sharp edge, a plane end or some more complex fabricated construction is a detail which is not usually considered in concept evaluation. Again the paradigm of extrapolation, dimensional addition, cellular union and deletion of unwanted cells is useful to provide a starting point for design detailing. The evaluation of the relative merits of different joint designs can be facilitated.

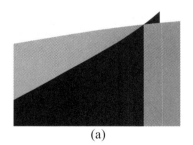
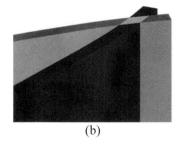

(a)   (b)

Figure 15: Local design of connection at aerofoil trailing edge.

3.6.3   Composite flap

The simplest possible representation is a single sandwich beam mesh, Figure 10. As ply details and edge details are specified, the contents of a cell representing the local substructure in the vicinity of the flap edge will be designed, Figure 10(b). Using mixed dimensional coupling to link the local 2D model to the beam model, the edge detail can be abstracted as an equivalent edge stiffener, Figure 10(c). The global effect of local changes in ply organisation can be economically assessed, as can the local stresses arising from global loading. If detailed non-linear analysis of local effects such as crack growth in the edge detail are of interest, the rest of the global model can be sub-structured at the mixed dimensional interface to provide appropriate drivers for the local model.

3.6.4   Major aircraft structures

Parametric modelling of aircraft wing structures has been described by Frey [21], who mention the use of "lines that represent the main datums of big bone structures" that are reminiscent of the 1D axes described above. Figure 16 shows the overlap of wing and fuselage solid models generated automatically from simple reduced dimensional models. Whilst parametric rules (perhaps using loading data from simple models) are probably sufficient for the layout of both wing and fuselage at points remote from their connection, detailed design of the area in the vicinity of the intersection will be necessary. Dimensional addition, the logical separation of the model into cells in the vicinity of and remote from the connection, and mixed dimensional coupling are all useful tools in assisting the evolution from simple concept to detailed geometry.

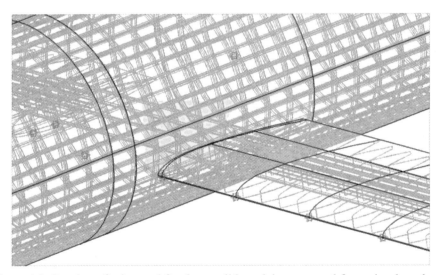

Figure 16: Overlap of wing and fuselage solid models generated from simple reduced dimensional models.

3.7   Limitations

Design by structural feature has many attractions. The full power of the CAD modelling functionality is available for the rapid creation and optimisation of candidate designs through variation of feature parameters. Feature suppression can be employed to produce models with different levels of engineering detail, from the coarse global behaviour to the detailed local evaluation. The key weakness of this approach is that different simulations will require suppression of different features, some of which will not be explicit in the feature tree. There is a need for automatic identification and treatment of generic features of interest to simulations.

4 Simulation Feature Identification

4.1 Introduction

Simulation features are typically either regions of geometric complexity that are small relative to the environment in which they exist, or regions which have one or more dimension which is small compared to the others which can be represented with reduced-dimensional elements. Mobley [22] used an analysis of edge meshes to identify proximity between adjacent edges, facilitating the detection of features such as near tangencies and face constrictions, as well as geometrical quality problems such as coincident edge precision discrepancies and poor intersection curve accuracy.

Whilst it is feasible to describe similar algorithms to identify adjacent mesh faces with no nearby edges i.e. thin sheets, the proximity information has no structure to facilitate reasoning about the model. There is however an alternative representation of geometry that does provide this structure, the medial axis transform.

4.2 Medial Axis Transform

The Medial Axis of a planar domain is the locus of the centre of all circles of maximal diameter that are contained within the object, Figure 17(a). The combination of the medial axis and the function describing the radius of the inscribed circle at every point on the medial axis is know as the Medial Axis Transform (MAT). It was proposed by Blum [23] for the characterisation of biological shape, but has since attracted interest in a number of areas including mesh generation [24, 25].

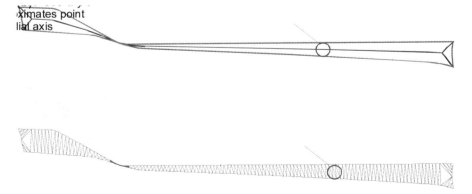

Figure 17: 2D medial axis and its approximation by a Delaunay triangulation.

Whilst the computation of exact medial geometry and topology is a challenging task, a useful approximation to the MAT can be obtained from a Delaunay triangulation of points distributed over the object boundary, Figure 17(b). The logic is that since the circumcircle of a Delaunay triangle contains no other point, then if the point distribution on the boundary is reasonably dense the triangle circumcircles approach inscribed circles in the domain.

This alternative representation provides a lower-dimensional skeleton of the object. The touching points of the inscribed disc centred on the Medial Axis identify parts of the object boundary which are in geometric proximity even if they are not topologically adjacent, Figure 17 (a). It facilitates the identification of relative sizes and aspect ratios for identifying features of structural significance such as long slender parts, thin sheets and small features embedded in a large amount of material.

The Delaunay triangulation approximation is also possible on 3D parametric surfaces where a metric based on the parametric derivatives [26] can be used to correct for length over the surfaces. Figure 18 shows a surface region where constrictions have been identified based on *relative* size – the area identified has an inscribed disc which is small compared to the largest inscribed disc in the surface. Whilst other approaches to finding proximity based on absolute distance are possible, the MAT provides a structure to this proximity information that is difficult to reproduce by alternative means.

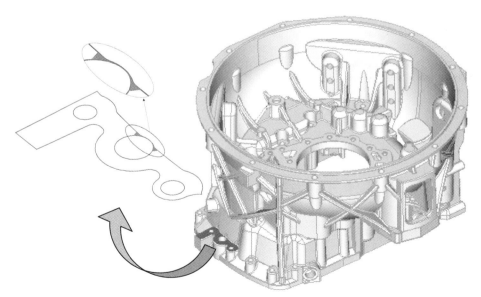

Figure 18: Shading on surface indicates a constriction exists, where the inscribed disc is small compared to the largest inscribed disc on the surface.

In 3D the robust computation of MAT topology is still an open problem [27, 28, 29]. The geometry of the MAT is controlled by the touching points of the inscribed disc. For curved boundaries a numerical solution may be required to find exact position of the inscribed sphere, but once the sphere touching points are found queries about the differential geometry of the medial surface such as edge tangent and curvature, surface normal and curvature can be evaluated directly from the differential geometry at the touching points [30]. This means that MAT geometry can be meshed using current technology [26] for curvature-sensitive meshing. Figure 19 shows a mid-surface mesh between two general spline surfaces, adaptively refined to capture the curvature induced by the mutually perpendicular waves in both bounding surfaces.

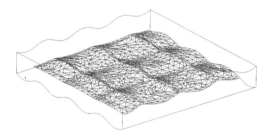

Figure 19: A medial surface mesh between a pair of general spline surfaces.

Figure 20 shows a medial surface mesh in a more complex solid – at every point the MAT identifies those parts of the object boundary which are in closest proximity and the distance between the boundary. Computation of derived information such the volume associated with each medial surface feature, minimum and maximum radii etc is straightforward. There are many opportunities for feature recognition and intelligent reasoning about solid shape with this technology.

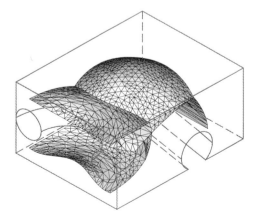

Figure 20: Medial surface of a more complex solid.

An unexpected side-effect of previous work on techniques to find small features has been to highlight "realism" errors, where the creation of the small features was unintentional and arose from the modelling technique used to create the part. While few of the errors discovered were detrimental, correcting them increased the quality of the CAD models, the manufacturing processes and the finished parts [31].

5   Simulation Model Derivation

Previous sections have highlighted the fundamental geometric modelling requirements to enable analysis driven design. However as the design evolves the CAD model necessarily becomes more complex and new simulations are required to update the design and performance data. The CAD data therefore has to be prepared appropriately to make it useful for downstream systems, particularly for analysis systems. This principally involves the generation of a mesh appropriate for the simulation concerned.

Models received into the CAD system may be modified in a number of generic ways. A set of transforming functions may be used to make the model more suitable for use by the down-stream system. These functions include the ability to collapse small features such as short edges and sliver faces, to join carpets of faces into simpler topologies for meshing, and remove small holes from the topology. Finally, a number of flavouring functions may be used to prepare the model for a particular downstream application. These functions include NURBS order reduction, selection of model or parameter space curves to be exported, and the preferred treatment of assembly components.

From a geometric modelling point of view low level (atomic) functions are required to carry out the necessary geometric modifications but from an analysis point of view these functions must further be orchestrated in an intelligent way to ensure suitable analysis models are generated.

The conditioning of geometric models for meshing will usually require the recognition of troublesome features and the substitution of these features with new topology or geometry which idealises the original and is more tractable for the meshing algorithms to be used. Such an idealisation could be the representation of a narrow step face with an edge to carry the element edge nodes, or the joining of several faces into a larger meshing patch of better size and shape for the size of the elements to be used in that region. Dimensional reduction to edges or faces may be used where beam or shell elements represent certain features of the model.

One aspect of the modelling interface is a set of find functions, which identify certain types of features in a CAD model. However once particular features have been found, they can either be replaced or have an attribute associated which indicates what should be done when meshing for analysis.

5.1   Replacement

The first approach is one of replacement. In the model which has been idealised for meshing, certain features are replaced by other features which are more appropriate. Thus the narrow step face mentioned above would be removed from the topology of the body it bounds, and replaced in that topology by the single new edge which represents it. This is not a simple task, as the short lines at the ends of the face may have to generated, and then collapsed to single vertices, and neighbouring faces have to be adjusted to use the new edges and vertices. More difficult still, the embedding geometry of these neighbouring faces has to be modified to extend to the new edges, and in some way to blend the geometry of the original faces with the step. Figure 21 below illustrate these procedures.

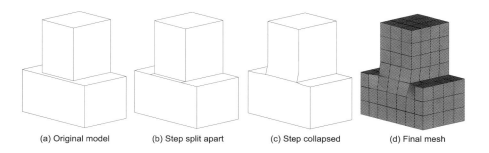

(a) Original model   (b) Step split apart   (c) Step collapsed   (d) Final mesh

Figure 21: Simplifying a model for analysis.

If the function of a software module is to undertake geometry cleanup and provide traditional geometries for downstream systems to perform the meshing, then this replacement process, however difficult, must be addressed. It is clear that significant model transformations will be made, and that an exported model idealised for one style or density of meshing may be unsuitable for a different meshing requirement.

Also, in such an environment it is desirable for the modelling software to retain information about the original geometry and topology, linked to the idealised features in such a way that the user can return at a later date and recover the original from the idealised model.

Replacement is of more importance if the downstream system requires hard geometry as input, in which case the preparation of a coherent solid is a key requirement. However in many systems this is not necessarily the case and the engineering model can be captured by the appropriate commands to regenerate the features, say through a log file. In such cases the use of modelling attributes may be more appropriate.

5.2   Attributes

If the same software module is to perform the meshing as well as the geometric idealisation, then an easier, more flexible approach may be considered. In this approach, the original geometry and topology is retained throughout, and the idealised geometry is held as meshing attribute properties of the original. This approach has been explored by Bridgett [32].

Bridgett describes a "logical model" for meshing where the new topologies of the idealised features for meshing are linked as attributes to the original topology, and the original embedding geometry is left unchanged. The idealised topology is used to control the topology of the mesh to be generated, while node coordinates are calculated by projection on to the original geometry. These concepts have evolved into the virtual topology ideas mentioned above [8].

This approach is very flexible, as these idealised attributes can be easily edited without sophisticated solid modelling operations. Also, as nodes are constrained to lie on the original geometry, the mesh model will reflect the original with optimum accuracy, and the level of detail picked up from the original geometry will be controlled automatically by the size of the elements generated. Furthermore, if the idealised topology is labelled as "soft" then the effectiveness of element smoothing can be enhanced by permitting nodes to drift off the idealised topology while remaining in the original geometry.

The principle weakness in this approach is in the node projection part of the algorithm, where powerful nearest point routines are required.

5.3   Atomic Functions

As described above, the modelling functions currently required from CAD systems include a range of atomic functions for building and merging topology, joining edges and faces, simplifying and cleaning complex geometries, particularly NURBS, and for conditioning models by finding and removing small features. The build functions are widely available but the functions for suppressing or removing feature not required for analysis are still only available in an ad-hoc fashion in a minority of commercial systems. The main functions required deal with fillets, holes, narrow regions and sub-partitions, each of which is discussed in turn in the following sections. These functions can be controlled by parameters specified by the user or automatically by the analysis package.

5.3.1   Fillet suppression

Many models contain fillets/blends/rounds where original sharp edges and vertices have been replaced by small faces which may have little structural significance, but make the topology of the part much more complex. This introduces needless difficulties for automatic mesh generators and prolongs analysis times.

The fixing functionality should replace all of the fillet faces in the given input set, and replace them with edges and vertices which represent their idealised positions. These idealised edges should lie on the fillet faces midway between the long edges, with idealised vertices at the centre of "suitcase corner" faces. This clearly suits the techniques of Attribute Suppression described above.

If replacement suppression is required, this could be achieved be generating edges and vertices as above on the fillet mid-faces, or alternatively, by removing the fillets completely and replacing them with the original sharp corners before the rounding. The latter is not recommended because it is both complex geometrically (requiring adjacent faces to be re-intersected with special case details to be considered), and dangerous, as it introduces stress raisers into the model by sharpening up internal angles.

5.3.2   Hole suppression

Small holes such as tapped bolt holes often cause considerable topological complexity without affecting the structural performance of the part significantly. A hole is considered to be an internal loop in a face, or a set of faces which connect such loops in a solid body. When a hole is suppressed its geometry is idealised to vertices and an edge defining the original axis of the hole. For Replacement Suppression these new vertices and centre-line edge will be "scratched" into the new definitions of the topology containing the hole. For Attribute Suppression they will be linked as attributes to this topology.

5.3.3   Sub-partitioning

In many analysis situations it is desirable to sub-partition a model so that detailed analysis or re-modelling can be concentrated on a selected feature. In such cases a boundary is often established at a distance beyond which the global effects of the changes to the detailed model are negligible. Several approaches exist for the transfer of engineering data across such boundaries.

In structural problems the stiffness properties of all the material beyond the boundary may be condensed (or reduced) into a matrix related to the freedoms of the nodes on the boundary. This part of the model is treated as a substructure, or "super-element", whose stiffness properties may be added efficiently to the stiffness of the elements in the detailed part in order to represent the boundary conditions on that part. In more sophisticated applications, super-elements may be reduced, assembled, and transformed many times to form a super-element assembly tree, in a manner very similar to a CAD assembly of components.

In other types of problem it may be convenient to use a very different mesh for the detailed part from that in the region beyond the partition boundary. The mesh near the detail may be a fine mesh of say, higher order hexahedral elements (for accurate stress recovery), while the mesh in the remainder of the model may be a coarse mesh of say, linear tetrahedral elements (which is adequate for displacements). Across the partition boundary where nodes do not coincide, constraint equations will have to be set up to ensure structural continuity between the detail and the rest of the model.

Often, for CAE modelling efficiency, there will be a dimension change in the analytical elements across the partition boundary. Examples include:

- Use of 2D elements to establish the overall pattern of air and pollution flow across a city, together with a 3D detail of the air in the vicinity of a house façade where the pollution is being deposited over time.
- Use of 1D beam elements and 2D membrane elements to represent the over-all structural load-path behaviour of an airframe, together with a detailed 3D stress analysis of an important stringer-frame junction.

At such multi-dimensional junctions it is important to provide the maximum amount of analysis field continuity. This can be partially achieved by introducing geometrical artefacts such as rigid links and spoked "cartwheels" to transfer loads from say, a cylindrical pipe modelled with shells, to the axial beam element which represents it. Often these artefacts are inadequate in their ability to transfer the solution state between the parts of different dimensionality, and this can cause undesirable anomalies near the boundary. The work in Section 3.4 provides a useful starting point to modelling in mixed dimensions - the coupling equations can be automatically specified from the meshes on the interface between dimensions.

5.3.4   Dimensional Reduction

Thin sheet objects can be efficiently analyses using shell elements. This approach gives the analyst the advantages of faster simulation time, less hardware resource requirements and consequently a faster design cycle. Dimensional addition to create solid models from shells was described in Section 3, but the inverse problem - dimensional reduction of thin solid sheets to the equivalent surface is often required. The medial axis described above provides an accurate representation of the mid-surface of a thin sheet.

From the pair of faces defining a given medial face, an accurate mid-surface mesh can be generated between them. All that is required is the differential geometry of the defining surfaces at the touching points of the inscribed sphere associated with that point on the medial axis [30]. The geometry of this mid-surface can be adaptively refined to within a user specified surface deviation limit, Figure 19 and Figure 20.

5.4   Patches for meshing

Complex topologies on parts can cause problems for meshing algorithms, where the complexity of the topology itself makes it unsuitable for mapped meshing strategies, or the small size and shape of the faces cause poorly dimensioned elements. In these cases it is desirable to provide larger, better shaped topological "patches" for meshing, by using selected existing or new topological edges while honouring the underlying geometric surfaces. The topology of the patches thus controls the topology of the mesh to be generated, while the nodes of the mesh are constrained to lie on the original geometry.

This patch generation needs to make use of any features which have been found previously by the atomic functions described above. In particular small gaps must have been healed by previous merging, and small fillets must previously have been idealised if necessary.

6   Analysis Modelling Errors

The assessment of simulation modelling errors is a key issue for delivering simulation results of known accuracy. One traditional technique for discretisation error estimation is the assessment of inter-element stress jumps. Stresses are discontinuous at element boundaries, and the amplitude of their jump at the discontinuity is a good estimator of stress accuracy. By measuring these stress jumps, one can obtain an assessment of the local accuracy of a solution, without reference to a previous analysis pass.

Bridgett [32] investigated the application of error estimates in appraising the validity of submodelling schemes. If the stiffness of the local submodel detail is similar to the stiffness of the equivalent region in the coarse representation, then stress values around the boundary of the submodel should be similar to those in corresponding positions of the simple coarse model. Summing up these stress jumps between coarse model and submodel around the boundary of the submodel provides an estimate of the accuracy of the simplification. Bridgett concluded that if these errors are high, then the detail should be reinstated and a further analysis performed.

As this is an *a posteriori* error estimation of the global-submodel combination, if the error is beyond the acceptable error specified by the user, then the offending details should be reinstated in the global model for reanalysis. This method facilitates an adaptive approach to model simplification.

McCune [18] studied errors produced at transitions between coupled dissimilar element types. The coupling procedure he suggested used constraint equations to impose compatibility of strain at the interface. Stress continuity requires that the stress distributions observed in the 2D continuum elements should correspond with the stresses implied by the 1D beam element forces and moments and so any mismatch between the two will indicate an error in the analysis that can be quantified. The differences between the 2D stresses at the interface and those computed from the 1D beam forces and moments were calculated to give the stress jumps at the interface. McCune showed that it was necessary to model a disturbed region in an effort to determine how close the transition can be to a disturbance in geometry, loading, or boundary conditions before the errors become unacceptably large.

Note that the creation of cells for local 3D design in shell or beam models adjacent to joints could be regarded as a form of *a priori* error estimate – the reduced dimensional models are fundamentally unreliable over lengths typical of the suppressed dimension.

7   Open Frameworks for Collaborative Design

There is a growing body of knowledge associated with the use of computer based methods; solutions are increasingly tending towards integration of standalone, 'best in class' software applications. Organisations are currently limited in their selection of resources because of cost or compatibility with existing systems. Furthermore, because these tools typically come from a variety of sources, they usually do not work well together, and employ different communication protocols.

The Object Management Group proposed an interface standard, known as CAD Services [16]. This standard enables exchange of geometry and topology data, the intent being to establish a series of high-level engineering interfaces that do not require low-level data structures to answer mechanical engineering queries. To avoid many of the problems associated with data translation, the standard provides CORBA interfaces with consistent functionality across native CAD implementations.

The US-based FIPER project [33] used this paradigm to establish a product development environment that can be distributed across multiple platforms, and exploits a variety of grid-based technologies. This internet-distributed framework provides geographically dispersed engineers and business partners with access to up-to-date resources including programs, files, data, computers, and networks, while facilitating integration and automation of existing and emerging design, simulation, and optimisation tools and processes.

Commercial tools are now emerging from the outcomes of the FIPER project. CADscript [15] for example, has the capability to open a CAD model over the internet, allowing the user to interact with it in a way that is almost indistinguishable from the original CAD system. As its implementation includes the OMG CAD Services standards and FIPER functionality, individual components from disparate CAD systems may be assembled, assessed and modified in a single environment. This new product development paradigm enables global communication of information and collaboration of tools, organisations, and business units.

As in CAD systems, CAE systems are tending towards the integration of individual software packages and data storage systems. Information is increasingly being channelled through multiple engineering applications. This facilitates the solution of multidisciplinary problems, such as optimisation or multi-physics analyses, using traditional single discipline simulation tools. The need to both, use a mix of computer systems, and to transport large amounts of data among many differing computer systems can result in a considerable data management problem.

XML, the extensible mark-up language, facilitates data integration from disparate sources and can deal with data from multiple applications and allows for local manipulation of structure. A US consortium has recently launched femML, the finite element modelling mark-up language [34]. Their objective is to address the problems of data interpretation and application interoperability in the Finite Element Modelling domain. While XML is emerging as a suitable format for small to medium sized models, technology related to the storage of large data sets has evolved. Utilities now provide support for the types of data and metadata commonly used by scientists, provide efficient storage of and access to large data sets and usually facilitate platform independent storage.

A current challenge in mainstream CAE is the simulation of multi-physics problems, those analyses that involve an interdependent combination of any two or more solution domains. While the capabilities of single-physics analysis will continue to grow, the progress of coupling between numerous physical phenomena will be even greater. Data translation and interpolation introduces errors not present in single physics problems, and so it is important to ensure the validity of analysis approach and results; benchmarking is essential. Tools such as MpCCI [35], a mesh-based parallel code-coupling interface, are emerging as a viable solution for coupled analysis via existing specialist single-physics codes.

In summary, distributed collaborations and outsourcing are now common in engineering developments, especially those activities involving disparate fields of enquiry. There are currently opportunities to integrate the above mentioned technologies in a way that provides interfaces that exhibit longevity, flexibility and scalability, all of which will be required for large-scale simulations of the future.

8   Conclusions

Multidimensional geometric modelling with a coherent, consistent treatment of line, surface and solid models is an essential requirement for simulation modelling. The ability to partition or subdivide the model into separate cells for the specification of analysis attributes would greatly facilitate linkage with design. Different analyses, for example of a FE analysis of a structure and a CFD analysis of the space surrounding it, could be specified by different cells of the same overall space. Analyses at different levels of detail may be specified by different partitioning or cellular subdivision of the same model with different meshing attributes.

Given a comprehensive modelling capability, design by structural feature can help facilitate conceptual CAD where the beam/shell models often used in preliminary design can be used to create the first iteration of the 3D solid geometry.

One of the most difficult issues is how to update the properties of high-level global models to reflect the results obtained from detailed local models. It has been suggested here that mixed dimensional modelling would be useful. Used in combination with cellular partitioning of geometry and substructuring for linear problems, this provides a way of managing the evolution of the design from a simple reduced dimensional model to a cellular solid. An updated global model could be recovered from dimensional reduction of the cells corresponding to thin sheets or slender bars and 3D models of connections substructured to equivalent joint stiffnesses using mixed dimensional coupling. The extension to nonlinear problems is obviously difficult but the general approach may have some merit.

Creating new 'hard' geometry of models for use by downstream applications raises some challenging geometric modelling problems. The Virtual Topology approach can be extended to the idea of attribute suppression. Thus, a blend could be suppressed by specifying a 1D mesh attribute, which would create a line of element edges for the adjacent surface meshes. A mid-surface could be identified as a meshing attribute to a solid cell. With this approach it is very much easier to keep geometric and simulation models in step. The range of potential simulations to be applied to any given model implies a generic requirement for tools for analysis feature recognition and modelling error estimation.

Advances in multi-disciplinary design would be greatly facilitated by the introduction of open frameworks that integrate design and simulation tools.

9   Acknowledgements

The authors gratefully acknowledge the kind assistance of their colleagues KY Lee, RW McCune and KW Shim in developing the ideas outlined above and in the preparation of this document.

References

[1]   M. Drela, "Integrated simulation model for preliminary aerodynamic, structural, and control-law design of aircraft", 40th AIAA SDM Conf., 1-13, St.Louis, USA, 1999.
[2]   A.J. Morris, R. Vignjevic, "Consistent finite element structural analysis and error control", Comput. Methods Appl. Mech, Engrg., 140, 87-108, 1997.
[3]   C. Armstrong, A. Bowyer, S. Cameron, J. Corney, G. Jared, R. Martin, A. Middleditch, M. Sabin, J. Salmon, J. Woodwark, "Djinn: Specification and Report", Information Geometers, 2000.
       http://www.bath.ac.uk/~ensab/GMS/Djinn/

[4]   R. Bidarra, K.J. de Kraker, W.F. Bronsvoort, "Representation and management of feature information in a cellular model", Computer-Aided Design, 30(4), 301–13, 1998.

[5]   http://www.spatial.com

[6]   http://www.parasolid.com

[7]   C.G. Armstrong, S.J. Bridgett, R.J. Donaghy, R.W. McCune, R.M. McKeag D.J. Robinson, "Techniques for Interactive and Automatic Idealisation of CAD Models", Proc. 6th Int. Conf. on Numerical Grid Generation in Computational Field Simulations, 643-662, London, 1998.

[8]   A. Sheffer, T.D. Blacker, M Bercovier, "Virtual topology operators for meshing", 6th International Meshing Roundtable, Sandia National Laboratories, 49-66, 1997.

[9]   S.J. Bridgett, "Detail Suppression of Stress Analysis Models", PhD thesis, The Queen's University of Belfast, N Ireland, 1997.

[10]   J. Steinbrenner, C. Fouts, N. Wyman, "Scripting Language as a Means of Automation in Gridgen", Proceedings of the 8th International Conference on Numerical Grid Generation, Honolulu, HW, June, 2002.

[11]   P. Benk, R. R. Martin and T Várady, "Algorithms for reverse engineering boundary representation models", Computer-Aided Design, 839-851, 1998.

[12]   P.V. Coveney, "Multiscale Modelling and Simulation", 2nd Annual Multiscale Modeling, Boston, MA, 2001.

[13]   C. Vaughan, "Product Data Management and the Engineering Analysis Environment", NAFEMS, 2000.

[14]   R. Bidarra, W.F. Bronsvoort, "Semantic feature modelling", Computer-Aided Design 32, 201–225, 2000.

[15]   D. Cheney, "Engineering Application Integration Using A Python-Based Scripting Approach", Proc. EuroConference on CAE Integration, University of Cambridge, England, 2001.

[16]   R.W. Claus, CAD Services V1.0, Joint Proposal to the OMG in Response to OMG Manufacturing Domain Task Force CAD Services RFP, http://mfg.omg.org/

[17]   M.J. Pratt, W.D. Anderson, "A shape modelling applications programming interface for the STEP standard", Computer-Aided Design 33, 531-543, 2001.

[18]   R.W. McCune, C.G. Armstrong, D.J. Robinson, "Mixed dimensional coupling in finite element models", International Journal for Numerical Methods in Engineering, 49, 725-750, 2000.

[19]   D.J. Monaghan, K.Y. Lee, C.G. Armstrong and H. Ou, "Mixed Dimensional Finite Element Analysis of Frame Models", Proc. 10th ISOPE conference, Seattle, USA, 2000.

[20]   K.W. Shim, D. J. Monaghan and C. G. Armstrong, "Mixed Dimensional Coupling in Finite Element Stress Analysis", Engineering with Computers, in press, 2002.

[21]   E. K. Frey, "Parametric modelling techniques for aircraft structures", AIAA Paper 99-1363, 1411-1415, 1999.

[22] A.V. Mobley, M.P. Carroll, S.A. Canann, "An object oriented approach to geometry defeaturing for finite element meshing", 7th International Meshing Roundtable, Sandia National Labs, 547-563, 1998.

[23] H.Blum, "Biological Shape and Visual Science", Journal of Theoretical Biology, 38, 205-287, 1973.

[24] M.A. Price, C.G. Armstrong, "Hexahedral Mesh Generation By medial Surface Subdivision: Part I. Solid with Convex Edges", International Journal for Numerical Methods in Eng., 40, 111-136, 1997.

[25] A. Sheffer, M. Etzion, A. Rappoport, M. Bercovier, "Hexahedral Mesh Generation Using the Embedded Voronoi Graph", Engineering with Computers, 15:248-262, 1999

[26] P.J. Frey, P.L. George, "Mesh Generation: Application to Finite Elements", Hemes Science, Oxford & Paris, 2000.

[27] D.J. Sheehy, C.G. Armstrong, D.J. Robinson, "Shape Description by Medial Surface Construction", IEEE Transaction on Visualisation & Comp. Graphics, 2 (1996)

[28] E.C. Sherbrooke, N.M. Patrikalakis, E. Brisson, "An algorithm for the medial axis transform of 3D polyhedral solids. IEEE Transactions on Visualization and Computer Graphics 2, 44-61, 1996.

[29] P. Sampl, "Semi-structured mesh generation based on medial axis", 6th International Meshing Roundtable, Sandia National Laboratories, 21-32, 2000.

[30] P.Y. Ang, C.G. Armstrong, "Adaptive curvature-sensitive meshing for medial axis", Engineering with Computers, in press, 2002.

[31] H Gu, T.R.Chase, D.C. Cheney, T. Bailey, D. Johnson, "Identifying, correcting and avoiding errors in computer-aided design models which affect intraoperability", Trans ASME Journal of Computing and Information Science in Engineering, 1, 156-166, 2001.

[32] S.J. Bridgett, "Detail Suppression of stress analysis models", PhD thesis, Queen's University of Belfast, 1997.

[33] P.J. Rohl, R.M. Kolonay, R.K. Irani, M. Sobolewski, K. Kao, M.W. Bailey, "A Federated Intelligent Product Environment", 8th AIAA/USAF/NASA/ISSMO Symposium on Multidisciplinary Analysis and Optimization, Long Beach, CA, September 6-8, 2000.

[34] J. Michopoulos, "femML: Finite Element Modeling Markup Language", ASME 2001 Design Engineering Technical Conferences & Computers and Information in Engineering Conference, Pittsburgh, PA, September 2001.

[35] M. G. Hackenberg, P. Post, R. Redler, B. Steckel, "MpCCI, Multidisciplinary Applications and Multigrid", Proc. ECCOMAS 2000, CIMNE, Barcelona, September 2000.

Chapter 5

©2002, Saxe-Coburg Publications, Stirling, Scotland
Engineering Computational Technology
B.H.V. Topping and Z. Bittnar, (Editors)
Saxe-Coburg Publications, Stirling, Scotland, 105-120.

Finite Element Methods for Electromagnetic Scattering

K. Morgan†, P.D. Ledger†, J. Peraire‡, O. Hassan† and N.P. Weatherill†
† Civil & Computational Engineering
  University of Wales, Swansea, Wales
‡ Aeronautics & Astronautics
  MIT, Cambridge, Massachusetts, USA

Abstract

An arbitrary order edge element approach for problems of electromagnetic wave scattering is described. The construction of an a–posteriori error estimator enables bounds to be placed on computed outputs of practical interest, such as the scattering width distribution. The use of adaptivity is investigated and the possibilities offered by the use of methods based upon reduced–order approximation are also outlined.

Keywords: electromagnetic scattering, arbitrary order edge elements, output error bounds, goal orientated adaptivity, reduced–order approximation.

1   Introduction

Edge element methods are popular in computational electromagnetics [1]–[5]. Demkowicz and co–workers [6, 7] developed a two dimensional hierarchical basis for edge elements, which enabled fully adaptive *hp* approximations to be computed. Our basic algorithm follows a similar approach, but employs the shape functions defined by Ainsworth and Coyle [8] and the chosen application area is the solution of electromagnetic wave scattering problems. An a–posteriori error estimation capability is added by employing the approach developed by the group of Patera and Peraire [9]–[12]. This enables inexpensive, sharp, rigorous and constant free bounds to be computed for the numerical error in outputs of engineering interest. The error estimation technique is incorporated within the high order element framework and the selected output is the scattering width. We will also demonstrate the use of reduced–order approximation for scattering simulations. Reduced–order models are constructed, from full finite element solutions, for a small set of problem parameters and enable the prediction of computational outputs for a new set of parameters [13, 14]. It is demonstrated how these models can also be supplemented by the calculation of rigorous error bounds on outputs [15, 16] and how they may be employed to enable the rapid calculation of outputs for multiple incident wave directions.

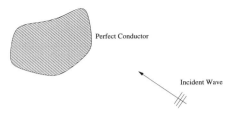

Figure 1: Schematic of the electromagnetic scattering problem

2   Problem Formulation

Consider problems involving the interaction between electromagnetic waves and a scattering obstacle. The waves are generated by a source in the far field and the obstacle, which is surrounded by free space, will be assumed to be a perfect electrical conductor (PEC). The problem, which is illustrated schematically in Figure 1, is governed by Maxwell's equations and the unknowns are the electric and magnetic field intensity vectors, E and H respectively. A scattered field formulation is employed, with Maxwell's equations expressed, in the frequency domain, in the dimensionless form

$$\text{curl } \boldsymbol{E}^s = -\mathrm{i}\omega\boldsymbol{\mu}_f \boldsymbol{H}^s \qquad\qquad \text{curl } \boldsymbol{H}^s = \mathrm{i}\omega\epsilon_f \boldsymbol{E}^s \tag{1}$$

$$\text{div}\,(\epsilon_f \boldsymbol{E}^s) = 0 \qquad\qquad \text{div}\,(\boldsymbol{\mu}_f \boldsymbol{H}^s) = 0 \tag{2}$$

Here the superscript s refers to the scattered field, $\mathrm{i} = \sqrt{-1}$ and $\omega = 2\pi/\lambda$, where λ is the wavelength of the incident wave. The quantities $\boldsymbol{\mu}_f$ and ϵ_f represent the relative permeability and relative permittivity tensors respectively for the propagation medium and both are equal to the unit tensor for free space. For two dimensional problems, $\boldsymbol{E}^s$ and $\boldsymbol{H}^s$ are functions of the cartesian coordinates x and y only. Then, in transverse electric (TE) simulations, $\boldsymbol{E}^s = (E_x^s, E_y^s, 0)^T$ and $\boldsymbol{H}^s = (0, 0, H_z^s)^T$, while $\boldsymbol{E}^s = (0, 0, E_z^s)^T$ and $\boldsymbol{H}^s = (H_x^s, H_y^s, 0)^T$ for the transverse magnetic (TM) case.

2.1   Boundary Conditions

At the surface of the scatterer, the condition

$$\boldsymbol{n} \wedge \text{curl } \boldsymbol{H}^s = -\boldsymbol{n} \wedge \text{curl } \boldsymbol{H}^i \tag{3}$$

is applied in TM simulations, where $\boldsymbol{n}$ is the unit outward normal vector to the surface, $\wedge$ is the vector product and the superscript i denotes the prescribed incident field. In

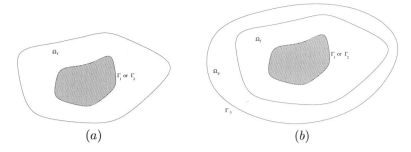

Figure 2: (a) Creation of the finite solution domain Ω_f and (b) the addition of Ω_p, the PML region, to the domain Ω_f

this case, the surface of the scatterer is Γ_1. The condition

$$n \wedge E^s = -n \wedge E^i \qquad (4)$$

is imposed for TE simulations and, in this case, the surface of the scatterer is Γ_2.

Far from the scattering obstacle, the scattered field consists of outgoing waves only. To simulate this condition, a finite solution domain, Ω_f, surrounding the scatterer is selected and a PML technique is employed [17]. To achieve this, an artificial material layer, Ω_p, is added to Ω_f and the outer surface of the PML is Γ_3. This process is illustrated in Figure 2. Within the PML, Maxwell's curl equations are considered in the form

$$\text{curl } E^s = -i\omega\mu_p H^s \qquad\qquad \text{curl } H^s = i\omega\epsilon_p E^s \qquad (5)$$

The thickness of the PML, and the variation in the material properties ϵ_p and μ_p through the PML, are prescribed in a manner designed to ensure that the scattered wave is completely absorbed within the PML layer, without reflection [18].

2.2 Variational Formulation

The governing curl equations (1) are combined and expressed as

$$\text{curl } \left(\Lambda_1^{-1}\text{curl } U^s\right) - \omega^2\Lambda_2 U^s = 0 \qquad (6)$$

where U^s denotes the scattered electric field for TE simulations and the scattered magnetic field in TM simulations. In Ω_f, Λ_1 and Λ_2 are both equal to the unit tensor while, in Ω_p, $\Lambda_1 = \mu_p$ and $\Lambda_2 = \epsilon_p$ for TE simulations while $\Lambda_1 = \epsilon_p$ and $\Lambda_2 = \mu_p$ in the TM case. The spaces

$$Z^D = \{v \mid v \in \mathcal{H}(\text{curl}; \Omega); \ n \wedge v = -n \wedge U^i \text{ on } \Gamma_2; \ n \wedge v = 0 \text{ on } \Gamma_3\} \qquad (7)$$

$$Z = \{v \mid v \in \mathcal{H}(\text{curl}; \Omega); \ n \wedge v = 0 \text{ on } \Gamma_2, \ n \wedge v = 0 \text{ on } \Gamma_3\} \qquad (8)$$

are introduced and a weak variational formulation of the problem is then [18]: find $U^s \in Z^D$, such that

$$A(U^s, W) = \ell(W) \qquad\qquad \forall W \in Z \qquad\qquad (9)$$

Here

$$A(U^s, W) = a(U^s, W) - \omega^2 m(U^s, W) \qquad\qquad (10)$$

and the bilinear forms that have been introduced are defined as

$$a(U^s, W) = \int_{\Omega_f + \Omega_p} \operatorname{curl} \overline{W} \cdot \Lambda_1^{-1} \operatorname{curl} U^s \, d\Omega \qquad\qquad (11)$$

$$m(U^s, W) = \int_{\Omega_f + \Omega_p} \overline{W} \cdot \Lambda_2 U^s \, d\Omega \qquad\qquad (12)$$

The overbar denotes the complex conjugate and, in equation (9),

$$\ell(W) = \int_{\Gamma_1} n \wedge \operatorname{curl} U^i \cdot \overline{W} \, d\Gamma \qquad\qquad (13)$$

Note that this is a valid formulation for scattering problems involving prescribed non zero values for ω [7].

2.3   Galerkin Approximation

With finite element subspaces Z_H of Z and Z_H^D of Z^D, the Galerkin approximate solution $U_H^s \in Z_H^D$ is such that

$$A(U_H^s, W) = \ell(W) \qquad \forall W \in Z_H \qquad\qquad (14)$$

When edge elements are used to discretise the solution domain, an approximation of the $\mathcal{H}(\operatorname{curl};\Omega)$ space is obtained in which the tangential component of the solution is continuous across element edges. The family of arbitrary order triangular and quadrilateral edge elements proposed recently by Ainsworth and Coyle [8] is adopted.

2.4   Computing the Scattering Width

In two dimensional scattering simulations, a non linear output of primary interest is the scattering width, or the radar cross section per unit length. The evaluation of this quantity requires the use of a near field to far field transformation, using solution information obtained on a collection surface, Γ_c, that totally encloses the scatterer and lies in free space. When an approximate solution U_H^s has been computed on the working mesh, the scattering width integral is approximated as

$$S(U_H^s; \phi) = \mathcal{L}^O(U_H^s; \phi) \overline{\mathcal{L}^O(U_H^s; \phi)} \qquad\qquad (15)$$

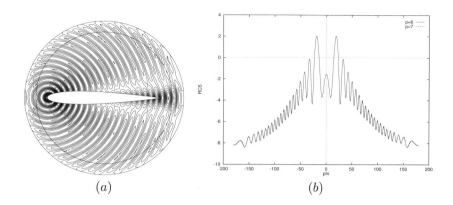

$$(a) \qquad\qquad (b)$$

Figure 3: Scattering of a plane TE wave by a perfectly conducting NACA0012 aerofoil of electrical length 10λ showing: (a) the contours of $Re(H_z^s)$ and (b) the computed distribution of the scattering width

where ϕ is the viewing angle,

$$\mathcal{L}^{O}\left(\boldsymbol{U}_H^s;\phi\right) = \int_{\Gamma_c} \left(\boldsymbol{n} \wedge \boldsymbol{U}_H^s \cdot \boldsymbol{V}\right) d\Gamma + \sum \int_k \left(\omega^2 \boldsymbol{U}_H^s \cdot \boldsymbol{y}_H - \operatorname{curl} \boldsymbol{U}_H^s \cdot \operatorname{curl} \boldsymbol{y}_H\right) d\Omega \tag{16}$$

and

$$\boldsymbol{V} = \begin{bmatrix} 0 \\ 0 \\ \delta \end{bmatrix} e^{-\mathrm{i}\omega(x'\cos\phi + y'\sin\phi)} \qquad \boldsymbol{Y} = \frac{\delta}{\omega} \begin{bmatrix} -\sin\phi \\ \cos\phi \\ 0 \end{bmatrix} e^{-\mathrm{i}\omega(x'\cos\phi + y'\sin\phi)} \tag{17}$$

The parameter δ is set equal to -1 for TE simulations and to $+1$ for the TM case. The summation in equation (16) extends over all elements $k \in \Omega_f$, such that $\partial k \cup \Gamma_c \neq \emptyset$, while the function $\boldsymbol{y}_H$ is zero at all interpolation points in Ω_f and equal to the edge element interpolation of $\boldsymbol{Y}$ on Γ_c. The simulation of scattering of a plane TE wave by a perfectly conducting NACA0012 aerofoil, of electrical length 10λ, is illustrated in Figure 3. The figure shows the computed contours of the real part of the magnetic field and distribution of the scattering width in decibels $\sigma = 10\log_{10}\{(\omega/4)\mathcal{S}(\boldsymbol{U}_H^s;\phi)\}$ plotted against the viewing angle ϕ. Mesh convergence is demonstrated by comparing the scattering width distributions computed on a mesh with elements of uniform order $p = 6$ and on the same mesh with elements of uniform order $p = 7$.

3 Error Estimation

Outputs computed from solutions obtained on discretisations with a sufficiently high p and small enough mesh spacing h will, in general, be indistinguishable from the exact.

However, such solutions can be expensive to compute. It is, therefore, important to be able to evaluate strict upper and lower bounds for specified outputs, such as the scattering width $\mathcal{S}(U_H^s, \phi)$, which are functions of the computed solution and an auxiliary variable ϕ, in the form

$$s^-(U_h^s, \phi) \leq \mathcal{S}(U_H^s, \phi) \leq s^+(U_h^s, \phi) \tag{18}$$

To accomplish this, an extension of the a–posteriori finite element error bound procedure proposed by Sarrate, Peraire and Patera [12] for the Helmholtz equation is employed. The method, which is capable of dealing with quadrilateral, triangular or hybrid discretisations [19, 20], reduces to a requirement for solving local Neumann sub–problems inside each element, with the balance between elements ensured by using Demkowicz's edge fluxes [21]. Linearisation must be employed when non–linear outputs, such as the scattering width, are considered, with the variable ϕ interpreted as the viewing angle. As an example, the bounds computed for the case of scattering of a plane TE wave by a perfectly conducting NACA0012 aerofoil of electrical length

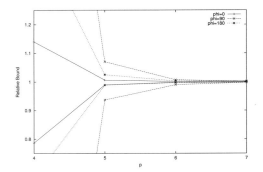

Figure 4: Scattering of a plane TE wave by a perfectly conducting NACA0012 aerofoil of electrical length 10λ: variation of the computed error bounds s^+/s_h and s^-/s_h, for the scattering width at $\phi = 0, 90\ 180$ degrees, with increase in p

10λ are shown in Figure 4. The viewing angles of interest are selected as $\phi = 0, 90$ and 180 degrees and the truth mesh, on which the output s_h is proposed to be indistinguishable from the exact, is computed on a mesh with uniform order $p = 7$ elements. The relative bounds s^+/s_h and s^-/s_h are seen to converge to unity as the polynomial order, p, is increased.

4 Adaptivity

With these procedures in place, an adaptive mesh procedure, based upon the computed error bound gap and with error indicator, Δ, defined as

$$\Delta = \frac{1}{2}\left(s^+ - s^-\right) \tag{19}$$

can now be proposed [20, 22]. The indicator is evaluated as the sum of element contributions, with the contribution from element k denoted by Δ_k. The element contributions are used in an adaptivity process which refines all elements k' for which $\Delta_{k'} \geq \theta \max_k \Delta_k$, where $\theta \in (0, 1)$ is a user specified parameter [23]. The process

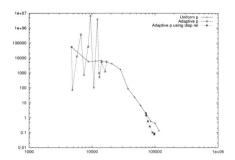

Figure 5: Scattering of a plane TM wave by a thin PEC cylindrical cavity, of diameter $D = 16\lambda$ and aperture 20 degrees, showing the convergence of the relative bound gap $(s^+ - s^-)/s_h$ with number of unknowns when uniform p and adaptive p refinement strategies are employed

is terminated when the error indicator falls below a prescribed value. Although mesh enrichment can be employed, we only illustrate here a process which locally increases the polynomial order without changing the underlying mesh. The technique of constrained approximation [6] is used to enforce continuity across each edge in the mesh. The performance of this adaptive procedure is illustrated by considering scattering of a plane TM wave by thin PEC cylindrical cavity of diameter $D = 16\lambda$ and with an aperture of 20 degrees. The truth mesh for this problem employs uniform order $p = 14$ elements. The convergence of the relative bound gap with the number of unknowns is shown in Figure 5 for (a) an adaptive procedure starting from a coarse discretisation; (b) an adaptive procedure starting from a mesh created so as to ensure that dispersion is already reduced to an acceptable level; and (c) for uniform refinement. In all cases, the objective is to terminate the adaptive procedure when the prescribed tolerance level $0.01 \times S(U_H^s, \phi)$ is reached for the specified angle of $\phi = 0$. For this example, it can be observed that, for case (a), with the adaptive strategy initiated on a coarse discretisation, the convergence behaviour is very erratic and is characterised by wild increases and decreases in the size of the bound gap. After 15 adaptive steps, no real convergence is apparent and the adaptive strategy is stopped. For case (c), with uniform refinement in p adopted, the solution convergence is uniform throughout. Following the adaptive p strategy of case (b), with an initial distribution of $p = 10$ elements deemed to be sufficient to reduce the dispersion to an acceptable level, we observe that the solutions produced require less unknowns than the uniform p strategy. These results indicate that it is appropriate to employ a meaningful starting mesh for the adaptive procedure. To complete this example, the contours of $Re(E_z^s)$ and

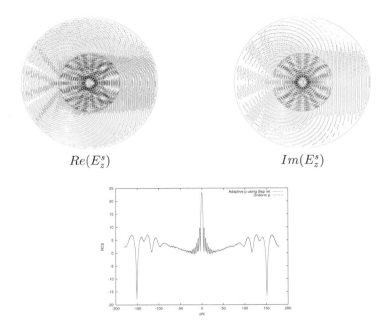

$$Re(E_z^s) \hspace{6cm} Im(E_z^s)$$

Figure 6: Scattering of a plane TM wave by a thin PEC cylindrical cavity, of diameter $D = 16\lambda$ and aperture 20 degrees, showing contours of $Re(E_z^s)$ and $Im(E_z^s)$ and the computed distributions of the scattering width

$Im(E_z^s)$ and a plot of the converged scattering width distributions following the adaptive procedures (b) and (c) above are shown in Figure 6.

5   Reduced Order Approximation

5.1   Bounding the Complete Scattering Width Distribution

Reduced–order, or low–order, models have been shown to provide a powerful method for computing outputs in the areas such as turbomachinery [13, 14]. These models are constructed from full finite element solutions for a small set of problem parameters and enable the prediction of outputs for a new set of parameters. Recent work [15, 16] has demonstrated the construction of rigorous constant free error bounds on the outputs of the reduced–order model. Here, we present an extension to the above error bounding capability by providing a method for the calculation of error bounds for the complete spectrum of viewing angles of the scattering width using a reduced–order modelling technique. The procedure that has been described focused on the point wise evaluation of error bounds, so this is an important extension. The procedure requires the computation of M adjoint problems, corresponding to viewing angles

$\phi_1, \phi_2, \cdots, \phi_M$. At other viewing angles, the error bounds are reconstructed using the reduced–order model.

On the working mesh, we again begin by obtaining the Galerkin approximate solution $\boldsymbol{U}_H^s \in Z_H^D$ from equation (14). Now, the residual

$$\mathcal{R}^U = \ell(\boldsymbol{W}) - \mathcal{A}(\boldsymbol{U}_H^s, \boldsymbol{W}) \tag{20}$$

is also defined and we note that $\mathcal{R}^U(\boldsymbol{W}) = 0$ for all $\boldsymbol{W} \in Z_H$. The output adjoints $\boldsymbol{\Psi}_i \in Z_H$, for $i = 1, 2 \cdots M$ and associated with the viewing angles $\phi_1, \phi_2, \cdots, \phi_M$, are obtained on the same mesh from

$$\mathcal{A}(\boldsymbol{W}, \boldsymbol{\Psi}_i) = -\ell^O(\boldsymbol{W}; \phi_i) \qquad \forall \boldsymbol{W} \in Z_H, \, i = 1, 2, \cdots, M \tag{21}$$

where, following a linearisation [20], for the scattering width

$$\ell^O(\boldsymbol{W}; \phi_i) = \mathcal{L}^O(\boldsymbol{U}_H^s; \phi_i)\overline{\mathcal{L}^O(\boldsymbol{W}; \phi_i)} + \mathcal{L}^O(\boldsymbol{W}; \phi_i)\overline{\mathcal{L}^O(\boldsymbol{U}_H^s; \phi_i)} \tag{22}$$

The adjoint residuals

$$\mathcal{R}_i^\Psi = -\overline{\ell^O(\boldsymbol{W}; \phi_i)} - \overline{\mathcal{A}(\boldsymbol{W}, \boldsymbol{\Psi}_i)} \qquad i = 1, 2, \cdots, M \tag{23}$$

are also defined and it may be observed that $\mathcal{R}_i^\Psi(\boldsymbol{W}) = 0$ for all $\boldsymbol{W} \in Z_H$ and all $i = 1, 2, \cdots, M$. A truth mesh is constructed by a refinement of each coarse mesh element T_H. The refinement is accomplished by either subdividing the element, or increasing the polynomial order on the element, or by a combination of both. Then, a coarse broken space is defined by

$$\hat{Z}^H = \{\boldsymbol{u}|\ \boldsymbol{u} \in \mathcal{H}(\text{curl}; \Omega), \boldsymbol{u}|_{T_H} \in Z_H(T_H), \forall T_H\} \tag{24}$$

and a corresponding fine broken space as

$$\hat{Z}^h = \{\boldsymbol{u}|\ \boldsymbol{u} \in \mathcal{H}(\text{curl}; \Omega), \boldsymbol{u}|_{T_H} \in Z_h(T_H), \forall T_H\} \tag{25}$$

When these broken spaces are introduced, edge fluxes are necessary to ensure that the solution computed on the broken elements remains in balance. This is achieved, following Demkowicz [21], by defining edge flux functionals λ^U and λ_i^Ψ as

$$\lambda^U(\boldsymbol{W}) = \int_{\partial T_H} (\boldsymbol{n} \wedge \boldsymbol{W}) \cdot \boldsymbol{f}_H^U \mathrm{d}s \tag{26}$$

$$\lambda_i^\Psi(\boldsymbol{W}) = \int_{\partial T_H} (\boldsymbol{n} \wedge \boldsymbol{W}) \cdot \boldsymbol{f}_H^\Psi|_i \mathrm{d}s \qquad i = 1, 2 \cdots, M \tag{27}$$

The quantities $\boldsymbol{f}_H^U$ and $\boldsymbol{f}_H^\Psi|_i \in \{\hat{Z}^H\}$ are determined from the requirement that

$$\lambda^U(\boldsymbol{W}) = \mathcal{R}^U(\boldsymbol{W}) \qquad \forall \boldsymbol{W} \in \hat{Z}^H \tag{28}$$

$$\lambda_i^\Psi(\boldsymbol{W}) = \mathcal{R}_i^\Psi(\boldsymbol{W}) \qquad \forall \boldsymbol{W} \in \hat{Z}^H, \, i = 1, 2, \cdots, M \tag{29}$$

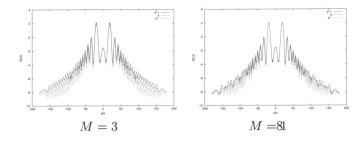

$$M = 3 \qquad\qquad\qquad\qquad M = 81$$

Figure 7: Scattering of a plane TE wave by a perfectly conducting NACA0012 aerofoil of electrical length 10λ showing the reconstructed bounds for the complete spectrum of viewing angles when a reduced–order model with $M = 3$ and $M = 81$ adjoints are computed on a working discretisation of $p = 5$.

For a prescribed viewing angle, ϕ, the output adjoint is reconstructed in the form

$$\mathbf{\Psi} = \sum_{i=1}^{M} \beta_i \mathbf{\Psi}_i \tag{30}$$

where the coefficients β_i, $i = 1, 2, \ldots, M$, are obtained from the requirement that

$$\mathcal{A}(\boldsymbol{W}, \boldsymbol{\Psi}) = -\ell^{\mathcal{O}}(\boldsymbol{W}; \phi) \qquad \forall \boldsymbol{W} \in \{\boldsymbol{\Psi}_1, \boldsymbol{\Psi}_2, \cdots \boldsymbol{\Psi}_M\} \tag{31}$$

Reconstructed errors are computed in the decoupled truth space according to

$$2\mathcal{B}^s(\hat{\boldsymbol{e}}^U, \boldsymbol{W}) = \mathcal{R}^U(\boldsymbol{W}) - \lambda^U(\boldsymbol{W}) \qquad \forall \boldsymbol{W} \in \hat{Z}^h \tag{32}$$

$$2\mathcal{B}^s(\hat{\boldsymbol{e}}_i^\Psi, \boldsymbol{W}) = \mathcal{R}_i^\Psi(\boldsymbol{W}) - \lambda^\Psi(\boldsymbol{W}) \qquad \forall \boldsymbol{W} \in \hat{Z}^h, \ i = 1, 2, \cdots, M \tag{33}$$

$$2\mathcal{B}^s(\hat{\boldsymbol{e}}_0^\Psi, \boldsymbol{W}) = -\ell^{\mathcal{O}}(\boldsymbol{W}; \phi) + \sum_{i=1}^{M} \beta_i \ell^{\mathcal{O}}(\boldsymbol{W}; \phi_i) \qquad \forall \boldsymbol{W} \in \hat{Z}^h \tag{34}$$

and, then,

$$\hat{\boldsymbol{e}}^\Psi = \hat{\boldsymbol{e}}_0^\Psi + \sum_{i=1}^{M} \beta_i \hat{\boldsymbol{e}}_i^\Psi \qquad\qquad \hat{\boldsymbol{e}}^\pm = \hat{\boldsymbol{e}}^U \mp \frac{1}{\kappa} \hat{\boldsymbol{e}}^\Psi \tag{35}$$

where κ is a suitably defined scaling parameter. To complete the process, the lower and upper bounds on the defined output are computed as

$$s^- = \mathrm{Re}\{S(\boldsymbol{U}_H^s; \phi)\} - \kappa \mathcal{B}^s(\hat{\boldsymbol{e}}^-, \hat{\boldsymbol{e}}^-) \tag{36}$$

$$s^+ = \mathrm{Re}\{S(\boldsymbol{U}_H^s; \phi)\} + \kappa \mathcal{B}^s(\hat{\boldsymbol{e}}^+, \hat{\boldsymbol{e}}^+) \tag{37}$$

and these bounds are optimized by appropriate choice of the scaling parameter [23]. To demonstrate an application of this procedure, consider the scattering of a plane TE wave by a perfectly conducting NACA0012 aerofoil of electrical length 10λ. The reconstructed bounds are initially obtained for the scattering width when a working discretisation consisting of $p = 5$ elements and a truth discretisation of $p = 7$ elements are used. Figure 7 shows the bounds obtained when the values $M = 3$ and $M = 81$ are chosen. We observe that, the bounds become small at locations corresponding to angles for which adjoint problems have been solved and that the bounds at other locations become tighter as the value of M is increased. In all cases, we it can be seen that the computational output is always bounded. Experience shows that refining the working discretisation has the effect of reducing the size of the bound gap.

5.2   Constructing the Scattering Width Distribution for Different Incident Angles

For design purposes, the engineer will often be interested in the rapid calculation of the scattering width distribution for all possible incident wave angles. To illustrate how this may be accomplished using a reduced–order model, we begin by returning to the standard Galerkin variational statement of equation (14), but expressed now in the form [15]: find $\boldsymbol{U}_H^s(\theta) \in Z_H^D$ such that

$$\mathcal{A}(\boldsymbol{U}_H^s(\theta), \boldsymbol{W}) = \ell(\boldsymbol{W}; \theta) \qquad \forall \boldsymbol{W} \in Z_H \qquad (38)$$

for a given incident wave direction, θ. In this case, the computed output $s_H(\theta; \phi) \in \mathbb{C}$ is defined to be

$$s_H(\theta, \phi) = \mathcal{L}^O(\boldsymbol{U}_H^s(\theta); \phi) \qquad (39)$$

The associated adjoint approximate solution, $\boldsymbol{\Psi}_H(\phi) \in Z_H$ satisfies

$$\mathcal{A}(\boldsymbol{W}, \boldsymbol{\Psi}_H(\phi)) = -\mathcal{L}^O(\boldsymbol{W}; \phi) \qquad \forall \boldsymbol{W} \in Z_H \qquad (40)$$

To construct reduced–order spaces, we select parameters $\{\theta_1, \cdots, \theta_{N_\theta}\}$, corresponding to incident wave angles, and parameters $\{\phi_1, \cdots, \phi_{N_\phi}\}$, corresponding to viewing angles of the scattering width. We then compute $\boldsymbol{U}_H^s(\theta_1), \cdots, \boldsymbol{U}_H^s(\theta_{N_\theta})$ and $\boldsymbol{\Psi}_H(\phi_1), \cdots, \boldsymbol{\Psi}_H(\phi_{N_\phi})$, and define

$$W_{N_\theta}^{\mathrm{pr}} = \mathrm{span}\{\boldsymbol{U}_H^s(\theta_i); i = 1, \cdots N_\theta\} \qquad W_{N_\phi}^{\mathrm{du}} = \mathrm{span}\{\boldsymbol{\Psi}_H(\phi_i); i = 1, \cdots N_\phi\} \qquad (41)$$

where the superscripts pr and du denote primal and dual problems respectively. When we are given a different incident angle, θ, we look for $\boldsymbol{U}_{N_\theta}^s(\theta) \in W_{N_\theta}^{\mathrm{pr}} \subset Z_H^D$ such that

$$\mathcal{A}(\boldsymbol{U}_{N_\theta}^s, \boldsymbol{W}) = \ell(\boldsymbol{W}) \qquad \forall \boldsymbol{W} \in W_{N_\theta}^{\mathrm{pr}} \qquad (42)$$

For this angle of incidence, we consider the sequence of viewing angles ϕ and compute, for each angle ϕ, $\boldsymbol{\Psi}_{N_\phi}(\phi) \in W_{N_\phi}^{\mathrm{du}} \subset Z_H$ such that

$$\mathcal{A}(\boldsymbol{W}, \boldsymbol{\Psi}_{N_\phi}) = -\mathcal{L}^0(\boldsymbol{W}) \qquad \forall \boldsymbol{W} \in W_{N_\phi}^{\mathrm{du}} \qquad (43)$$

and $s_N(\theta, \phi) \in \mathbb{C}$ according to

$$s_N = \mathcal{L}^{\mathcal{O}}(\boldsymbol{U}^s_{N_\theta}) - \left[\ell\left(\boldsymbol{\Psi}_{N_\phi}\right) - \mathcal{A}(\boldsymbol{U}^s_{N_\theta}, \boldsymbol{\Psi}_{N_\phi})\right] \qquad (44)$$

Here, s_N represents an adjoint enhanced reduced–order approximation to the output, with the unenhanced version given by $\mathcal{L}^{\mathcal{O}}(\boldsymbol{U}^s_{N_\theta})$. It should be noted that $\boldsymbol{U}^s_{N_\theta}$ and $\boldsymbol{\Psi}_{N_\phi}$ are obtained from a linear combination of the predetermined data given in $W^{\mathrm{pr}}_{N_\theta}$ and $W^{\mathrm{du}}_{N_\phi}$. The matrices which result from equations (42) and (43) can be precomputed to reduce the computational cost of computing subsequent outputs for new θ values. Inversion of the resulting system for $\boldsymbol{U}^s_{N_\theta}, \boldsymbol{\Psi}_{N_\phi}$ is at most, $O(\max(N^3_\theta, N^3_\phi))$ and we deduce that, for small N_θ and N_ϕ, s_N will be much less expensive to compute than s_H.

It is also of interest to consider the construction of certainty bounds for this reduced–order model approximation to the output, where the bounds are measured with respect

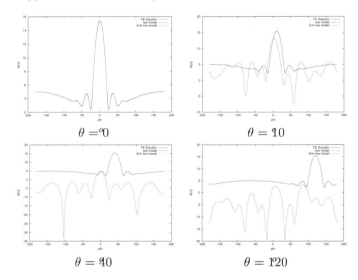

Figure 8: Scattering of a plane TM wave by a PEC circular cylinder of diameter $D = 2\lambda$ showing a comparison between the scattering width distributions computed using the enhanced and unenhanced reduced–order model solutions and the finite element solution for waves incident at different angles, θ

to the output given by the finite element solution. To do this, we demonstrate that equation (44) provides an accurate reduced–order model. Initially, equations (38)–(40) are expressed in the matrix form

$$\boldsymbol{A}\boldsymbol{U}^s_H = \boldsymbol{L} \qquad (45)$$
$$s_H = \boldsymbol{g}^T\boldsymbol{U}^s_H \qquad (46)$$
$$\boldsymbol{A}^T\boldsymbol{\Psi}_H = -\boldsymbol{g} \qquad (47)$$

and, given an approximate solution $U^s_{N_\theta}$ to equation (45) and a corresponding approximate adjoint solution Ψ_{N_ϕ} to equation (47), we can write

$$s_H = g^T U^s_H = g^T U^s_{N_\theta} - (\Psi_H - \Psi_{N_\phi})^T A(U^s_H - U^s_{N_\theta}) - \Psi^T_{N_\phi} A(U^s_H - U^s_{N_\theta}) \quad (48)$$

Furthermore, if we recognise that

$$A(U^s_H - U^s_{N_\theta}) = L \qquad A U^s_{N_\theta} = R^U \quad (49)$$

we obtain

$$s_H = \left[g^T U^s_{N_\theta} - \Psi^T_{N_\phi}(L - A U^s_{N_\theta}) \right] - (\Psi_H - \Psi_{N_\phi})^T R^U \quad (50)$$

where the term in square brackets is computable without prior knowledge of U^s_H or Ψ_H. In fact, this term is identical to the expression for s_N given in equation (44) and represents the adjoint enhanced reduced–order approximation to the output. The other term represents the error between the reduced–order model approximation to the output and the output given by the finite element solution. This error term may be further manipulated by writing

$$s_H = s_N - (\Psi_H - \Psi_{N_\phi})^T A A^{-1} R^U \quad (51)$$

and, after defining

$$R^\Psi = g \qquad A^T \Psi_H = A^T \Psi_H - A^T \Psi_{N_\phi} \quad (52)$$

it follows that

$$s_H = s_N - R^{\Psi^T} A^{-1} R^U \quad (53)$$

The error between the reduced–order model approximation to the output and the output given by the finite element solution is, therefore, bounded by

$$|s_H - s_N| \leq \| R^\Psi \| \, \| A^{-1} \| \, \| R^U \| \quad (54)$$

where $\| \cdot \|$ denotes the Euclidean norm. It follows from this result [24] that

$$|s_H - s_N| \leq \frac{\| R^\Psi \| \, \| R^U \|}{\min \mu_i} \quad (55)$$

where μ_i denote the singular values of the matrix A.

To demonstrate the numerical performance of this process, consider the case of scattering of a plane TM wave by a PEC circular cylinder of diameter $D = 2\lambda$. The computational domain is in the form of circular annulus of inner radius λ and outer radius 2λ and the mesh employed consists of 200 triangular elements of polynomial order $p = 3$. The choice $N_\theta = 3$ is made, with corresponding incident angles $\theta_i = \{-0, 0, 90\}$ degrees, while we choose $N_\phi = 81$ with $\phi_i = 91 + 20(i-1)$ degrees for $i = 1, 2, \ldots, 18$. Using the computed results, the reduced–order model is invoked to predict the scattering width distributions for waves incident at angles of

$\theta = \{0, 10, 40, 120\}$ degrees in turn. A comparison of the resulting distributions obtained by the adjoint enhanced reduced–order model, the unenhanced reduced–order model and the finite element solution is shown in Figure 8. It can be observed that for the case $\theta = 0$, which corresponds to one of the primal problems, both the adjoint enhanced reduced–order model and the unenhanced reduced–order model produce results that are in excellent agreement with the finite element distribution. For incident angles $\theta = \{10, 40, 120\}$, which do not correspond to incident angles for the primal problems, the adjoint enhanced reduced–order model continues to be in excellent agreement with the output given by the full finite element solution. However, the scattering width distributions obtained for these angles of incidence using the unenhanced reduced–order model are seen to be inaccurate. From this initial numerical investigation, it is clear that the adjoint enhanced reduced–order model provides a fast and accurate prediction of scattering width over a wide range of angles of incidence. The use of an unenhanced reduced–order model is seen to be inaccurate for small values of N_θ.

6 Conclusions

An *hp* edge element procedure for the simulation of electromagnetic wave scattering problems in the frequency domain on general triangular meshes has been presented. Arbitrary order edge elements have been employed, with the computational domain truncated by using the PML approach. Bounds are produced on outputs of electromagnetic scattering problems and, of particular interest to aerospace engineers, is the demonstration of the practicality of using the method to bound the computed scattering width at prescribed viewing angles. It has also been demonstrated how the bounding procedure can be used to drive an adaptive procedure based upon element contributions to the error bound gap. The possibilities offered by the application of reduced–order modelling techniques to this problem area have also been addressed. It has been demonstrated how reduced–order models can be employed to bound the complete computed scattering width distribution and to construct the scattering width distribution for different incident wave angles.

7 Acknowledgements

Paul Ledger acknowledges the support of the UK Engineering and Physical Sciences Research Council (EPSRC) in the form of a PhD studentship under grant GR/M59112. Jaime Peraire acknowledges the support of EPSRC in the form of a visiting fellowship award under grant GR/N09084.

References

[1] J. C. Nédélec, *Computation of eddy currents on a surface in* $\mathbb{R}^3$ *by finite element methods*, SIAM Journal of Numerical Analysis, 15, 580–594, 1978

[2] J. C. Nédélec, *Mixed finite elements in* $\mathbb{R}^3$, Numerische Mathematik, 35, 315–341, 1980

[3] J. C. Nédélec, *A new family of mixed finite elements in* $\mathbb{R}^3$, Numerische Mathematik, 50, 57–81, 1986

[4] P. Monk, *An analysis of Nédélec's method for the spatial discretisation of Maxwell's equations*, Journal of Computational and Applied Mathematics, 47, 101–121, 1993

[5] P. Monk, *On the p– and hp–extension of Nédélec's curl conforming elements*, Journal of Computational and Applied Mathematics, 53, 117–137, 1994

[6] L. Demkowicz and W. Rachowicz, *A 2D hp–adaptive finite element package for electromagnetics (2Dhp90em)*, TICAM Report 98–15, University of Texas at Austin, 1998

[7] L. Demkowicz and L. Vardapetyan, *Modeling of electromagnetic/scattering problems using hp–adaptive finite elements*, Computer Methods in Applied Mechanics and Engineering, 152, 103–124, 1998

[8] M. Ainsworth and J. Coyle, *Hierarchic hp–edge element families for Maxwell's equations on hybrid quadrilateral/triangular meshes*, Computer Methods in Applied Mechanics and Engineering, 190, 6709–6733, 2001

[9] M. Paraschivoiu, J. Peraire and A. T. Patera, *A–posteriori finite element bounds for linear functional output of elliptic partial differential equations*, Computer Methods in Applied Mechanics and Engineering, 150, 289–312, 1997

[10] Y. Maday, J. Peraire and A. T. Patera, *A general formulation for a–posteriori bounds for output functionals of partial differential equations: application to the eigenvalue problem*, C. R. Acad. Sci. Paris, Ser. I, 328, 823–828, 1999

[11] J. Peraire and A. T. Patera, *Asymptotic a–posteriori finite element bounds for the outputs of noncoercive problems: the Helmholtz and Burgers equations*, Computer Methods and Applied Mechanics, 171, 77–86, 1999

[12] J. Sarrate, J. Peraire and A. T. Patera, *A–posteriori error bounds for nonlinear outputs of the Helmholtz equation*, International Journal for Numerical Methods in Fluids, 31, 17–36, 1999

[13] K. E. Wilcox, J. D. Paduano and J. Peraire, *Low order aerodynamic models for areoelastic control of turbomachines*, AIAA Paper 99–1467, 1999

[14] K. E. Wilcox, J. Peraire and J. White, *An Arnoldi approach for generation of reduced order models for turbmachinery*, Computers and Fluids, 31, 369–389, 1999

[15] L. Machiels, Y. Maday and A. T. Patera, *Output bounds for reduced–order approximations of partial differential equations*, Computer Methods in Applied Mechanics and Engineering, 190, 3413–3426, 2001

[16] Y. Maday, A. T. Patera and D. V. Rovas, *A reduced–basis output bound method for noncoercive linear problems* Technical Report, Laboratoire d'Analyse

Numérique, Universite Paris VI, 2000

[17] J.–P. Berenger, *A perfectly matched layer for the absorption of electromagentic waves*, Journal of Computational Physics, 114, 185–200, 1994

[18] P. D. Ledger, O. Hassan, K. Morgan and N. P. Weatherill, *Arbitrary order edge elements for electromagnetic scattering simulation using hybrid meshes and a PML*, accepted for publication in International Journal of Numerical Methods in Engineering, (2002)

[19] P. D. Ledger, *An hp–adaptive finite element procedure for electromagnetic scattering problems*, PhD Thesis, University of Wales Swansea, 2002

[20] P. D. Ledger, K. Morgan, J. Peraire, O. Hassan and N. P. Weatherill, *Efficient, highly accurate hp–adaptive finite element computations of the scattering width output of Maxwell's equations*, submitted to International Journal for Numerical Methods in Fluids, 2002

[21] L. Demkowicz, *A–posteriori error analysis for steady state Maxwell's equations*, in P. Ladaveze and J. T. Oden (editors), New Advances in Adaptive Computational Methods in Mechanics, Elsevier, New York, 1998

[22] P. D. Ledger, K. Morgan, J. Peraire, O. Hassan and N. P. Weatherill, *The development of an hp–adaptive finite element procedure for electromagnetic scattering problems*, submitted to Finite Elements in Analysis and Design, 2002

[23] J. Peraire and A. T. Patera, *Bounds for linear functional outputs of coercive partial differential equations: local indicators and adaptive refinement*, in P. Ladeveze and J.T. Oden (editors), Proceedings of the Workshop on New Advances in Adaptive Computational Methods in Mechanics, Elsevier, New York, 1998

[24] J. T. Oden and L. Demkowicz, *Applied Functional Analysis*, CRC Press, Bacon Raton, 1996

©2002, Saxe-Coburg Publications, Stirling, Scotland
Engineering Computational Technology
B.H.V. Topping and Z. Bittnar, (Editors)
Saxe-Coburg Publications, Stirling, Scotland, 121-146.

Chapter 6

Finite Element Non-Linear Dynamic Soil-Fluid-Structure Interaction

R.S. Crouch
Computational Mechanics Unit
Department of Civil and Structural Engineering
University of Sheffield, United Kingdom

Abstract

A strategy for simulating the dynamic response and permanent deformation of concrete structures sited on unbounded domains (and those bounded by fluids) is outlined here. In this review paper, the focus is on the constitutive model and development of a coupled continuum FE approach able to handle the dynamic far-field. A generalised elasto-plasticity constitutive model is described in some detail. The far-field dynamic radiation condition is satisfied analytically using the novel Scaled Boundary Finite Element method. Another issue touched upon is the identification of instability indicators; these potentially offering valuable engineering performance measures. Several different FE structural analyses are presented, including the response of a soil-embedded fluid-filled concrete tank and the soil-structure interaction of a pre-stressed concrete containment vessel (PCCV) hit by an aircraft.

Keywords: dynamic, far-field, acoustic, interaction, NLFEA, concrete, structures.

1 Introduction

Unstructured FE continuum modelling of reinforced and pre-stressed concrete engineering facilities has for some time been the preferred analytical method when simulating the deformation response of complex structural geometries under extreme loading. While non-linearity of the material behaviour is the result of fracture growth, pore collapse and sliding (all discontinuous, local processes), an *equivalent continuum* approach is often adopted. This rather uncomfortable idealisation is a consequence of the current limits of computational power, which prevents a detailed local tracing of the evolution of each crack in civil-engineering-scale structures.

Despite much time having been invested worldwide on the development of a generalised constitutive model for this heterogeneous duplex material (concrete), we remain quite some way from realising a robust, accurate model. Of the more advanced models that do exist, the issues of stability, uniqueness, and physical realism continue to provide significant challenges to the computational mechanics researcher. Outside of constitutive modelling, another area of engineering mechanics where further work is required is that of providing efficient solutions to modelling large coupled dynamic systems where far-field (infinite) boundaries exist. As ever, the problem is one of finding accurate methods that do not impose an excessive computational burden.

This paper presents a personal view of some recent tools advanced in both areas (constitutive modelling and the far-field treatment), together with a discussion on some structural integrity indicators linked to physical and numerical instability. The text is split into 4 sections. First, the constitutive modelling is addressed. Here, a description of a recently developed isothermal inviscid frameworks is given. The need for regularisation is discussed and a viscous extension to the elasto-plasticity model sketched. At the end of section 4, some remarks on material complexity are made. Addressing the need for new laboratory data, some results from multi-axial compression tests conducted at elevated temperature are presented, together with examples of uniaxial tension tests designed to reveal the material *characteristic length*.

In section 3, the discrete FE matrix form of the incremental non-linear dynamic equation of equilibrium is presented. Here an implicit approach is adopted, which is solved using a Newton-Raphson scheme with an element-by-element (GMRES) stabilised Bi-Conjugate Gradient non-symmetric iterative solver [1], [2]. The FE system includes terms coupling an inviscid compressible fluid to the deformable solid. The background to the technique simulating the dynamic far-field is explained and its introduction into a general FE code described. Section 4 of this paper makes use of a hierarchical approach to assessing FE instability.

Within sections 3 and 4, there are FE examples illustrating the issues described above. Inevitably, these analyses are simplified in the sense of using unsatisfactorily over-coarse meshes.

2   An equivalent continuum model for concrete

This is not the paper to justify the use of an isotropic plasticity model (some discussion of this is given in [3]). Rather, a description of the hardening/softening formulation is given in some detail.

2.1   Isotropic plasticity

The peak nominal stress (PNS) achievable by an isotropic concrete specimen under multi-axial loading may be expressed in terms of the 3 normalised cylindrical invariants $\bar{\xi}$, $\bar{\rho}$ and θ. The Lode angle, θ, equals $-\frac{\pi}{6}$, 0 and $+\frac{\pi}{6}$ on the extension, shear and compression meridians respectively.

Modified generalised, Hoek-Brown [4] expressions are proposed. This criterion has been adopted as it appears to offer a good balance between physical relevance, ease of calibration and agreement with experimental results (also see [5], [6], [7]). The compression and extension meridians are as follows

$$\bar{\rho}_c = (\frac{1}{6})^\gamma \sqrt{\frac{2}{3}} \left(-m + \sqrt{m^2 - 12\sqrt{3}m\bar{\xi} + 36}\right)^\gamma \tag{1}$$

$$\bar{\rho}_e = (\frac{1}{3})^\gamma \sqrt{\frac{2}{3}} \left(-m + \sqrt{m^2 - 3\sqrt{3}m\bar{\xi} + 9}\right)^\gamma \tag{2}$$

where $0 < \gamma < 1$ and M is a function of f_t. These meridians provide a continuous, smooth surface, intersecting the hydrostatic axis normally in the tensile region. Adoption of (1) and (2) implies the following relationship between the ratio B_0 (that is, the ratio of *extensional* deviatoric strength to *compressive* deviatoric strength, at the same $\bar{\xi}$, ρ_e/ρ_c) and $\bar{\xi}$ for a given m and γ.

$$B_0 = \left(\frac{-2m + 2\sqrt{m^2 - 3\sqrt{3}m\bar{\xi} + 9}}{-m + \sqrt{m^2 - 12\sqrt{3}m\bar{\xi} + 36}}\right)^\gamma \tag{3}$$

Here, the generalised smooth convex elliptic expression, first employed by Bhowhmik and Long [8], is used. Surface convexity ensures a stable material behaviour according to Drucker's stability postulate. Expressed in the first sextant of principal stress space $(-\frac{\pi}{6} \le \theta \le \frac{\pi}{6})$, we have

$$r = \frac{2d_0}{d_1 - \sqrt{((d_1)^2 - 4d_0 d_2)}} \tag{4}$$

where

$$\begin{aligned} d_0 &= c_1\cos^2\theta - c_2\sin^2\theta + c_3\sin\theta\cos\theta \\ d_1 &= 2(c_4\sqrt{3}\cos\theta - c_5\sin\theta) \\ d_2 &= B_0(4 - 3B_0 c_0) \end{aligned} \tag{5}$$

$$\begin{aligned} c_0 &= \frac{(2-\sqrt{3}B_1)(2B_0-\sqrt{3}B_1)}{(B_1(1+B_0)-\sqrt{3}B_0)^2} & c_1 &= 3 - c_0(1+B_0)^2 \\ c_2 &= 1 + 3c_0(1-B_0)^2 & c_3 &= 2c_0\sqrt{3}(1-B_0^2) \\ c_4 &= (1+B_0)(1-B_0 c_0) & c_5 &= (1-B_0)(1-3B_0 c_0) \end{aligned} \tag{6}$$

subject to the following restrictions

$$\frac{1}{2} < B_0 \le 1 \qquad \frac{\sqrt{3}B_0}{1+B_0} < B_1 < \frac{2B_0}{\sqrt{3}} \qquad B_1 = \frac{\rho_s}{\rho_c} \tag{7}$$

Given $\bar{\rho}_c = \bar{\rho}\, r$, and using (1) one obtains

$$F = (\sqrt{\frac{3}{2}})^{\frac{2}{\gamma}}(\bar{\rho}r)^{\frac{2}{\gamma}} + \frac{m}{3}(\sqrt{\frac{3}{2}})^{\frac{1}{\gamma}}(\bar{\rho}r)^{\frac{1}{\gamma}} + \frac{m}{\sqrt{3}}\bar{\xi} - 1 = 0 \qquad (8)$$

B_1 may be expressed as a function of B_0 in terms of its upper and lower limits

$$B_1 = \frac{\sqrt{3}B_0}{1 + B_0} + \alpha(\frac{2B_0}{\sqrt{3}} - \frac{\sqrt{3}B_0}{1 + B_0}) \qquad \text{where} \quad 0 < \alpha < 1 \qquad (9)$$

The complete surface is described by just four parameters f_c, f_t, γ and α. Furthermore, it appears that α may be fixed at 0.6 for most structural concretes, thus the active number of PNS parameters reduces to 3.

In hardening and softening plasticity, the yield function F depends not only on the stress tensor σ_{ij} but also a number of internal (or state) variables. These variables (collected in the tensor q_k) control the evolution of the surfaces.

The proposed function introduces constraints to ensure that the yield surace meridians always intersect the hydrostatic axis normally in the compression and tension quadrants. This is achieved by means of a C_2 continuous factor $\hat{k}$ operating on the PNS meridians.

$$F = \left(\sqrt{\frac{3}{2}}\,\bar{\rho}r\right)^{\frac{2}{\gamma}} + \frac{m}{3}\left(\sqrt{\frac{3}{2}}\,\hat{k}\bar{\rho}r\right)^{\frac{1}{\gamma}} + \frac{m\hat{k}^{\frac{2}{\gamma}}}{\sqrt{3}}\bar{\xi} - \hat{k}^{\frac{2}{\gamma}} = 0 \qquad (10)$$

$(\hat{k})^2$ is defined as a quadratic function of $\bar{\xi}$ and the strain hardening variable k.

$$\hat{k}^2 = k^p\left(1 - \left(\frac{\bar{\xi}}{\bar{\xi}_h}\right)^2\right) \qquad (11)$$

where p is a new material constant which controls the proximity of the yield surfaces to the PNS surface. The intersection point on the hydrostatic axis $\bar{\xi}_h$ in the compression domain is defined as a function of k, $\bar{\xi}_h = 1/(k - 1)$, allowing $\bar{\xi}_h$ to approach infinity as k tends to 1.

The isotropic-hardening hypothesis describes the manner in which the loading surface grows with respect to a scalar-valued measure of plastic strain. k is expressed in terms of both the equivalent plastic strain ϵ^p and a measure of confinement, $\bar{\xi}$, $\dot{\bar{\epsilon}}^p = \sqrt{\frac{2}{3}\dot{\epsilon}_{ij}^p.\dot{\epsilon}_{ji}^p}$ The confinement dependent hardening variable k_h is given as

$$\dot{k}_h = \frac{\dot{\bar{\epsilon}}^p}{\zeta} \qquad \text{where} \qquad \zeta = -A_h + \sqrt{(A_h)^2 - B_h\bar{\xi} + C_h} \qquad (12)$$

ζ represents the maximum plastic strain, which depends on the level of confinement ($\bar{\xi}$). Equation (12) implies that the material continues to undergo plastic straining as the confinement level increases. A_h, B_h and C_h are new material constants determined from triaxial compression experiments. The relationship between k and k_h is given as $k = k_0 + (1 - k_0)\sqrt{k_h(2 - k_h)}$ where k_0 identifies the initial yield surface.

Softening has been defined as a gradual decrease of the mechanical resistance during a continuously increasing deformation forced upon a material specimen or structure. It is widely accepted that concrete softening is a macroscopic process which results from changes occurring on the microlevel as a result of spatially discrete crackgrowth [9]. Due to the heterogeneity of concrete, a uniformly distributed boundary stress or deformation can result in a highly non-uniform distribution of internal stresses and strains. These gradients cause local concentrations of critical stresses (or strains) leading to rupture of internal bonds. The material is gradually weakened because of the diminishing number of internal bonds which remain to resist the externally applied load.

The principal task when developing a softening formulation is to link the evolution of the yield surface with the degradation of tensile strength and identify the internal variables that control the process. In this formulation, softening of the surface is controlled by the loss of cohesion c. The model introduces a form of kinematic softening whereby the PNS surface translates along the hydrostatic axis (away from the tensile region) to arrive at a *residual* PNS state. With this approach, some minor softening exists even for high levels of confinement.

$$F = \left(\sqrt{\frac{3}{2}}\,\bar{\rho}r\right)^{\frac{2}{\gamma}} + \frac{m}{3}\left(\sqrt{\frac{3}{2}}\,\bar{\rho}r\right)^{\frac{1}{\gamma}} + \frac{m}{\sqrt{3}}\bar{\xi} - c = 0 \tag{13}$$

where c lies in the range $0 \leq c \leq 1$. Note that $(\hat{k})^{\frac{2}{\gamma}}$ does not appear in (13) as $\hat{k} = 1$ when softening is taking place. When $c = 1$, the material is intact and (13) describes the PNS meridians. Whereas, when $c = 0$, the material is considered to be completely fractured; possessing only its residual frictional strength which may be zero (in the case of tensile states) or close to the PNS values (for high confinement compressive states).

The relation between the loss of cohesion and the uniaxial tensile strength f_t is simply $c = \frac{\sigma_t}{f_t}$ where σ_t is the residual, softened, tensile strength. The cohesion c decays exponentially as a function of the *effective crack width* w

$$c = \exp\left(-a_s(w)^{\gamma_s}\right) \tag{14}$$

subject to the restrictions that at $w = 0$, $c = 1$ and the slope $\frac{\partial c}{\partial w}$ is equal to zero. a_s controls the rapidity of the decay and γ_s ensure continuity of the stress-strain curve.

The fracture energy, $G_f = \int_0^{w_r} \sigma_t dw$, is generally treated as being a material constant, independent of the specimen length (although there is not universal agreement about this). Therefore, mapping between the *discrete* crack opening w and the *smeared* tensile strain (used at the macroscopic constitutive level) leads to the introduction of the size ℓ of the region associated with the fracture, normal to the crack $w = \ell \epsilon^f$, where ϵ^f is referred to as the fracture strain. ϵ^f is a measure of tensile plastic strains which have occurred once the PNS surface has been reached.

$$\dot{\epsilon}_f = |\langle \dot{\epsilon}_i^p \rangle| = \sqrt{\langle \dot{\epsilon}_1^p \rangle^2 + \langle \dot{\epsilon}_2^p \rangle^2 + \langle \dot{\epsilon}_3^p \rangle^2} \tag{15}$$

where $\langle \rangle$ are the Macauley brackets which extract the (positive) tensile components of the principal plastic strain rates.

The fracture model for Mode I type tensile cracking is extended to encompass Mode II/III type shear fractures. It is known that distributed micro-cracking occurs under increasing confinement as Mode II or Mode III fracturing appears. Mixed mode failure can be introduced by enforcing plastic softening in compression to be also based on a fracture energy concept which is influenced by the principal stress state. The general crack model can be interpreted as a multiple tensile crack approach. Mixed mode fracturing introduces the idea of the number, N, of cracks which are formed in a specimen under a given state of stress. The resulting fracture energy is $N.G_f$. This approach provides a smooth transition from brittle (abrupt softening) to ductile (mild softening) behaviour on the basis of micro-crack density through inclusion of the stress ratio ξ/ρ and the Lode angle θ.

$$\begin{aligned} N &= \sqrt{2}(-\tfrac{\xi}{\rho} + \tfrac{1}{\sqrt{2}})(1 - \cos(\theta + \tfrac{\pi}{6}))(N_{uc} - 1) + 1 \quad \text{for} \quad \tfrac{\xi}{\rho} < \tfrac{1}{\sqrt{2}} \\ N &= 1 \qquad\qquad\qquad\qquad\qquad\qquad\qquad\qquad \text{for} \quad \tfrac{\xi}{\rho} \geq \tfrac{1}{\sqrt{2}} \end{aligned} \tag{16}$$

Finally, the expression describing the equivalent mixed-mode crack opening displacement rate is given by

$$\dot{w} = \ell \frac{\dot{\lambda}}{N} \left| \left\langle \frac{\partial G}{\partial \sigma_i} \right\rangle \right| \tag{17}$$

No discussion of non-linear constitutive models is complete without mentioning the problems attached to deriving an accurate integration of the rate equations. Here much work has been done by Simo, Hughes and other workers [10], [11], [12], [13], [14]. Tahar [15] produced an automatic sub-incrementation scheme in conjunction with the Closest Point Projection method to cater for predictor-trial stress states well away from the yield surface. Figure 2(b) shows an example of the iteration counts required to return 6400 trial states (starting within the deviatoric plane at the tensile closure in the upper contour plot, or deviatoric plane at the compressive closure in the lower plot) onto the hardening-softening surfaces. The significant point is that a routine has been found which provides a stable corrector to a converged state.

Figure 1(e) illustrates a serious, but apparently un-reported, difficulty resulting from using yield function definitions of the common form (10). Outside the surfaces (where the hydrostatic stress state is beyond the intersection points, the function is undefined. This is problematic since trial stress states will invariable fall into these regions as part of a predictor-corrector return algorithm. With F undefined, the gradients cannot be determined and so the method is inoperative. A further difficulty with (10) is that even with the *closure zone*, F can be highly irregular leading to serious convergence difficulties near the hydrostatic compression closure. Tahar employed a sub-incrementation scheme to handle such cases, but it's use is inefficient for trial states close to the hydrostatic axis. To remedy this, an alternative approach is proposed, whereby the yield function takes the form

$$F = \frac{\bar{\rho}}{g.\hat{k}.\bar{\rho}_c|_\eta} - 1 \tag{18}$$

where $\bar{\rho}_c|_\eta$ identifies the deviatoric stress invariant at the intersection of the current yield surface with a *radial projection* sharing the same ρ/ξ ratio as the trial stress state.

A non-associated plastic potential can be introduced by modifying the volumetric component of the yield surface function. That is, the gradient of the potential function need only differ from the yield function gradient in terms of its volumetric component. As an example, the direction of the plastic strain $\frac{\partial G}{\partial \sigma_{ij}}$ may be obtained by means of the fourth order transformation tensor A_{ijkl}.

$$\frac{\partial G}{\partial \sigma_{ij}} = A_{ijkl} \frac{\partial f}{\partial \sigma_{kl}} \qquad \text{where} \qquad A_{ijkl} = \delta_{ik}\delta_{jl} - \frac{1}{3}(1 - \eta)\delta_{ij}\delta_{kl} \tag{19}$$

The constant η defines the degree of volumetric non-associativity. When $\eta = 1$, an associated flow rule is recovered, whereas when $\eta = 0$ incompressible plastic flow occurs.

There is, however, a serious price to pay for the introduction of non-associativity, in that it results in a non-symmetric tangential stiffness matrix. The latter is not convenient from a numerical point of view. It necessitates greater storage and a more general solution technique for large systems of equation as generated in typical finite element analyses. There is also a second, physically unacceptable consequence of a non-associated rule; the uniqueness of a solution is not guaranteed and energy may be created during a load-unload cycle.

Experimentalists will agree that the shapes of the pre-peak yield surfaces in stress space are not so straightforward to detect from laboratory tests, whereas the PNS surface is. Given the lack of high quality data, many researchers have assumed a fan-like geometric form for the yield surfaces in concrete. The natural consequences of these surfaces is that plastic dilation is predicted at the onset of yielding. As noted above, this leads to excessive volumetric expansion which has been countered-acted by introducing a non-associated flow rule. However, if the yield surfaces are closed in the compression region (rather than fan-like open), then some plastic compaction is obtained in the early stages of deviatoric loading, with plastic dilation only being realised as the PNS surface is approached. Recent data suggest his to be the case for moderate to high confinement conditions (see Figure 1(a), where the incremental flow directions from cylindrical triaxial tests are shown). However, close to the tensile

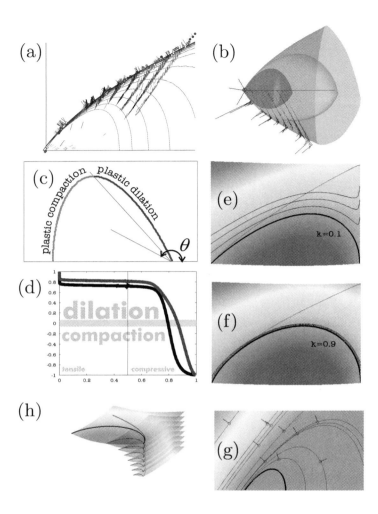

Figure 1: Some details of the plasticity model: (a) incremental flow directions from triaxial compression tests, (b) three nested yield surfaces (showing experimental flow vectors and stress paths), (c) yield surface showing zones of dilation and compaction, (d) dilation/deviatoric plastic strain ratio versus stress ratio, (e) difficulty associated with standard yield function definition (F highly irregular near compression closure and undefined outside closed regions), (f) alternative (*radial*) definition of yield function, (g) further example of alternative yield function showing random stress states with their fully defined flow directions, (h) image of successive yield surfaces using the alternative yield function definition.

zone, associated flow rule will significantly over-estimate the plastic dilation. This is not such a surprise. Material nonlinearity under tensile states could hardly be thought of as resulting from *flow processes*. Plasticity fundamentally describes slip processes and not loss of cohesion. It appears that, if one is to stick with a plasticity model, then in order to achieve reasonable agreement with laboratory data, the degree of non-associativity should increase as tensile states are approached.

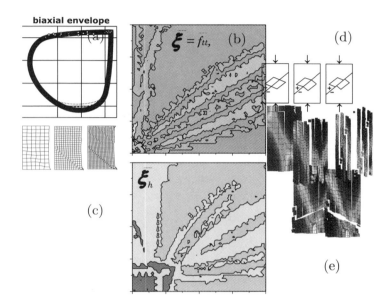

Figure 2: Simulations from the elasto and visco-plastic constitutive model: (a) biaxial envelope, (b) stress return iteration contours (red indicates 1 iteration, light green > 5, (c) kinematic categorisation of displacement field localisation, (d) effect of changing mesh density and mesh alignment on shear-band prediction (flaw introduced in bottom right corner of mesh).

2.2 Viscoplastic regularisation

Complexity arises in many mechanical systems. The term has often been invoked in connection with all kinds of obstacles encountered in the advancement of science. Yet certain challenging engineering problems, which appear at first unrelated, may share similar mathematical structures influenced by numerous interacting events occurring on different geometric scales. Recognition and ordering of such a hierarchical system can lead to new insights in large-scale simulations. Unfortunately, in the case of multi-phase engineering materials, identification of the key constituents and understanding the dominant physio-chemical actions operating on the different scales are not suffi-cient to rid the *complex* label and provide accurate simulations, as the level of resolu-tion needed to capture such processes lie well beyond current and near-future compu-tational capabilities. To overcome this, non-local, probabilistic-based tools could offer a solution which operates on a higher (simpler) scale. As an example, the concept of a periodic, equivalent continuum is regularly employed by constitutive modelers to describe media and events which are inherently aperiodic and discontinuous (such as geomaterial deformation and diffuse micro-cracking in solids). Such an idealisation looses its predictive realism when localisation appears (through material instability) on a scale where significant changes to the topology occurs. By adopting an interact-ing, non-local approach, constructed in a thermodynamically consistent framework, these difficulties may be avoided.

The use of a mesh-dependent *length scale* in the specific fracture energy approach of the plasticity model described above is a common, but fundamentally this is a flawed way of handling the difficulties surrounding strain softening and localisation issues in an equivalent continuum. Over the past 15 years, standard, local continuum constitutive models have been *enriched* by introducing an internal length scale either via some form of non-local approach, a micro-polar method or a rate dependency (for example, see [22]). The latter has been applied to the elasto-plasticity model [16]. The following Duvaut-Lions (D-L) method was used. Starting from the rate expression for the additive decomposition (using matrix notation rather than subscript indices) and the hypoelastic relationship

$$\{\dot{\varepsilon}\} = \{\dot{\varepsilon}^e\} + \{\dot{\varepsilon}^{vp}\} \qquad\qquad \{\dot{\sigma}\} = [D]\{\{\dot{\varepsilon}\} - \{\dot{\varepsilon}^{vp}\}\} \qquad (20)$$

Expressing in incremental form

$$\{\Delta\sigma\} = [D]\{\{\Delta\varepsilon\} - \{\Delta\varepsilon^{vp}\}\} \qquad (21)$$

Defining incremental values corresponding to the visco-plastic time increment Δt by weighting the visco-plastic strain rate at the beginning of the visco-plastic time-step $\{_t\dot{\varepsilon}^{vp}\}$ and at the end of the visco-plastic time-step $\{_{t+\Delta t}\dot{\varepsilon}^{vp}\}$ one may write

$$\{\Delta\varepsilon^{vp}\} = ((1-\theta)\{_t\dot{\varepsilon}^{vp}\} + \theta\{_{t+\Delta t}\dot{\varepsilon}^{vp}\})\,\Delta t \qquad (22)$$

where θ is an interpolation parameter which lies in the range $0 \leq \Theta \leq 1$. For $\theta = 0$, $\theta = \frac{1}{2}$, and $\theta = 1$ we obtain (i) an explicit scheme, (ii) a scheme corresponding to the implicit trapezoidal rule and (iii) a Backward Euler scheme, respectively. Construct-ing a truncated Taylor series expansion of the visco-plastic strain rate at the end of the interval

$$\{_{t+\Delta t}\dot{\varepsilon}^{vp}\} = \{_t\dot{\varepsilon}^{vp}\} + \left\{\frac{\partial\dot{\varepsilon}^{vp}}{\partial\sigma}\right\}^T\{\Delta\sigma\} + \left\{\frac{\partial\dot{\varepsilon}^{vp}}{\partial\sigma^{ep}}\right\}^T\{\Delta\sigma^{ep}\} \qquad (23)$$

or, after substituting the generalised incremental form into the above

$$\{_{t+\Delta t}\dot{\varepsilon}^{vp}\} = \{_t\dot{\varepsilon}^{vp}\} + \frac{[D]^{-1}}{\hat{\tau}}(\{\Delta\sigma\} - \{\Delta\sigma^{ep}\}) \qquad (24)$$

With the approximated value for the visco-plastic strain rate at time $t + \Delta t$ (24) and with the aid of (22) and (21), the incremental stress may be written as

$$\{\Delta\sigma\} = [D_{cons}^{vp}]\{\Delta\varepsilon\} - \{\Delta S\} \tag{25}$$

where $[D_{cons}]$ is the consistent tangent operator for the visco-plastic model, which for the Backward Euler case is

$$[D_{cons}^{vp}] = \frac{\hat{\tau}}{\hat{\tau} + \Delta t}\left[[D] + \frac{\Delta t}{\hat{\tau}}[D_{cons}^{ep}]\right] \tag{26}$$

and $[D_{cons}^{ep}]$ is the consistent tangent operator for the *backbone* inviscid model. $\{\Delta S\}$ identifies the pseudo-stress

$$\{\Delta S\} = \frac{\Delta t}{\hat{\tau} + \Delta t}\{\{_t\sigma\} - \{_t\sigma^{ep}\}\} \tag{27}$$

Figure 2(e) shows strain contours obtained from using the D-L elasto-visco-plasticity model when simulating compression and shearing in a prismatic specimen with an artificially introduced defect in one corner (lower right, Figure 2(d)). The figures show the influence of altering the FE mesh alignment on the predicted form of the localisation bands. Preliminary studies have not been conclusive. It is evident that introducing a rate dependency into this model has the effect of spreading-out the zone where deformations concentrate, however, determination of a functional form of the viscosity that yields results independent of the mesh orientation and density has yet to be realised. One should not expect objectivity across all meshes, as sufficient detail always is required to capture properly the evolving morphology of the localisation bands.

2.3 The need for further experimental testing

There remains the need for more high quality experimental data on the time-dependent multi-axial behaviour of concrete. Much of the early work in this field suffered from loading devices which prevented the specimen from moving lateral because of platen friction, and with little, or no attention being given to the role of pore fluid pressures. For further progress, greater care is needed in designing the boundary conditions and capturing the response. The new multi-axial test facility (mac^{2t}) at Sheffield University (Figure 3(a) and (b)) operates on finely ground cubes, loaded via thin-film PTFE tiles and employs a system of 6 lasers to detect specimen deformations down to nanometers. A further feature of this device is its ability to undertake compression testing to $400MPa$ (with the 3 principal stresses controlled independently) at temperatures up to $300^{o}\,C$. The necessity for such work arises from the nuclear industry where the integrity of reactor pressure vessels is of major importance.

The mac^{2t} rig has been used in conjunction with another apparatus which determines the residual tensile properties (F_t, G_f, E, Fig 3(e) and (c)) of specimens exposed to higher temperatures and pre-compression. This device also attempts to provide a uniform displacement field throughout the tensile specimen to allow smeared fracture tests to be performed and the material characteristic length to be calculated (Fig 3(f) and (d)).

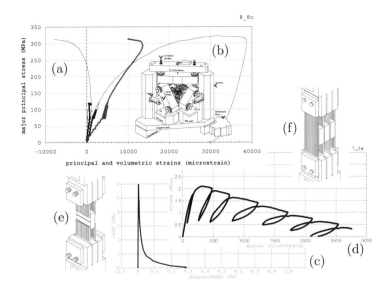

Figure 3: Experimental compression and tensile test results from Sheffield University: (a) triaxial compression at $70MPa$ confinement, (b) mac^{2t}, (c) discrete Mode I tension test on a notched prism as performed in (d), (e) *smeared* Mode I test as performed on a 3-slice specimen in (f).

3 Non-linear FE fluid-soil-structure interaction

Section 3.1 shows how the incremental implicit form of non-linear dynamic force balance for a deformable solid coupled to a slow moving fluid domain is constructed (for further details see [17]).

In section 3.2, alternative numerical methods of treating the dynamic far-field when undertaking a soil-structure interaction analyses are briefly reviewed. Both direct and sub-structure-based methods are considered. This paper concentrates on the use of the novel Scaled Boundary Finite Element Method (SBFEM) developed by Wolf and Song. The time-domain solution is described, both in terms of its theoretical background and numerical application within a 3D non-linear dynamic Finite Element code *ya*FE*c*.

3.1 Dynamic equilibrium: Implicit HHT scheme

The Hilbert, Hughes and Taylor time-integration scheme (α method) is adopted here. This popular scheme introduces damping (of the erroneously over represented higher frequency modes) into the classical Newmark method without significantly degrading the accuracy. If $\alpha = 0$, the method becomes identical to the Newmark scheme. An unconditionally stable second-order algorithm ([18] and the references therein). is obtained if $-\frac{1}{3} \leq \alpha \leq 0$, $\gamma = \frac{1}{2} - \alpha$ and $\beta = \frac{1}{4}(1 - \alpha)^2$. The following *sub* and *super* script notation applies $_{time-step}^{iteration}d$ (over-dots indicate time differentiation).

The standard Newmark approximation to the up-dated displacements is

$$\{_{t+\Delta t}^{k+1}d\} = \{_{t+\Delta t}^{k}d\} + \{_{t+\Delta t}^{k+1}\delta d\} \tag{28}$$

The up-dated accelerations are given by

$$\{_{t+\Delta t}^{k+1}\ddot{d}\} = \frac{1}{\beta(\Delta t)^2}\left\{ \overbrace{\{_{t+\Delta t}^{k}d\} + \{_{t+\Delta t}^{k+1}\delta d\}}^{\{_{t+\Delta t}^{k+1}d\}} - \{_t d\} - \Delta t\{_t \dot{d}\} \right\} - \frac{1-2\beta}{2\beta}\{_t \ddot{d}\} \tag{29}$$

Finally, the up-dated velocity may be written as

$$\{_{t+\Delta t}^{k+1}\dot{d}\} = \left(1 - \frac{\gamma}{\beta}\right)\{_t \dot{d}\} + \Delta t\left(1 - \frac{\gamma}{2\beta}\right)\{_t \ddot{d}\} + \frac{\gamma}{\beta\Delta t}\left(\{_{t+\Delta t}^{k}d\} + \{_{t+\Delta t}^{k+1}\delta d\} - \{_t d\}\right) \tag{30}$$

The incremental equation of motion may be expressed in the form $[K^*]\{_{t+\Delta t}^{k+1}\delta d\} = \{\delta f^*\}$. This is obtained from the HHT dynamic equilibrium expression

$$[M]\{_{t+\Delta t}^{k+1}\ddot{d}\} + (1+\alpha)[C]\{_{t+\Delta t}^{k+1}\dot{d}\} - \alpha[C]\{_t \dot{d}\}+$$

$$(1+\alpha)[_{t+\Delta t}^{k+1}\bar{K}]\{_{t+\Delta t}^{k+1}d\} - \alpha[_t\bar{K}]\{_t d\} \qquad = (1+\alpha)\{_{t+\Delta t}f_{ext}\} - \alpha\{_t f_{ext}\} \tag{31}$$

where $[\bar{K}]$ indicates a secant stiffness matrix (from the origin to $\{_{t+\Delta t}^{k+1}d\}$ in the case of $[_{t+\Delta t}^{k+1}\bar{K}]$, or from the origin to $\{_t d\}$ in the case of $[_t\bar{K}]$).

Substituting $\{_{t+\Delta t}^{k+1}d\} = \{_{t+\Delta t}^{k}d\} + \{_{t+\Delta t}^{k+1}\delta d\}$ into the above, one arrives at

$$[M]\{_{t+\Delta t}^{k+1}\ddot{d}\} + (1+\alpha)[C]\{_{t+\Delta t}^{k+1}\dot{d}\} - \alpha[C]\{_t \dot{d}\}+$$

$$(1+\alpha)[_{t+\Delta t}^{k+1}\bar{K}]\left\{\{_{t+\Delta t}^{k}d\} + \{_{t+\Delta t}^{k+1}\delta d\}\right\} - \alpha[_t\bar{K}]\{_t d\} \qquad = (1+\alpha)\{_{t+\Delta t}f_{ext}\} - \alpha\{_t f_{ext}\} \tag{32}$$

$$[_{t+\Delta t}^{k+1}\bar{K}]\left\{\{_{t+\Delta t}^{k}d\} + \{_{t+\Delta t}^{k+1}\delta d\}\right\} = [_{t+\Delta t}^{k}\bar{K}]\{_{t+\Delta t}^{k}d\} + [_{t+\Delta t}^{k+1}\tilde{K}]\{_{t+\Delta t}^{k+1}\delta d\} \tag{33}$$

where $[_{t+\Delta t}^{k+1}\tilde{K}]$ is the *chord* stiffness.

We may also write

$$\{_{t+\Delta t}^{k}f_{int}\} = [_{t+\Delta t}^{k}\bar{K}]\{_{t+\Delta t}^{k}d\} \qquad \text{and} \qquad \{_t f_{int}\} = [_t\bar{K}]\{_t d\} \tag{34}$$

$$[M]\{_{t+\Delta t}^{k+1}\ddot{d}\} + (1+\alpha)[C]\{_{t+\Delta t}^{k+1}\dot{d}\} - \alpha[C]\{_t \dot{d}\}+$$

$$(1+\alpha)[_{t+\Delta t}^{k+1}\tilde{K}]\{_{t+\Delta t}^{k+1}\delta d\} + (1+\alpha)\{_{t+\Delta t}^{k}f_{int}\} - \alpha\{_t f_{int}\} \qquad = (1+\alpha)\{_{t+\Delta t}f_{ext}\} - \alpha\{_t f_{ext}\} \tag{35}$$

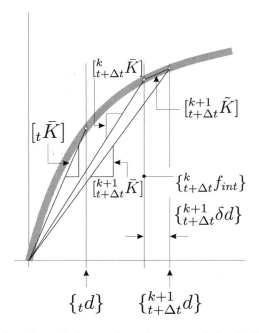

Figure 4: Schematic illustration of secant and chord stiffnesses

However, $[^{k+1}_{t+\Delta t}\tilde{K}]$ is unknown. In it's place, the tangent stiffness at time-step $t + \Delta t$ and iteration k (that is, $[^{k}_{t+\Delta t}K]$) is used within a Newton-Raphson iterative solution strategy. This approach will result in an *out-of-balance* residual force (which one aims to reduce within a pre-defined tolerance) during successive iterations.

$$[M]\{^{k+1}_{t+\Delta t}\ddot{d}\} + (1+\alpha)[C]\{^{k+1}_{t+\Delta t}\dot{d}\} - \alpha[C]\{_t\dot{d}\}+$$

$$(1+\alpha)[^{k}_{t+\Delta t}K]\{^{k+1}_{t+\Delta t}\delta d\} + (1+\alpha)\{^{k}_{t+\Delta t}f_{int}\} - \alpha\{_t f_{int}\} \quad = (1+\alpha)\{_{t+\Delta t}f_{ext}\} - \alpha\{_t f_{ext}\} \tag{36}$$

Substituting (29) and (30) into (36), and re-arranging

$$\left[\frac{1}{\beta(\Delta t)^2}[M] + \frac{\gamma(1+\alpha)}{\beta\Delta t}[C] + (1+\alpha)[^{k}_{t+\Delta t}K]\right]\{^{k+1}_{t+\Delta t}\delta d\} =$$

$$- \left[\frac{1}{\beta(\Delta t)^2}[M] + \frac{\gamma(1+\alpha)}{\beta\Delta t}[C]\right]\{^{k}_{t+\Delta t}d\} + \left[\frac{1}{\beta(\Delta t)^2}[M] + \frac{\gamma(1+\alpha)}{\beta\Delta t}[C]\right]\{_t d\}$$

$$+ \left[\frac{1}{\beta\Delta t}[M] + \left(\frac{\gamma(1+\alpha)}{\beta} - 1\right)[C]\right]\{_t\dot{d}\} + \left[\frac{1-2\beta}{2\beta}[M] + \Delta t(1+\alpha)(\frac{\gamma}{2\beta} - 1)[C]\right]\{_t\ddot{d}\} \tag{37}$$

$$- (1+\alpha)\{^{k}_{t+\Delta t}f_{int}\} + \alpha\{_t f_{int}\} + (1+\alpha)\{_{t+\Delta t}f_{ext}\} - \alpha\{_t f_{ext}\}$$

Thus, we may identify $[K^*] = \left[\frac{1}{\beta(\Delta t)^2}[M] + \frac{\gamma(1+\alpha)}{\beta\Delta t}[C] + (1+\alpha)[^k_{t+\Delta t}K] \right]$ and

$$\{\delta f^*\} = \left[\frac{1}{\beta(\Delta t)^2}[M] + \frac{\gamma(1+\alpha)}{\beta\Delta t})[C] \right] \overbrace{\{\{_td\} - \{^k_{t+\Delta t}d\}\}}^{-\{^k_{t+\Delta t}\Delta d\}} +$$

$$\left[\frac{1}{\beta\Delta t}[M] + \left(\frac{\gamma(1+\alpha)}{\beta} - 1 \right)[C] \right] \{_t\dot{d}\} + \left[\frac{1-2\beta}{2\beta}[M] + \Delta t(1+\alpha)(\frac{\gamma}{2\beta} - 1)[C] \right] \{_t\ddot{d}\} -$$

$$(1+\alpha)\{^k_{t+\Delta t}f_{int}\} + \alpha\{_tf_{int}\} + (1+\alpha)\{_{t+\Delta t}f_{ext}\} - \alpha\{_tf_{ext}\}$$

(38)

Allowing for Rayleigh damping ($[C] = A[M] + B[_0K]$) in parts of the structure

$$[K^*] = \left[\overbrace{\frac{1 + A\gamma\Delta t(1+\alpha)}{\beta(\Delta t)^2}}^{c_2}[M] + \overbrace{\frac{B\gamma(1+\alpha)}{\beta\Delta t}}^{c_1}[_0K] + \overbrace{(1+\alpha)}^{c_0}[^k_{t+\Delta t}K] \right]$$

(39)

$$\{\delta f^*\} = [c_2[M] + c_1[_0K]] \overbrace{\{\{_td\} - \{^k_{t+\Delta t}d\}\}}^{-\{^k_{t+\Delta t}\Delta d\}} +$$

$$\left[\overbrace{\frac{1 + A\Delta t(\gamma(1+\alpha) - \beta)}{\beta\Delta t}}^{c_4}[M] + \overbrace{B\left(\frac{\gamma(1+\alpha)}{\beta} - 1 \right)}^{c_3}[_0K] \right] \{_t\dot{d}\} +$$

(40)

$$\left[\overbrace{\frac{A\Delta t(1+\alpha)(\gamma - 2\beta) + (1-2\beta)}{2\beta}}^{c_6}[M] + \overbrace{\frac{B\Delta t(1+\alpha)(\gamma - 2\beta)}{2\beta}}^{c_5}[_0K] \right] \{_t\ddot{d}\} -$$

$$(1+\alpha)\{^k_{t+\Delta t}f_{int}\} + \alpha\{_tf_{int}\} + (1+\alpha)\{_{t+\Delta t}f_{ext}\} - \alpha\{_tf_{ext}\}$$

Now consider the introduction of a fluid domain bounding the deformable solid. Using a pressure-based (p) formulation for the compressive, inviscid fluid, one has

$$\begin{bmatrix} K^*_s & -(1+\alpha)[Q] \\ -(1+\alpha)[Q]^T & \bar{K}^*_f \end{bmatrix} \left\{ \begin{array}{c} ^{k+1}_{t+\Delta t}\delta d \\ ^{k+1}_{t+\Delta t}\delta p \end{array} \right\} = \left\{ \begin{array}{c} \sum^{VI}_{i=I}\delta f^i_s \\ \sum^{VI}_{i=I}\delta f^i_f \end{array} \right\}$$

(41)

where

$$[K^*_s] = \left[c_2[M_s] + c_1[^0K_s] + \overbrace{(1+\alpha)}^{c_0}[^{t+\Delta t}K^k_s] \right]$$

$$[\bar{K}^*_f] = -\left[\frac{c_0}{\rho_f}[M_f] + \overbrace{\frac{1}{\rho_f}\frac{1}{\beta\Delta t^2}(1+\alpha)^2}^{c_7}[K_f] \right]$$

(42)

$$\{\delta f^I_s\} = c_0[Q]\{_{t+\Delta t}p^k\} - \alpha[Q]\{_tp\}$$
$$\{\delta f^{II}_s\} = -[c_2[M_s] + c_1[^0K_s]]\{_{t+\Delta t}d^k\}$$
$$\{\delta f^{III}_s\} = [c_2[M_s] + c_1[^0K_s]]\{_t d\}$$
$$\{\delta f^{IV}_s\} = [c_4[M_s] + c_3[^0K_s]]\{_t\dot{d}\}$$
$$\{\delta f^V_s\} = [c_6[M_s] + c_5[^0K_s]]\{_t\ddot{d}\}$$
$$\{\delta f^{VI}_s\} = c_0\{_{t+\Delta t}f^{ext}_s\} - \alpha\{_tf^{ext}_s\} - c_0\{^k_{t+\Delta t}f^{int}_s\} + \alpha\{_tf^{int}_s\}$$

and

$$\{\delta f_f^I\} \;=\; c_0[Q]^T\{_{t+\Delta t}d^k\}$$

$$\{\delta f_f^{II}\} \;=\; -c_0[Q]^T\{_t d\} - \overbrace{(1+\alpha)\Delta t}^{c_9}[Q]^T\{_t\dot{d}\} - \overbrace{\frac{(\Delta t)^2(1+\alpha)(1-2\beta)}{2}}^{c_8}[Q]^T\{_t\ddot{d}\}$$

$$\{\delta f_f^{III}\} \;=\; \left[\frac{c_0}{\rho_f}[M_f] + \frac{1}{\rho_f}\overbrace{\beta\Delta t^2(1+\alpha)^2}^{c_7}[K_f]\right]\{_{t+\Delta t}^{k}p\}$$

$$\{\delta f_f^{IV}\} \;=\; -\left[\frac{1}{\rho_f}c_0[M_f] + \frac{1}{\rho_f}\overbrace{\beta(\Delta t)^2(1+\alpha)\alpha}^{c_{10}}[K_f]\right]\{_t p\} \tag{43}$$

$$\{\delta f_f^{V}\} \;=\; -\left[\frac{1}{\rho_f}\overbrace{\Delta t(1+\alpha)}^{c_9}[M_f]\right]\{_t\dot{p}\}$$

$$\{\delta f_f^{VI}\} \;=\; -\left[\frac{1}{\rho_f}\overbrace{(\Delta t)^2(1+\alpha)(1-2\beta)}^{c_8}[M_f]\right]\{_t\ddot{p}\}$$

3.2 Dynamic soil-structure interaction: Scaled Boundary Finite-Element Method

The search for a rigorous, accurate and computationally efficient technique which is able to simulate the behaviour of an un-bounded anisotropic soil domain posses a further challenge within engineering mechanics. Such a model forms an essential ingredient of any dynamic soil-structure analysis. When comparing the response of a structure embedded in flexible soil with that of the same structure founded on rigid rock, one notes that (i) the presence of the soil makes the dynamic system more flexible and (ii) the radiation of energy of the propagating waves away from the structure (if occurring) increases the damping. If seismic motion is applied at the base of the structure then the free-field motion will differ from the *bed-rock* control motion; usually being amplified towards the free surface. The excavation of the soil and insertion of a stiff base results in some averaging of the translation plus a *rotation* of the foundation given horizontal earthquake motion and the inertial loads resulting from the motion of the structure will further modify the seismic motion along the base of the structure.

The degree of this interaction depends not only on the foundation stiffness, but also the stiffness and mass properties of the structure and the nature of the applied excitation. Ignoring DSSI effects could lead to an overly conservative design.

To illustrate the consequences of using an extended mesh with different dimensions, consider the vertical pressure loading of an elastic soil half-space. Symmetry allows only one quarter of the mesh to be examined (this is of course an axi-symmetric problem but the subsequent structural analysis is not, so this problem serves as a useful benchmark example). The next two figures (5) and (6) show the stress contours illustrating wave propagation (including reflections). A graph of the vertical displacement history is also shown for each mesh.

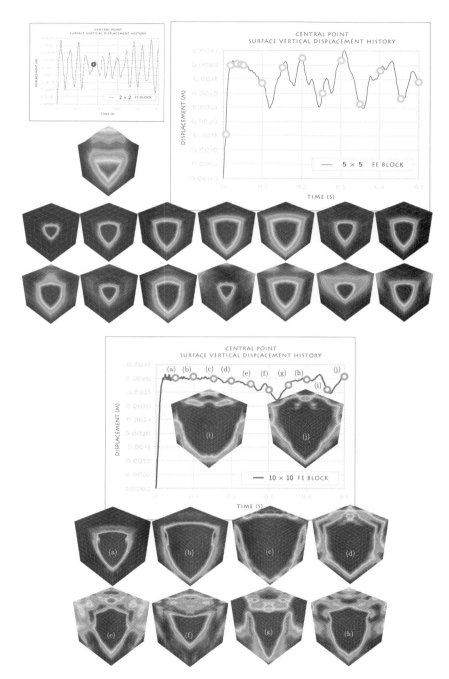

Figure 5: Halfspace problem: Vertical displacement versus time with stress contours showing wave propagation in the extended FE meshes. Top left 2 × 2 × 2, top right 5 × 5 × 5, bottom 10 × 10 × 10 FE mesh.

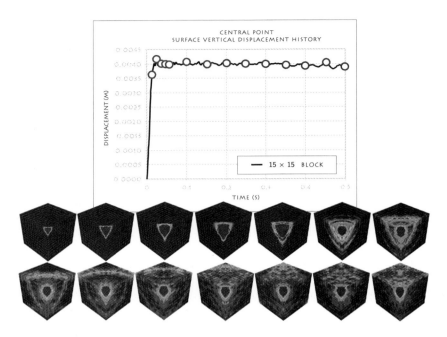

Figure 6: Halfspace problem: Vertical displacement versus time with stress contours showing wave propagation in the extended 15 × 15 FE mesh.

Notice the very noisy signal (due to multiple stress wave reflections) for the smallest mesh. The conventional classification of DSSI methods to model the dynamic elastic far-field (direct and substructure methods) is not so useful when performing truly non-linear time-domain analyses, which must be undertaken in the time domain). Perhaps a clearer classification is one which distinguishes between methods that are *local in space and time* from those that are *non-local in space and time*. Direct methods could fall into the former category and sub-structure methods generally lie in the latter.

A great variety of local techniques exist. These include lumped parameter or cone models based on relatively simple strength-of-materials analytical concepts, using discrete springs, masses and dashpots. Basic models of this kind often are constructed on the assumption that the foundation lies on the surface of a halfspace (that is, it is not embedded). Much effort has been expended on developing efficient transmitting, or silent, boundaries. However, these (like cone or lumped parameter idealisations) are invariably only silent for planar waves striking the boundary normally. Infinite elements have proved effective in static soil-structure interaction problems. However, they fail to provide a rigorous general solution in dynamic SSI. The difficulty with this approach is the necessity to predefine the form of the decay shape functions a priori, but this is not possible since the behaviour differs depending on whether the excitation lies above or below the cut-off frequency.

In non-local approximations the solution is generally formulated in terms of the dynamic interaction forces on a fictitious boundary marking the limit of the conventional FE discretisation. The derivation is usually constructed in the frequency domain. The time domain relationship following from the convolution integral. This leads to the notion of a unit impulse response matrix. The consequences of this non-local formulation are that unit impulse response matrices need to be calculated at each time step. These matrices, in general, will be fully populated (but symmetric). Respecting the fundamental concept of conservation of momentum, employing a strain-displacement relationship and introducing an elastic constitutive relationship allows the governing PDEs to be derived for the far-field. For practical cases, the shape of the boundary and variation of the material properties preclude an analytical solution of the governing PDEs of elastodynamics. These PDEs may be reformulated into integral expressions using weighted residual principles. It is possible to arrive at a set of integral equations which represent the medium's behaviour involving only boundary integrals. Such an approach implicitly incorporates the Sommerfield radiation condition (the far-field is a sink, not a source). However, solutions have not yet be found for all classes of material anisotropy and complicated singular integrals may appear in the solution. The numerical treatment of these integrals is referred to as the Boundary Element Method which has proved to be extremely useful in rigorous DSSI analyses.

Over the past decade a new technique based on FE methods has emerged [19]. The approach has undergone a number of developments, reflected in the changing names: sub-structure cloning, similarity-based far-field, consistent infinitesimal boundary finite element and the scaled boundary finite element method. The approach recognisees that if the physical problem can be represented by ODEs, classical methods can lead to an exact analytical solution. To offer this possibility when modelling the dynamic far-field, the governing PDEs may be transformed from the global Cartesian coordinate system to a scaled boundary coordinate system. In the circumferential direction the boundaries are discretised using surface finite elements, reducing the PDEs to ODEs in the radial coordinate. The method is semi-analytical. In the circumferential directions the FE weighted residual approximations (using isoparametric concepts) apply, leading to convergence in the FE sense.

Displacements along a ray emanating from the scaling centre are functions of the radial dimension alone. Formulated in the frequency domain, one obtains a system of 2nd-order ODES for the (frequency dependent) displacement amplitude in terms of the radial coordinate. The dynamic stiffness matrix for the unbounded soil may be obtained from the interaction force-displacement relationship at a surface with a constant radial dimension. An asymptotic expansion of the dynamic stiffness matrix at high frequency permits the radiation damping condition to be satisfied rigorously. This asymptotic expansion of the dynamic stiffness corresponds to the early-time asymptotic expansion of the unit impulse response matrix in the time domain.

Introduction of SBF elements on the structure-unbounded medium interface results in the following effective internal force appearing on the left side of the dynamic

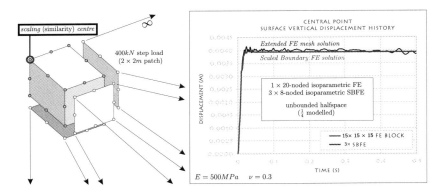

Figure 7: Halfspace problem: Vertical displacement versus time using a single FE and three SBFEs.

equation of equilibrium

$$[_1M^\infty]\left\{\{^{k+1}_{t+\Delta t}\dot{d}\} - \{_t\dot{d}\}\right\} + \underbrace{\sum[_{t+\Delta t \to 2\Delta t}M^\infty]\left\{\{_{\Delta t \to t}\dot{d}\} - \{_{0 \to t-\Delta t}\dot{d}\}\right\}}_{convolution} \tag{44}$$

$[M^\infty]$ represents a piecewise constant unit impulse matrix holding the same dimensions as a damping matrix. The numerical approximation to the *convolution* integral may be moved to the (*known*) right side of the equilibrium equation, as can the $[_1M^\infty]\{_t\dot{d}\}$ term. This leaves the $[_1M^\infty]\{^{k+1}_{t+\Delta t}\dot{d}\}$ term on the left. However, using the Newmark approximation (30), one may express this contribution as

$$[_1M^\infty]\left\{\frac{\gamma}{\beta\Delta t}\left\{\{^k_{t+\Delta t}d\} + \{^{k+1}_{t+\Delta t}\delta d\} - \{_t d\}\right\} + \left(1 - \frac{\gamma}{\beta}\right)\{_t\dot{d}\} + \Delta t\left(1 - \frac{\gamma}{2\beta}\right)\{_t\ddot{d}\}\right\} \tag{45}$$

So, moving the *known* terms to the right and adding to

$$[_1M^\infty]\{_t\dot{d}\} - \sum[_{t+\Delta t \to 2\Delta t}M^\infty]\left\{\{_{\Delta t \to t}\dot{d}\} - \{_{0 \to t-\Delta t}\dot{d}\}\right\} \tag{46}$$

On the left we obtain

$$[_1M^\infty]\frac{\gamma}{\beta\Delta t}\{^{k+1}_{t+\Delta t}\delta d\} \tag{47}$$

On the right we get

$$\begin{aligned}&[_1M^\infty]\left\{\frac{\gamma}{\beta\Delta t}\left\{-\{^k_{t+\Delta t}d\} + \{_t d\}\right\} - \left(1 - \frac{\gamma}{\beta}\right)\{_t\dot{d}\} - \Delta t\left(1 - \frac{\gamma}{2\beta}\right)\{_t\ddot{d}\}\right\} + \\ &[_1M^\infty]\left\{_t\dot{d}\right\} - \sum[_{t+\Delta t \to 2\Delta t}M^\infty]\left\{\{_{\Delta t \to t}\dot{d}\} - \{_{0 \to t-\Delta t}\dot{d}\}\right\}\end{aligned} \tag{48}$$

which may be re-arranged as follows

$$
\overbrace{\tfrac{\gamma}{\beta \Delta t}[{}_1 M^\infty]\{-{}^k_{t+\Delta t}\Delta d\}}
$$
$$
-\frac{\gamma}{\beta \Delta t}[{}_1 M^\infty]\{{}^k_{t+\Delta t}d\} + \frac{\gamma}{\beta \Delta t}[{}_1 M^\infty]\{{}_t d\} + \tfrac{\gamma}{\beta}[{}_1 M^\infty]\{{}_t \dot d\} - \Delta t \left(1 - \tfrac{\gamma}{2\beta}\right)[{}_1 M^\infty]\{{}_t \ddot d\} \qquad (49)
$$
$$
- \sum [{}_{t+\Delta t \to 2\Delta t} M^\infty]\left\{\{{}_{\Delta t \to} {}_t \dot d\} - \{{}_{0 \to} {}_{t-\Delta t}\dot d\}\right\}
$$

Substituting the above into (39) and (40), we get (for an unbounded solid domain)

$$
[K^*] = \left[c_2[M] + c_1[{}_0 K] + c_0[{}^k_{t+\Delta t}K] + \frac{\gamma}{\beta \Delta t}[{}_1 M^\infty]\right] \qquad (50)
$$

$$
\{\delta f^*\} = \left[c_2[M] + c_1[{}_0 K] + \frac{\gamma}{\beta \Delta t}[{}_1 M^\infty]\right] \overbrace{\left\{\{{}_t d\} - \{{}^k_{t+\Delta t}d\}\right\}}^{-\{{}^k_{t+\Delta t}\Delta d\}} +
$$
$$
\left[c_4[M] + c_3[{}_0 K] + \tfrac{\gamma}{\beta}[{}_1 M^\infty]\right]\{{}_t \dot d\} + [c_6[M] + c_5[{}_0 K]]\{{}_t \ddot d\} - \qquad (51)
$$
$$
\Delta t \left(1 - \tfrac{\gamma}{2\beta}\right)[{}_1 M^\infty]\{{}_t \ddot d\} - \sum[{}_{t+\Delta t \to 2\Delta t} M^\infty]\left\{\{{}_{\Delta t \to} {}_t \dot d\} - \{{}_{0 \to} {}_{t-\Delta t}\dot d\}\right\} -
$$
$$
(1 + \alpha)\{{}^k_{t+\Delta t}f_{int}\} + \alpha\{{}_t f_{int}\} + (1 + \alpha)\{{}_{t+\Delta t}f_{ext}\} - \alpha\{{}_t f_{ext}\}
$$

Figures 7 and 8 give an indication of the power of the method. In Figure 7 the same halfspace problem as treated earlier by the extended mesh, is analysed. Not only is the response smoother, it is some 2 order of magnitude faster in terms of CPU-time.

Figure 8 ilustrates the effect of embedding a reinforced concrete tank within a stiff soil and leaving empty, or filling with water, before striking with a pressure impulse on the top slab. Very different *bounce-back* results are seen depending on the nature of the ground support and presence of the fluid.

Figure 9 shows the deformation contours resulting from an aircraft impacting a $61m$ high nuclear containment vessel. It is interesting to note that the maximum reinforcement stresses reduce when soil-structure interaction is included.

4 Instability monitors: A hierarchical approach

Large (many degrees-of-freedom) nonlinear FE analyses are prone to instabilities for all but the simplest constitutive models. Rapid strain softening in brittle behaviour can impose a severe test on the convergence of a Newton-Raphson scheme. Indicators of when, and the reasons why, FE codes crash, are of real importance. The following approach, although not novel [20] is helpful. By examining the behaviour on, and consequences of, non-linearity on several levels, greater understanding may be gained.

Recall that material instabilities are associated with an acute intensification of the spatial strain gradient. Depending on the smoothness of the displacement field we can distinguish three kinematic descriptions ([21] and references therein)

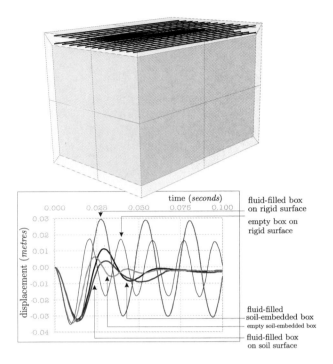

Figure 8: Example of a dynamic fluid-structure interaction analysis: Water-filled reinforced concrete box embedded within a soft elastic soil. A square-wave (step) impulse was applied to the central element within the top slab of this simplified analysis. The presence of the water and the different foundation conditions significantly influence the degree of *bounce-back* in the top slab. Graph shows the central vertical displacement history for 5 cases.

1. *diffuse localisation* identifies the case where high strains are concentrated in a narrow band while much lower strains exist outside this region and the strain field remains continuous throughout. This state is attained when

$$det|^{sym}D^{ep}| \leq 0 \qquad (52)$$

2. *weak discontinuities* describe a finite band where localised deformations occur. The displacement field remains continuous but the strain field exhibits a jump (only in the direction normal to the localisation band). In physical terms a weak discontinuity corresponds to a damage process zone with a near constant high density of micro-defects. Weak localisation is initiated when

$$det|Q^{ep}| = 0 \qquad (53)$$

where the acoustic tensor $Q_{jk} = n_i D_{ijkl} n_l$.

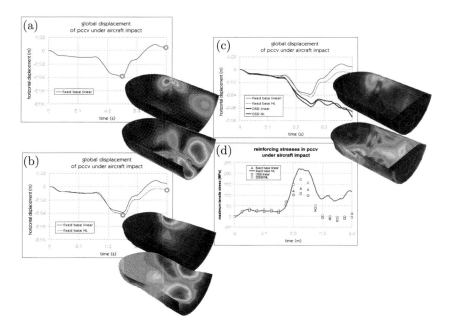

Figure 9: Displacement contours, displacement histories and reinforcement stresses in PCCV under aircraft impact: (a) elastic response for fully-fixed (rigid) base case, (b) elasto-plastic analysis, (c) elastic and elasto-plastic analysis for case where vessel is embedded $(5m)$ in stiff clay and (d) reinforcement stress histories for different analyses.

3. *strong discontinuities* describe where jumps occur in the displacement field across the discontinuity surface. In this case the strain field consists of a regular part obtained by standard differentiation of the displacement field, and a singular part, having the character of a multiple of the Dirac delta distribution. In physical terms a strong discontinuity corresponds to a sharp crack.

The above classification refers to the *material* (mesoscopic) level. Within a Finite Element approach, two further levels are worth investigating (i) the elemental level, where the element frequency spectrum, or element determinant may be calculated and (ii) the global structural level, where the eigenvalues may also be examined and the diagonal monitored. It is evident that the lowest levels inform the higher levels. Instability at a material point level may not lead to instability on an elemental or structural level.

The upper half of Figure 10 illustrates some results from a split Hopkinson bar type experiment on a rectangular, prismatic concrete specimen. In this experiment, a tensile stress wave is transmitted through the specimen via steel bars. These bars should be sufficiently long so as to prevent stress wave reflections occurring at the end

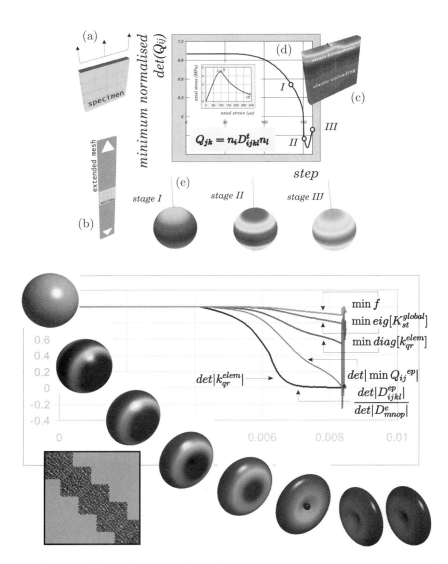

Figure 10: Instability indicators: Upper figure shows response of a central sampling point within a concrete split Hopkinson bar simulation. The main graph indicates the change in the minimum determinant of the acoustic tensor with time. The lower figure shows similar (but more detailed) results for a beam under dynamic flexure. The evolution in the acoustic tensor is illustrated by the initially spherical surface (undamaged), becoming disc-like then self inverting as the structure is on the point of collapse.

within the time of interest. Using two similarity centres placed at opposite infinite poles, only the boundary to the specimen need be modelled (figure 10(a)) rather than an extended mesh (only a portion of which is shown in figure 10(b)). Figure 10(d) shows the drop in the normalised minimum determinate of the acoustic tensor at a particular Gauss point at the top of the specimen. Stages I, II and III are identified by the corresponding stress-strain curve in the inset graph. Interestingly, at this point, the determinate begins to recover a more positive value prior to the analysis failing to reach convergence. The three spheres in part (e) of figure 10 indicate the directional variation in the acoustic tensor determinant.

Finally, in the lower part of figure 10, the results from a simple reinforced concrete beam analysis are shown. The decay in six different measures of instability are reported. It appears in this example that change in the minimum value of the element stiffness matrix diagonal, in the most distressed element, predicts the impending trouble as the rate of decay accelerates like that of $det|Q|$. Monitoring the minimum diagonal value is far less computational involved than determining the minimum acoustic tensor determinant at each Gauss point. The 3-dimensional object from top left to bottom right, chart the change in the directional properties of the acoustic tensor determinant at a particular sampling point. Note that the 6th *ball* has inverted itself as the determinant falls below zero in that direction.

5   Conclusions

This review paper has introduced are few important components of a FE method to analyse the nonlinear dynamic fluid-soil-structure interaction of reinforced concrete structures. It is shown that the SBFE method can offer an accurate and efficient treatment for the dynamic far-field provided the number of time-steps is limited (the convolution integral is time-consuming, and memory intensive as all past velocities must be stored) and the number of SBF elements is modest. Incorporation of the SBFE method into a fully coupled fluid-structure interaction analysis code using an element-by-element approach leads to considerable time-saving when tackling 3-dimensional problems. Several different instability indicators have been examined. This work is ongoing, but preliminary results suggest monitoring the diagonal terms in the element stiffness matrices may provide simple, rapid measures of the impending collapse.

References

[1] R Barrett, M Berry, T F Chen, J Demmel, J M Donato, J Dongarra, V Eijkhout, R Pozo, C Romine and H Van der vorst, "Templates for the solution of linear systems", SIAM, 1994.

[2] I M Smith, "General purpose parallel finite element programming", 7th ACME conference, Durham, 1999.

[3] J Lubliner, "Plasticity theory", Macmillan Publishing Company, 1990.

[4] E Hoek, E and E T Brown, Empirical strength criterion for rock masses, ASCE, "Journal of Geotechnical Engineering", 106(9), 1013-1035, 1980.

[5] K J Willam and E P Warnke, Constitutive model for the triaxial behaviour of concrete, "IABSE Seminar on Concrete Structures Subjected to Triaxial Stresses III-1", 184, 1974.

[6] G Etse and K J Willam, Fracture energy formulation for inelastic behaviour of plain concrete, ASCE, "J Eng Mech", 120(9), 1983-2011, 1994.

[7] E Pramono and K J Willam, Fracture energy based plasticity formulation of plain concrete, ASCE, "J Eng Mech", 106(9), 1013-1203, 1989.

[8] S K Bhowmik, and J H Long, A general formulation for the cross-sections of yield surfaces in octahedral planes, "NUMENTA 90", Pande and Middleton (Eds.), Elsevier, 795-803, 1990.

[9] H E Read and G A Hegemier, Strain softening of rock, soil and concrete: A review article, "Mechanics of Materials", 3, 271-294, 1984.

[10] J C Simo and T J R Hughes, "Computational Inelasticity", Springer-Verlag New York, 1998.

[11] M Ortiz and E P Popov, Accuracy and stability of integration algorithms for elasto-plastic constitutive equations, "International Journal for Numerical Methods in Engineering", 21, 1561-1576, 1985.

[12] J C Simo and R L Taylor, Consistent tangent operators for rate independent elasto-plasticity, "Computer Methods in Applied Mechanics and Engineering", 48, 101-118, 1985.

[13] C J Pearce, "Computational plasticity in concrete failure mechanics", PhD Thesis, University of Wales, Swansea, 1996.

[14] K Runesson and S Sture and K Willam, Integration in computational plasticity, "Computers and Structures", 30, 119-130, 1988.

[15] B Tahar, "C_2 continuous hardening/softening elasto-plasticity model for concrete", PhD Thesis, University of Sheffield, 2000.

[16] S Mesmar, "On the use of Duvaut-Lions viscosity as a regularisation technique for hardening/softening constitutive models", PhD Thesis, University of Sheffield, 2000.

[17] H S Wu, "3D non-linear dynamic fluid-structure interaction analysis of reinforced concrete structures", PhD Thesis, Department of Civil & Structural Engineering, University of Sheffield, UK, 2000.

[18] M Géradin and A Cardona, Time integration of the equations of motion in mechanical analysis, "Computers & Structures", 33(3), 801-820, 1989.

[19] J P Wolf and C Song, "Finite-Element modelling of unbounded media", John Wiley, 1996.

[20] K Willam, Recent issues in computational plasticity, "Complas II", Owen, Hinton and Onate (Eds), Volume 2, Pineridge Press, 1353-1377, 1989.

[21] N Bićanić, R de Borst, W Gerstle, D W Murray, G Pijaudier-Cabot, V Saouma, K J Willam and J Yamazaki, Computational Aspects of Structures (Chapter 7), "Finite element analysis of reinforced concrete structures II", Proceedings of the International Workshop, J Isenberg (Ed), ASCE, 1991.

[22] L J Sluys, "Wave propagation, localisation and dispersion in softening solids", PhD Thesis, University of Delft, 1992.

©2002, Saxe-Coburg Publications, Stirling, Scotland
Engineering Computational Technology
B.H.V. Topping and Z. Bittnar, (Editors)
Saxe-Coburg Publications, Stirling, Scotland, 147-165.

Chapter 7

Projection Techniques embedded in the PCGM for handling Hanging Nodes and Boundary Restrictions

A. Meyer
Faculty of Mathematics
Technical University of Chemnitz, Germany

Abstract

We consider an adaptive finite element method using so called 'hanging nodes' from subdividing an edge of the actual finite element mesh and subdividing only one of the adjacent triangles (or quadrilaterals) at this edge. The solution method of the resulting linear system requires a solution belonging to a special subspace that generates a conformal finite element function. We discuss the use of a special projection operator within the preconditioned conjugate gradient method and a favorable implementation of this projection in combination with the hierarchical finite element basis for linear and quadratic elements. Analogously, projections onto boundary restrictions are incorporated. The final adaptive run is demonstrated by means of a Signorini contact problem.

Keywords: finite elements, linear equation solver, conjugate gradients, adaptive meshes

1 Introduction

We consider an adaptive finite element method for the numerical solution of partial differential equations, given in a weak formulation

$$\text{find } u \in \mathbb{V}_0 \text{ with } a(u, v) = \langle f, v \rangle \ \forall v \in \mathbb{V}_0. \tag{1}$$

In the simplest example of a Laplace equation, the space $\mathbb{V}_0$ is a subspace of $H^1(\Omega)$ with zero–Dirichlet type boundary conditions on $\partial\Omega$ and

$$a(u, v) = \int_\Omega (\nabla u) \cdot (\nabla v) \, d\Omega \,.$$

We start with this example with a scalar function u to approximate in the first three Chapters of this paper, where the projection technique with hanging nodes are described. Later on we generalize the problem (1) to the case of linear elasticity, where a vector function $\vec{u}$ (the displacement vector) is approximated. Then we use instead of (1) the similar formulation

$$\text{find } \vec{u} \in \mathbb{V}_0 \text{ with } a(\vec{u}, \vec{v}) = \langle \vec{f}, \vec{v} \rangle \; \forall \vec{v} \in \mathbb{V}_0. \tag{2}$$

when

$$a(\vec{u}, \vec{v}) = \int_\Omega \epsilon(\vec{v}) : C : \epsilon(\vec{u}) \; d\Omega$$

An adaptive finite element solution for such a problem starts with a coarse conformal triangulation of Ω, where (1) is approximated from piecewise linear functions w.r.t. the given mesh. After having an approximate solution some error estimator (see i.e. [12, 1] for an overview) leads to an adaptive refinement, i.e. some of the actual triangles are subdivided into 4 equal sub-triangles ('red' subdivision) due to the local large estimated error contribution.

This procedure can disturb the consistency of the mesh, if one triangle is refined on one side of an edge and on the other side not (compare fig.1). There are two possibilities to overcome this difficulty:

1. Usually a 'green' refinement is used for all these un-consistent triangles (τ_1 in fig.1). These 'green' triangles have to be removed before the next refinement steps are done, otherwise some angles could become very small. A generalization to quadrilaterals or bricks is complicate.

2. We accept the so called 'hanging nodes', but have to ensure that the finite element ansatz functions used on this non-conformal triangulation remain continuous piecewise linear (or piecewise quadratic) functions. This idea works on quadrilaterals or bricks as well.

This paper shall be concerned with the case 2. We will show, how the properties of the preconditioned conjugate gradient method can be used in such a way that a solution within this continuous subspace of the non-conformal (discontinuous) function space is guaranteed.

The advantage of such a proceeding is the following:
Either we assemble a stiffness matrix $\tilde{K}$ in the usual way:

> *for each element do*
> > *1) generate element matrix and right hand side*
> > *2) add these into the allocated arrays*

or we work without any assembly of $\tilde{K}$ and carry out a matrix vector multiply within the PCGM element–by–element.
In both cases the effective stiffness matrix $\tilde{K}$ is defined from a non-conformal finite element space in the presence of at least one 'hanging node'. This space, called $\mathbb{V}^{(non)}$ is spanned by the basis functions $\tilde{\varphi}_1, \ldots, \tilde{\varphi}_n$ (on a mesh with n nodes), which are

collected into a row vector $\tilde{\Phi} = (\tilde{\varphi}_1, \ldots, \tilde{\varphi}_n)$ in the following. Each function $\tilde{\varphi}_i$ (nonzero only in triangles that contain the node i as vertex) is as usually the sum of the shape functions defined in the single triangles. Hence, if at least one 'hanging node' occurs in the actual mesh, some of the functions $\tilde{\varphi}_i$ are discontinuous (not in $H^1(\Omega)$).

Usually the 'hanging nodes' do not carry a degree of freedom, their values are defined from the values at neighboring nodes. This is equivalent to the definition of a continuous subspace $\mathbb{V}^{(con)} \subset \mathbb{V}^{(non)}$ with the smaller dimension $(n - \#\text{'hanging nodes'})$ and we look for the finite element solution $u \in \mathbb{V}^{(con)}$ with

$$a(u, v) = \langle f, v \rangle \ \forall v \in \mathbb{V}^{(con)} \cap \mathbb{V}_0. \tag{3}$$

On the other hand, the stiffness matrix $\tilde{K}$ is defined with the basis $\tilde{\Phi}$ of $\mathbb{V}^{(non)}$ from $\tilde{K} = (a(\tilde{\varphi}_j, \tilde{\varphi}_i))_{i,j=1}^n$, so

$$u = \sum_{i=1}^{n} u_i \tilde{\varphi}_i = \tilde{\Phi}\underline{u}$$

is a continuous function in $\mathbb{V}^{(con)}$ only if some restrictions on the vector $\underline{u} \in \mathbb{R}^n$ are fulfilled. (Vectors in $\mathbb{R}^n$ are column vectors and are underlined to distinguish them from functions. From the definition of the basis $\tilde{\Phi}$ as row vector of the functions $\tilde{\varphi}_i$, a short abbreviation of the linear combination $u = \tilde{\Phi}\underline{u}$ is used throughout this paper.)

Note that $\underline{u} = (u_1, \ldots, u_n)^T$ contains expanding coefficients of u w.r.t. the basis considered. These coefficients u_i coincide with the values of the function u at node i only in the case of the nodal basis (e.g. $\tilde{\Phi}$). Later on we consider hierarchical basis functions as another basis in $\mathbb{V}^{(non)}$, then this is no longer true.

Using the basis $\tilde{\Phi}$, we transform (1) into a linear system

$$\tilde{K}\underline{u} = \tilde{\underline{b}},$$

but we have to solve

$$P^T \tilde{K} P \underline{u} = P^T \tilde{\underline{b}}$$

with $\underline{u} \in \mathbb{U} = im\, P \subset \mathbb{R}^n$ and P the projection onto the subspace $\mathbb{U}$ that leads to continuous functions:

$$u = \tilde{\Phi}\underline{u} \in \mathbb{V}^{(con)} \iff \underline{u} \in \mathbb{U}$$

For using these ideas within an adaptive finite element method we have to investigate two basic features:

1. A special variant of PCGM has to be designed, that guarantees a solution within a prescribed subspace $\mathbb{U} \subset \mathbb{R}^n$ working with the larger $(n \times n)$–matrix $\tilde{K}$. This is the goal of Chapter 2.

2. The projector P requires a cheap implementation which is simple for linear elements, but more complicate for quadratic ones. This is discussed in Chapters 3 and 4.

2   The Projected PCGM

2.1   The Basic Conjugate Gradient Method

Both algorithms, the PCGM and the projected PCGM are nothing but variants of the basic CG method, which is well–known from HESTENESS/STIEFEL [5] for a long time, if we replace the symmetric matrix A and the EUCLIDIAN inner product in $\mathbb{R}^n$ by a symmetric operator $\mathcal{A}: \underline{\mathbb{U}} \to \underline{\mathbb{U}}$ with respect to another inner product $\langle \cdot, \cdot \rangle$. (Nothing but this symmetry is used in the proofs for the basic CG [5]).
So, let $\mathcal{A} : \underline{\mathbb{U}} \to \underline{\mathbb{U}}$ be symmetric and positive definite with respect to the inner product

$$\langle \cdot, \cdot \rangle : \underline{\mathbb{U}} \times \underline{\mathbb{U}} \to \mathbb{R}^1 .$$

Then the CG method for solving $\mathcal{A}\underline{u} = \underline{\tilde{b}}$ reads as

$$
\begin{aligned}
\text{Start:} \quad & \underline{u} \in \underline{\mathbb{U}} \text{ arbitrary,} \\
& \underline{w} := \mathcal{A}\underline{u} - \underline{\tilde{b}} \\
& \underline{q} := \underline{w}, \quad \gamma := \langle \underline{w}, \underline{w} \rangle
\end{aligned}
$$

$$
\begin{aligned}
\text{Iteration:} \quad & 1. \quad \delta := \langle \mathcal{A}\underline{q}, \underline{q} \rangle, \quad \alpha := -\gamma/\delta \\
& 2. \quad \underline{\hat{u}} := \underline{u} + \alpha\underline{q} \\
& 3. \quad \underline{\hat{w}} := \underline{w} + \alpha\mathcal{A}\underline{q} \\
& 4. \quad \hat{\gamma} := \langle \underline{\hat{w}}, \underline{\hat{w}} \rangle, \quad \beta := \hat{\gamma}/\gamma \\
& 5. \quad \underline{\hat{q}} := \underline{\hat{w}} + \beta\underline{q}
\end{aligned}
$$

with $(\underline{\hat{u}}, \underline{\hat{q}} \ \hat{\gamma})$instead of $(\underline{u}, \underline{q}, \gamma)$ *goto* 1.

As is well–known, the rate of convergence depends on the eigenvalues of $\mathcal{A}$: If

$$\underline{\gamma} \le \lambda_i(\mathcal{A}) \le \bar{\gamma}, \tag{4}$$

then the k–th step of the iteration has

$$\langle \mathcal{A}(\underline{u} - \underline{u}^*), (\underline{u} - \underline{u}^*) \rangle \le \eta^{2k} \cdot const \tag{5}$$

with $\eta = \frac{1-\sqrt{\xi}}{1+\sqrt{\xi}}$, $\xi = \underline{\gamma}/\overline{\gamma}$ and $\underline{u}^*$ the exact solution.

2.2   Preconditioned Conjugate Gradient Method

If we try to solve a linear $(n \times n)$ system

$$K\underline{u}^* = \underline{b},$$

with an ill–conditioned s.p.d. matrix K, we have to introduce a preconditioner C with 'good' eigenvalues of $C^{-1}K$. Then PCGM follows directly from 2.1 by replacing

$$\mathcal{A} = C^{-1}K, \ \tilde{b} = C^{-1}b$$

and
$$\langle \underline{u}, \underline{v} \rangle = (C\underline{u}, \underline{v})$$

with $(\cdot, \cdot)$ the EUCLIDIAN inner product in $\underline{U} = \mathbb{R}^n$, because $\mathcal{A}$ is s.p.d. w.r.t. $\langle \cdot, \cdot \rangle$.

So, the rate of convergence depends on the spectral bounds of $\mathcal{A} = C^{-1}K$:

$$\gamma \le \lambda_i(C^{-1}K) \le \bar{\gamma} \tag{6}$$

(which is : $\gamma(C\underline{x}, \underline{x}) \le (K\underline{x}, \underline{x}) \le \bar{\gamma}(C\underline{x}, \underline{x}) \ \forall \underline{x} \in \mathbb{R}^n$)

and (5) reads as
$$(K(\underline{u} - \underline{u}^*), \underline{u} - \underline{u}^*) \le \eta^{2k} \cdot const.$$

In the implementation, the step 3. is often replaced by

$$
\begin{array}{lll}
3a) & \hat{\underline{r}} & := & \underline{r} + \alpha\, K\underline{q} \\
3b) & \hat{\underline{w}} & := & C^{-1}\hat{\underline{r}}
\end{array}
$$

with the residuum $\underline{r} = K\underline{u} - \underline{b}$ of the original linear system.

2.3   Projected PCGM

Now we try to solve the linear system

$$P^T \tilde{K} P \underline{u}^* = P^T \tilde{\underline{b}}$$

with $\underline{u}^* = P\underline{u}^* \in \underline{U} \subset \mathbb{R}^n$ and $\underline{U}$ a proper subspace of dimension $n_0 < n$.
Let $\underline{U} = im\, P$ and $\underline{V} = im\, P^T$, where P is a projector $\mathbb{R}^n \to \underline{U}$, then $K = P^T \tilde{K} P$ is a unique mapping $\underline{U} \to \underline{V}$, so there exists a unique solution $\underline{u}^* \in \underline{U}$ for each $P^T \tilde{b} \in \underline{V}$.
Let $\tilde{K}$ be a symmetric positive definite $(n \times n)$–matrix. For each $\underline{u} \in \underline{U}$, the residuum

$$\underline{r} = P^T \tilde{K} P \underline{u} - P^T \tilde{\underline{b}} = P^T(\tilde{K}\underline{u} - \tilde{\underline{b}})$$

is a vector in $\underline{V}$. This means that a basic CGM without preconditioning $(\mathcal{A} = P^T \tilde{K} P)$ is impossible $(\hat{\underline{u}} := \underline{u} + \alpha\underline{r}$ makes no sense).

So, we have to define a preconditioner $C : \underline{U} \to \underline{V}$ and especially $C^{-1} : \underline{V} \to \underline{U}$ is required to form $\underline{w} := C^{-1}\underline{r} \in \underline{U}$.
In order to obtain a well–defined CGM from the basics in 2.1, we consider the following restriction of the EUCLIDIAN inner product in $\mathbb{R}^n$ to $\underline{U}$ and $\underline{V}$ as a dual pairing:

$$(\underline{u}, \underline{v})_D = (\underline{u}, \underline{v}) \quad \text{for each } \underline{u} \in \underline{U} \text{ (first argument)}$$
$$\text{and } \underline{v} \in \underline{V} \text{ (second argument)}.$$

Then $K = P^T \tilde{K} P$ and C are symmetric positive definite operators w.r.t. $(\cdot, \cdot)_D$ in the following sense:

$$
\begin{array}{lll}
(\underline{u}_1, K\underline{u}_2)_D & = & (\underline{u}_2, K\underline{u}_1)_D \quad \forall \underline{u}_1, \underline{u}_2 \in \underline{U} \\
(\underline{u}_1, C\underline{u}_2)_D & = & (\underline{u}_2, C\underline{u}_1)_D \quad \forall \underline{u}_1, \underline{u}_2 \in \underline{U} \\
(C^{-1}\underline{v}_1, \underline{v}_2)_D & = & (C^{-1}\underline{v}_2, \underline{v}_1)_D \quad \forall \underline{v}_1, \underline{v}_2 \in \underline{V}.
\end{array}
$$

Now, we define $\mathcal{A} = C^{-1}K : \underline{\mathbb{U}} \to \underline{\mathbb{U}}$ and $\langle \underline{u}_1, \underline{u}_2 \rangle = (\underline{u}_1, C\underline{u}_2)_D \; \forall \underline{u}_1, \underline{u}_2 \in \underline{\mathbb{U}}$. Here, $\langle \cdot, \cdot \rangle$ is an inner product $\underline{\mathbb{U}} \times \underline{\mathbb{U}} \to \mathbb{R}^1$ and $\mathcal{A}$ is symmetric positive definite w.r.t. $\langle \cdot, \cdot \rangle$:

$$
\begin{aligned}
\langle \mathcal{A}\underline{u}_1, \underline{u}_2 \rangle &= (C^{-1}K\underline{u}_1, C\underline{u}_2)_D = \left(\underline{u}_2, C(C^{-1}K\underline{u}_1) \right)_D = (\underline{u}_2, K\underline{u}_1)_D \\
&= (\underline{u}_1, K\underline{u}_2)_D = (\underline{u}_1, CC^{-1}K\underline{u}_2)_D = \langle \underline{u}_1, \mathcal{A}\underline{u}_2 \rangle.
\end{aligned}
$$

Hence, all presuppositions of (2.1) are fulfilled and the CGM for solving $\mathcal{A}\underline{u}^* = \underline{b}$ (with $\underline{b} = C^{-1}P^T\tilde{\underline{b}} \in \underline{\mathbb{U}}$) reads as:

Start: $\underline{u} \in \underline{\mathbb{U}}$ arbitrary
calculate $\underline{r} = \tilde{K}\underline{u} - \tilde{\underline{b}}$,
note: $P^T\underline{r} \in \underline{\mathbb{V}}$ is the original residuum
$\underline{q} := \underline{w} := C^{-1}P^T\underline{r} \in \underline{\mathbb{U}}$
$\gamma := \langle \underline{w}, \underline{w} \rangle = (\underline{w}, C\underline{w})_D$
$= (\underline{w}, P^T\underline{r})_D$
$= (\underline{w}, \underline{r})$ (from $\underline{w} \in \underline{\mathbb{U}}$)

Iteration: 1. $\delta := \langle \mathcal{A}\underline{q}, \underline{q} \rangle = (C^{-1}K\underline{q}, C\underline{q})_D = (\underline{q}, K\underline{q})_D =$
$= (\underline{q}, \tilde{K}\underline{q})$, $\alpha := -\gamma/\delta$
2. $\hat{\underline{u}} := \underline{u} + \alpha\underline{q}$ (update in $\underline{\mathbb{U}}$)
3a) $\hat{\underline{r}} := \underline{r} + \alpha\tilde{K}\underline{q}$
3b) $\hat{\underline{w}} := C^{-1}P^T\hat{\underline{r}}$
4. $\hat{\gamma} := \langle \hat{\underline{w}}, \hat{\underline{w}} \rangle = (\hat{\underline{w}}, \hat{\underline{r}}), \; \beta := \hat{\gamma}/\gamma$
5. $\hat{\underline{q}} := \hat{\underline{w}} + \beta\underline{q}$ (update in $\underline{\mathbb{U}}$)

REMARK 1: The projection $P^T\underline{r}$ in step 3a) is not explicitly done, if the preconditioner is chosen as

$$C^{-1} = P\tilde{C}^{-1}P^T : \underline{\mathbb{V}} \to \underline{\mathbb{U}}. \tag{7}$$

In this case $P^T\underline{r}$ (and $P^T\hat{\underline{r}}$) never occur in the iteration but $\hat{\underline{w}}$ and $\hat{\gamma}$ are well–defined from the structure of C^{-1}.

REMARK 2: The rate of convergence depends on the eigenvalues of the operator $\mathcal{A}$:

$$\underline{\gamma} \le \lambda_i(P\tilde{C}^{-1}P^T\tilde{K}P) \le \bar{\gamma}. \tag{8}$$

We have for the k–th step

$$
\begin{aligned}
\langle \mathcal{A}(\underline{u} - \underline{u}^*), \underline{u} - \underline{u}^* \rangle &= (\underline{u} - \underline{u}^*, K(\underline{u} - \underline{u}^*))_D \\
&= \left(\tilde{K}(\underline{u} - \underline{u}^*), \underline{u} - \underline{u}^* \right) \le \eta^{2k} \cdot const
\end{aligned}
$$

with η as in 2.1 . For investigating these eigenvalues for our application the Ficticious Space Lemma [8] has to be used (see Chapter 4).

REMARK 3: The special structure of the projectors P and P^T occur only once within the preconditioning step 3b), so we have the usual PCGM with a special projected preconditioner running within the subspace $\underline{\mathbb{U}}$.

3   Implementing the Projection

From Chapter 2 we conclude that the PCGM solution of the linear system

$$P^T \tilde{K} \underline{u} = P^T \underline{\tilde{b}}$$

for $\underline{u} \in \underline{\mathbb{U}}$ is a typical PCGM, if the projection, introduced into the preconditioner

$$C^{-1} = P \tilde{C}^{-1} P^T$$

can be cheaply implemented. To clarify the structure of the matrix P, we have to consider the finite element spaces as defined in the Introduction.

For sake of simplicity, we require for our actual mesh the following
MESH ASSUMPTIONS:
(a) At most one 'hanging node' per edge in the linear case as in fig. 1 (resp. one pair of 'hanging nodes' in the quadratic case as in fig. 2).
(b) Each triangle contains at most one edge with a 'hanging node' (i.e. the adaptive mesh generator subdivides each triangle 'red' if more than one of its edges are subdivided).

3.1   The Case of (Bi–, Tri–) Linear Elements

Without loss of generality we can suppose that exactly one 'hanging node' (the last one, node n) has been produced by a 'red' subdivision of one triangle without subdividing the other one sharing an edge (k, k').
So, the node n is 'son' of the 'fathers' k and k' in the hierarchical generation of the actual fine mesh and we have the situation as in fig. 1.

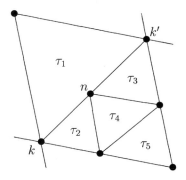

Figure 1: The Case of (Bi–, Tri–) Linear Elements

Then, the non-conformal basis $\tilde{\Phi}$ contains 3 discontinuous piecewise linear functions:

$\tilde{\varphi}_n$ (nonzero in τ_2, τ_4 and τ_3; zero in τ_1)
$\tilde{\varphi}_k$ (the support contains τ_1 and τ_2) and
$\tilde{\varphi}_{k'}$ (the support contains τ_1 and τ_3).

A function $u = \tilde{\Phi}\underline{u}$ (with $\underline{u} = (u_1, \ldots, u_n)^T$) is continuous if and only if

$$u_n = \frac{1}{2}(u_k + u_{k'}), \tag{9}$$

because this implies a linear dependence on u_k and $u_{k'}$ from both sides of this edge (k, k').
The matrix representation of the projector

$$P : \mathbb{R}^n \to \underline{\mathbb{U}} = \left\{ \underline{u} : u_n = \frac{1}{2}(u_k + u_{k'}) \right\}$$

is obviously $P = I - e_n e_n^T + \frac{1}{2}e_n e_k^T + \frac{1}{2}e_n e_{k'}^T$.

A much more simple formula is found with respect to the hierarchical finite element basis $\tilde{\Psi}$. This basis was used for efficient hierarchical preconditioning 2D f.e. systems by YSERENTANT[14].

Let $\tilde{\Psi} = (\tilde{\psi}_1, \ldots, \tilde{\psi}_n)$ the hierarchical basis of the same (discontinuous) f.e. space $\mathbb{V}^{non} = span\tilde{\Phi}$. Then, before subdividing the edge (k, k') we had no 'hanging node', so

$$\mathbb{V}^{con} = span(\tilde{\psi}_1, \ldots, \tilde{\psi}_{n-1})$$

and here the functions $\tilde{\psi}_k$ and $\tilde{\psi}_{k'}$ are continuous with a support containing τ_1 and $\bigcup_{i=2}^{5} \tau_i$. So, only $\tilde{\psi}_n = \tilde{\varphi}_n$ is the discontinuous basis function in the basis $\tilde{\Psi}$ and we have

$$u = \tilde{\Psi}\underline{v} \in \mathbb{V}^{con} \iff v_n = 0, \tag{10}$$

a much more simple representation of $\mathbb{V}^{con}$.
For representing $\underline{\mathbb{U}}$ (subspace of $\mathbb{R}^n$ of coefficient vectors with respect to the nodal basis functions $\tilde{\Phi}$), we compare

$$u = \tilde{\Psi}\underline{v} = \tilde{\Phi}\underline{u}.$$

With the well–known transformation matrix Q mapping the nodal basis to the hierarchical one:

$$\tilde{\Psi} = \tilde{\Phi}Q, \tag{11}$$

we obtain

$$u = \tilde{\Psi}\underline{v} = \tilde{\Phi}Q\underline{v} = \tilde{\Phi}\underline{u} \iff Q\underline{v} = \underline{u}.$$

This leads to another (hierarchical) representation of the matrix P:

$$\underline{u} \in \underline{\mathbb{U}} \iff \underline{u} = P\underline{u} \iff \underline{v} = Q^{-1}\underline{u} \text{ fulfills } v_n = 0,$$

hence

$$P = Q\hat{P}Q^{-1} \tag{12}$$

with

$$
\begin{aligned}
\hat{P} &= diag(1,1,\ldots,1,0) \\
&= I - e_n e_n^T.
\end{aligned}
$$

The implementation of P (and P^T) within the preconditioning step 3b) of the PCG algorithm in 2.3 is then best combined with the hierarchical preconditioner $\tilde{C}^{-1} = QQ^T$ in $\mathbb{R}^n$: From (12) and (7) we have

$$
\begin{aligned}
C^{-1} = P\tilde{C}^{-1}P^T &= Q\hat{P}Q^{-1}\,QQ^T\,Q^{-T}\hat{P}Q^T \\
&= Q\hat{P}Q^T.
\end{aligned}
$$

So, the projection into the proper subspace is done after transforming the residual $\underline{r} = \tilde{K}\underline{u} - \underline{b}$ into the hierarchical basis.

For better convergence and for 3D–calculations this can be generalized to the BPX–preconditioner in a straight forward manner.

3.2   The Case of Quadratic Elements

Here, we consider 6–node triangles or 8–node quadrilaterals in 2D (resp. 10–node tetrahedrons or 20–node bricks in 3D).

Again, for simple description, we consider only one subdivided edge, that produced 2 'hanging nodes', following fig.2:

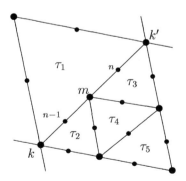

Figure 2: The Case of Quadratic Elements

Let the edge (k, m, k') be subdivided into the two smaller edges

$$(m, n-1, k) \text{ and } (m, n, k')$$

(again the two last nodes $n - 1$ and n are 'hanging nodes').

Now, the subspace $\mathbb{V}^{con}$ is again represented by restrictions of u_n and u_{n-1} depending on u_k, u_m and $u_{k'}$ but this is no more a local information on the two sub-edges. A cheap implementation of P (and P^T) is again possible in considering the hierarchical basis $\tilde{\Psi}$.

For quadratic elements, the hierarchical basis is defined as follows: On the finest mesh, we forget the edge–mid-nodes and define the hierarchical basis as in the linear case for the functions belonging to vertex nodes. Then the quadratic edge bubbles are added to the basis for completing $\tilde{\Psi}$. Obviously $span\tilde{\Psi} = span\tilde{\Phi}$ and $\tilde{\Psi} = \tilde{\Phi}Q$ remain valid with the same matrix Q as in the linear case.

From the element by element definition of the basis functions as sums of shape functions, we obtain the following 3 discontinuous functions in the hierarchical basis:

$$\tilde{\psi}_m = \begin{cases} \text{quadratic bubble} & \text{from } \tau_1\text{–side} \\ \text{piecewise linear} & \text{from } \tau_2/\tau_3\text{–side} \end{cases}$$

$$\tilde{\psi}_n = \begin{cases} 0 & \text{from } \tau_1\text{–side} \\ \text{quadratic bubble} & \text{in } \tau_2 \end{cases}$$

$$\tilde{\psi}_{n-1} = \begin{cases} 0 & \text{from } \tau_1\text{–side} \\ \text{quadratic bubble} & \text{in } \tau_3 \end{cases}$$

In the hierarchical basis $\tilde{\Psi}$, the functions $\tilde{\psi}_k$ and $\tilde{\psi}_{k'}$ are continuous (the same as in the linear case).

So, $u = \tilde{\Psi}\underline{v} \in \mathbb{V}^{con}$ is a continuous function, when $v_m\tilde{\psi}_m + v_n\tilde{\psi}_n + v_{n-1}\tilde{\psi}_{n-1}$ is continuous over the edge (k, m, k'), which is equivalent to

$$v_n = v_{n-1} = \frac{1}{4}v_m.$$

($v_n\tilde{\psi}_n + v_{n-1}\tilde{\psi}_{n-1}$ corrects the jump in $v_m\tilde{\psi}_m$ at the edge).

Now, the projection $\hat{P}$ w.r.t. the hierarchical basis is non–symmetric

$$\hat{P} = I - e_n e_n^T - e_{n-1}e_{n-1}^T + \frac{1}{4}e_n e_m^T + \frac{1}{4}e_{n-1}e_m^T$$

and the same calculation as in 3.1 yields

$$C^{-1} = Q\hat{P}\hat{P}^T Q^T$$

as a generalization of the hierarchical preconditioner to 'hanging nodes' for quadratic elements. In the general case, the implementation of $\hat{P}$ and $\hat{P}^T$ is a simple edge–oriented algorithm:

$\underline{v} := \hat{P}^T\underline{v}$: for each edge (m, i, j) do
                  if node i is 'hanging' then
                  $v_m := v_m + \frac{1}{4}v_i, v_i := 0$

$\underline{w} := \hat{P}\underline{w}$: for each edge (m, i, j) do
                  if node i is 'hanging' then
                  $w_i := \frac{1}{4}w_m$

4   The Estimation of the Eigenvalue Bounds

For complete use of the ideas above, we have to prove that

$$\kappa(P\tilde{C}^{-1}P^T \; P^T\tilde{K}P)$$

is bounded independent on h (or growing as $\sim |\ln h|$ for the simple hierarchical preconditioner $\tilde{C}^{-1} = QQ^T$). Here the Ficticious Space Lemma [8] has to be used:

FICTICIOUS SPACE LEMMA: If we have

1. a symmetric p.d. operator $\tilde{A} : \tilde{\mathbb{H}} \to \tilde{\mathbb{H}}$
   (Hilbert space with inner product $\langle \cdot, \cdot \rangle_\sim$, the 'ficticious space')

2. $\tilde{C}^{-1}$ a 'good' preconditioner for $\tilde{A}$, i.e.
   $$\gamma_1 \langle \tilde{A}, \tilde{u}, \tilde{u} \rangle_\sim \leq \langle \tilde{A}\tilde{C}^{-1}\tilde{A}\tilde{u}, \tilde{u} \rangle_\sim \leq \gamma_2 \langle \tilde{A}\tilde{u}, \tilde{u} \rangle_\sim \; \forall \tilde{u} \in \tilde{\mathbb{H}}.$$

3. Let $A : \mathbb{H} \to \mathbb{H}$ s.p.d. w.r.t. $\langle \cdot, \cdot \rangle$ – inner product in $\mathbb{H}$.

4. Let $\mathcal{R} : \tilde{\mathbb{H}} \to \mathbb{H}$ a restriction operator with

   $$\langle A\mathcal{R}\tilde{u}, \mathcal{R}\tilde{u} \rangle \leq c_R \langle \tilde{A}\tilde{u}, \tilde{u} \rangle_\sim \; \forall \tilde{u} \in \tilde{\mathbb{H}}$$

5. Let $\mathcal{Q} : \mathbb{H} \to \tilde{\mathbb{H}}$ a prolongation with $\mathcal{R}\mathcal{Q}u = u \; \forall u \in \mathbb{H}$ and

   $$\langle \tilde{A}\mathcal{Q}u, \mathcal{Q}u \rangle_\sim \leq c_Q^{-1} \langle Au, u \rangle \; \forall u \in \mathbb{H}$$

then : $\mathcal{C}^{-1} = \mathcal{R}\tilde{C}^{-1}\mathcal{R}^*$ is a 'good' preconditioner for $\mathcal{A}$ with

$$\underline{\gamma}\langle Au, u \rangle \leq \langle A\mathcal{C}^{-1}Au, u \rangle \leq \bar{\gamma}\langle Au, u \rangle$$

and

$$\underline{\gamma} \geq \gamma_1 c_Q , \qquad \bar{\gamma} \leq \gamma_2 c_R.$$

We would like to use this Lemma with

$\mathbb{H} = \mathbb{V}^{con}$ (A is defined from the underlying bilinear form $a(\cdot, \cdot)$
with the conformal f. e. basis $span\Phi$ and
has K as matrix representation)

$\tilde{\mathbb{H}} = \mathbb{V}^{non}$ ($\tilde{A}$ is defined from $a(\cdot, \cdot)$
w.r.t. the basis $\tilde{\Phi}$,
belonging to the stiffness matrix $\tilde{K}$)

Then $\mathcal{R} : \mathbb{V}^{non} \to \mathbb{V}^{con}$ has the previous matrix representation P, and from $\mathbb{V}^{con} \subset \mathbb{V}^{non}$ we can choose $\mathcal{Q}$ as the identity.

The F.S.L. could be applied yielding our preconditioner $P\tilde{C}^{-1}P^T$ as matrix representation of $\mathcal{C}^{-1}$, but for the definition of $\tilde{C}^{-1}$ acting on the non-conformal f.e. space, the spectral bounds γ_1, γ_2 are unclear. So, we consider another auxiliary ficticious space $\tilde{\mathbb{H}} = \mathbb{V}^{green}$.

Let $\mathbb{V}^{green} = span\tilde{\Phi}$ the f.e. space on the same triangulation for which $\mathbb{V}^{con}$ and $\mathbb{V}^{non}$ are defined but instead of letting 'hanging nodes', the triangles (τ_1 in the examples) are subdivided 'green' into 2 parts. Then, in the example of 3.1 we have

$$\tilde{\varphi}_i = \hat{\varphi}_i \ \forall i \neq k, k', n,$$

when

$$span\tilde{\Phi} = span(\tilde{\varphi}_1, \ldots, \tilde{\varphi}_n) = \mathbb{V}^{non}$$
$$span\hat{\Phi} = span(\hat{\varphi}_1, \ldots, \hat{\varphi}_n) = \mathbb{V}^{green}$$

Now, $\hat{\varphi}_k$, $\hat{\varphi}_{k'}$ and $\hat{\varphi}_n$ are continuous functions from the usual conformal mesh. If for a function $u = \hat{\Phi}\underline{u} \in \mathbb{V}^{green}$, we define $\mathcal{R}u = \tilde{\Phi}P\underline{u}$ with the same projection matrix as in 3.1 or 3.2, the arising function coincides with the analogous definition from chapters 3.1/3.2 using the non-conformal basis:

$$\mathcal{R}u = \tilde{\Phi}P\underline{u} = \hat{\Phi}P\underline{u} \in \mathbb{V}^{con}.$$

Hence, although the stiffness matrices $\tilde{K}$ and $\hat{K} = (a(\hat{\varphi}_j, \hat{\varphi}_i))_{i,j=1}^n$ do not coincide the projections $P^T\tilde{K}P = P^T\hat{K}P$ do.

Now, we can use the F.S.Lemma with
$\tilde{\mathbb{H}} = \mathbb{V}^{green}$, $\hat{K}$, preconditioner $\tilde{C}^{-1}$ and
$\mathbb{H} = \mathbb{V}^{con}$, $K = P^T\tilde{K}P = P^T\hat{K}P$ and $\mathcal{R} : \tilde{\mathbb{H}} \to \mathbb{H}$ represented by P.
For completing this chapter the constant c_R has to be estimated:

$$a(\mathcal{R}u, \mathcal{R}u) \leq c_R a(u, u) \ \forall u \in \mathbb{V}^{green}.$$

Knowing the fact that the dimension of $\mathbb{V}^{con} = \mathcal{R}\mathbb{V}^{green}$ is (n–#hanging nodes) and these 'hanging nodes' can occur in the (locally) finest level only, the angle between the subspaces $\mathbb{V}^{con}$ and $\mathbb{W}$ (when $\mathbb{V}^{green} = \mathbb{V}^{con} + \mathbb{W}$) is 'good':

$$a(u, v) \leq \gamma \left(a(u, u)a(v, v)\right)^{1/2} \quad \forall u \in \mathbb{V}^{con} \quad \forall v \in \mathbb{W}$$

with $0 < \gamma < 1$ (independent on h, see i.e. [2, 6]).
So, c_R follows with $u = \mathcal{R}u + v \ \forall u \in \mathbb{V}^{green}, v \in \mathbb{W}$ and

$$\begin{aligned}
a(u, u) &= a(\mathcal{R}u, \mathcal{R}u) + 2a(\mathcal{R}u, v) + a(v, v) \\
&\geq a(\mathcal{R}u, \mathcal{R}u) - 2\gamma \left(a(\mathcal{R}u, \mathcal{R}u) \cdot a(v, v)\right)^{1/2} + a(v, v) \\
&\geq (1 - \gamma)\left(a(\mathcal{R}u, \mathcal{R}u) + a(v, v)\right)
\end{aligned}$$

yielding $c_R = (1 - \gamma)^{-1}$.

Using the Ficticious Space Lemma we obtain for

$$\kappa(C^{-1}K) \le \bar{\gamma}/\underline{\gamma} \le \gamma_2/\gamma_1 \cdot c_R.$$

So, the preconditioner $C^{-1} = P\tilde{C}^{-1}P^T$ for $K = P^T\tilde{K}P$ is as good as $\tilde{C}^{-1}$ for $\hat{K}$.

REMARK 4: We have
$\gamma_2/\gamma_1 = \mathcal{O}(|\ln h|^2)$ for the hierarchical preconditioner in 2D [14] or
$\gamma_2/\gamma_1 = \mathcal{O}(1)$ for BPX preconditioners [4, 13, 9].

5   Other Subspaces from Boundary Conditions

The technique described above is not restricted to the case of handling hanging nodes. Each subspace problem is easily attacked with the same technique if the projectors P and P^T for the preconditioning step are easy to implement.

For example consider so called slip condition on a fixed wall without friction. Here, we have the following boundary value problem from usual linear elasticity.
Find $\vec{u} \in \mathbb{V}_0$ with

$$a(\vec{u}, \vec{v}) = \langle f, \vec{v} \rangle \quad \forall \vec{v} \in \mathbb{V}_0.$$

Here, $\vec{u}$ is fixed to zero (or to a given displacement) on a boundary part Γ_D and a "slip condition" $\vec{u} \cdot \vec{n} = 0$ has to be fulfilled on another part of the boundary Γ_S say, so

$$\mathbb{V}_0 = \left\{ \vec{u} \in (H^1(\Omega))^2 : \vec{u} = 0|_{\Gamma_D}, \vec{u} \cdot \vec{n} = 0|_{\Gamma_S} \right\}.$$

(Implicitly, we have zero tangential stresses along Γ_S as natural boundary conditions

$$\sigma \cdot \vec{t} = 0|_{\Gamma_S},$$

when $\vec{t} \perp \vec{n}$ is the tangential vector.)

Both restrictions on $\vec{u}$ are absent in the stiffness matrix, which is defined from assemblying the element matrices

$$K_e = \left(a(\vec{\varphi}_j, \vec{\varphi}_i) \right)_{i,j}$$

with $\vec{\varphi}_j$ the form functions of the element e, the restrictions of the ansatz functions

$$\vec{\Phi} = (\vec{\varphi}_1, \dots, \vec{\varphi}_{2n})$$

to the element.

The assembly of K_e over all elements would be a singular matrix, so again we have to work in a subspace of vectors $\underline{u}$ such that $\vec{u} = \vec{\Phi}\underline{u}$ defines a valid displacement function in $\mathbb{V}_0$. Let $\underline{u} \in \mathbb{R}^{2n}$ be written as

$$\underline{u} = (\underline{u}_1^T, \underline{u}_2^T, \dots, \underline{u}_n^T)^T$$

with sub-vectors $\underline{u}_i$ of nodal displacements. Then $\vec{u} \in \mathbb{V}_0$ ($\Leftrightarrow \underline{u} \in \underline{\mathbb{U}}$) means

for boundary nodes $\quad a_i \in \Gamma_D \quad : \quad \underline{u}_i = 0$

but for $\qquad\qquad\qquad a_i \in \Gamma_S \quad : \quad \vec{n}_i \cdot \underline{u}_i = 0$

which is $\qquad\qquad\qquad\qquad\qquad \underline{u}_i \in span(\vec{t}_i)$.

That is, the projector onto $\underline{\mathbb{U}} \subset \mathbb{R}^{2n}$ is the following block-diagonal matrix of (2×2)–blocks P_i:

$$P = diag(P_1, \ldots, P_n)$$

with

$$P_i = \begin{cases} I & \text{if } a_i \text{ is interior node} \\ & \text{or } a_i \notin (\Gamma_D \cup \Gamma_S) \\ \mathbb{O} & \text{if } a_i \in \Gamma_D \\ I - \vec{n}_i \vec{n}_i^T & \text{if } a_i \in \Gamma_S \end{cases}$$

Note that $I - \vec{n}_i \vec{n}_i^T = \vec{t}_i \vec{t}_i^T$ in 2D ($\|\vec{n}\| = \|\vec{t}_i\| = 1$).

In 3D the P_i are (3×3)–blocks and we have to use $\vec{n}_i$ as outer normal vector of a boundary face on Γ_S. Again as in the hanging nodes case, we include this projection into the preconditioning step of PCGM ensuring that $\underline{w} \in \underline{\mathbb{U}}$ for each iteration, so the PCGM solve arrises in the desired subspace, if the starting vector does.

6   Generalization to Contact Problems

The same "slip condition" as in Chapter 4 occurs in a contact simulation for nodes which are in contact.

For simplicity lets us consider Signorini's contact in the linear elastic case without friction. That is the following problem [11, 10]:

minimize $\qquad J(\vec{u}) = \frac{1}{2}a(\vec{u}, \vec{u}) - \langle f, \vec{u} \rangle$

w.r.t. $\quad \vec{u} \in \mathcal{K} = \{$ $\vec{u} \in (H_0^1(\Omega))^d$: with non-penetration of a given obstacle $\}$.

For an obstacle $\{\vec{x}_0 + \lambda \vec{r} : \lambda \in \mathbb{R}\}$ as straight line in 2D we can use

$$\mathcal{K} = \left\{ \vec{u} \in (H_0^1(\Omega))^2 : (\vec{x} + \vec{u}(\vec{x}) - \vec{x}_0) \cdot \vec{n} \geq 0 \right\}$$

if $\vec{n} = \vec{r}^\perp$ is the desired direction into the half-space of allowed abidance of the body Ω after deformation.

From the unknown nature of the convex set $\mathcal{K}$ many authors solve such contact problems by nonlinear iterations using penalty terms or similar ideas. In any case the codes watch, if the deformed body "penetrates" the obstacle ($\vec{u} \notin \mathcal{K}$ and correct this error.

In our adaptive strategy this can be done easily from defining the starting vector $\underline{u}^{(0)}$ of the PCGM as admissible (correct $\underline{u}^{(0)}$ for some "penetrating nodes" such that $\vec{\Phi}\underline{u}^{(0)} \in \mathcal{K}$) and switch on a projector $P_i = \vec{r}\vec{r}^T$ for this node a_i. This means we calculate for the new solution the deformed position $a_i + \underline{u}_i$ at its energy minimal position along the obstacle. The projector $P_i = \vec{r}\vec{r}^T$ (the same as in Chapter 4) is maintained at this node for all succeeding finer meshes, except a negative contact pressure would occur. Then P_i is switched back to I.

From this simple "contact control" the true contact area arrises automatically during the adaptive refinement. The error estimators follow the stress intensities, so we have fine meshes near the contact ends.

7 Example

Let us demonstrate all these projection solves in one example of a Signorini contact simulation of a rotational symmetric body Ω (a ball in 3D) with contact on the plane $\{x_3 = 0\}$ and forces $\vec{f} = (0, 0, -f_0)^T$.
The coarse mesh is shown in Figure 3 containing a very few number of triangles. (Note that outer edges carry the information of being arcs, which is used in the refinement procedure). In the beginning one node is in contact, so the matrix $P^T K P$ in the coarsest mesh is non–singular. After refining the projector P works on the contact nodes as in Chapter 5 as well as on the hanging nodes as in Chapter 3. The Figures 4 and 5 show the finer meshes with about 500 and 5000 nodes.

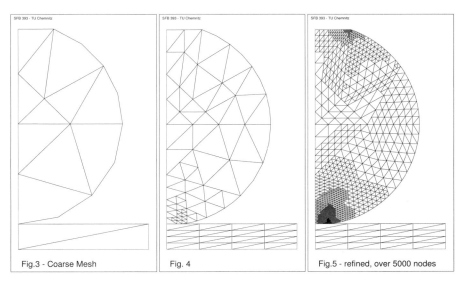

Fig.3 - Coarse Mesh   Fig. 4   Fig.5 - refined, over 5000 nodes

Figure 3,4 and 5: The development of the meshes in the adptive refinement

Up to more than 40 000 nodes we have run this adaptive refinement, yielding the estimated error plot of Fig. 8 (the error estimated is proportional to the elastic modulus $2 \cdot 10^5$) and the running times of the following Table 1.

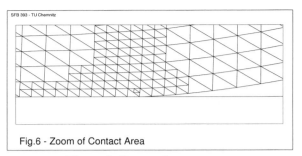

Figure 6: Zoom of contact area

Figure 6 is a zoom of the contact area and its very fine elements around and Figure 7 contains the contact pressure versus r-axis.

```
        NetFine :          |  ElMat : |    PCGM        |est.Err.
#Nodes/#Elem's /#Edges|  time[s]  | It time[s]  | (square)
----------------------|----------|-------------|--------
     41 /      13 /     26 |  0.000   | 22   0.000  | 3.8E+04
    108 /      40 /     93 |  0.008   | 29   0.008  | 8.9E+04
    187 /      67 /    158 |  0.008   | 25   0.008  | 1.4E+04
    236 /      82 /    195 |  0.008   | 28   0.008  | 1.3E+04
    261 /      85 /    204 |  0.000   | 25   0.016  | 1.1E+04
    319 /     103 /    246 |  0.008   | 35   0.016  | 6.1E+03
    368 /     118 /    283 |  0.008   | 31   0.016  | 5.5E+03
    420 /     136 /    323 |  0.000   | 21   0.016  | 4.8E+03
    524 /     178 /    417 |  0.016   | 26   0.031  | 3.6E+03
    552 /     184 /    431 |  0.000   | 37   0.047  | 3.4E+03
    615 /     205 /    480 |  0.008   | 35   0.047  | 3.0E+03
    688 /     232 /    541 |  0.016   | 21   0.039  | 2.5E+03
    768 /     262 /    609 |  0.016   | 25   0.047  | 2.3E+03
    865 /     301 /    694 |  0.016   | 31   0.078  | 1.9E+03
   1549 /     619 /   1376 |  0.109   | 51   0.258  | 9.7E+02
   1992 /     826 /   1813 |  0.078   | 49   0.336  | 7.1E+02
   3347 /    1459 /   3166 |  0.219   | 33   0.422  | 4.1E+02
   4843 /    2185 /   4662 |  0.258   | 40   0.828  | 2.7E+02
   7724 /    3580 /   7543 |  0.477   | 44   1.664  | 1.6E+02
  11952 /    5638 /  11771 |  0.703   | 43   2.727  | 1.0E+02
  17642 /    8428 /  17461 |  0.953   | 52   4.969  | 7.0E+01
  27787 /   13399 /  27606 |  1.695   | 52   7.906  | 4.4E+01
  43958 /   21364 /  43777 |  2.711   | 41   9.883  | 2.8E+01
```

Table 1: The adptive history, number of nodes, elements and edges, solver times and error estimators

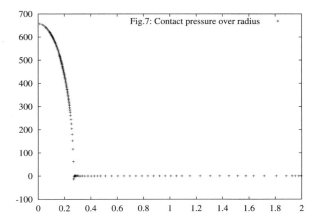

Figure 7: The contact pressure versus r-axis

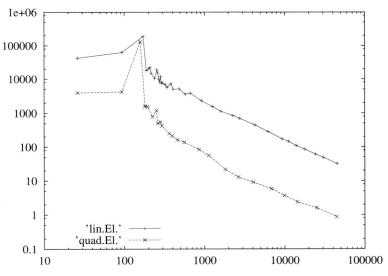

Figure 8: Typical decrease of error estimation

8   Conclusion

An efficient implementation of adaptive finite element solvers should use special data structures for holding the nodes, the edges and the elements (in 3D: additionally the faces). In the data structure "element" we have included the edges, the nodes and the element stiffness matrix so we avoid the assembly of K and we generate an element matrix only for new elements in the adaptive procedure. Moreover, the data structure "edges" contains three nodes (end points and the midpoint) which coincides with the necessary hierarchy of nodes during adaptive refinement of the mesh. In the resulting PCG–solver from element by element matrix multiply and hierarchical precondition-ers no additional information is required than contained in these data structures.

The finest edges contain the information on possibly hanging nodes and on bound-ary restrictions, which are non existent in the actual stiffness matrix.

Hence, the new preconditioner

$$P\tilde{C}^{-1}P^T$$

with the projector P onto the desired subspace solves this problem very efficiently and implicitly within the PCGM run.

We have given the special structure of P for hanging nodes in linear and quadratic elements as well as for "Dirichlet type" boundary conditions such as prescribed values or "slip conditions". This is simply generalized to contact problems.

References

[1] M. Ainsworth, J. Oden, *"A Posteriori Error Estimation in Finite Element Anal-ysis"*, Comp.Meth.Appl.Mech.Eng., 142,1-2, 1-88,1997.

[2] O.Axelsson, I.Gustafsson, *"Preconditioning and Two–level Multigrid Methods of Arbitrary Degree of Approximation"*, Math. Comp., 40, 219-242,1983.

[3] J. H. Bramble, J. E. Pasciak, A. H. Schatz, *"The Construction of Preconditioners for Elliptic Problems by Substructuring I – IV"*, Mathematics of Computation, 47,103–134,1986,   49, 1–16,1987, 51, 415–430,1988   and 53, 1–24,1989.

[4] I. H. Bramble, J. E. Pasciak, J. Xu, *"Parallel Multilevel Preconditioners"*, Math. Comp., 55, 191, 1-22,1990.

[5] M.R.Hestenes, E.Stiefel, *"Methods of Conjugate Gradients for Solving Linear Systems"*, J. Res. Nat. Bur. Stand., 49, 409-436,1952.

[6] M. Jung and J. F. Maitre, *"Some Remarks on the Constant in the Strengthened C.B.S. Inequality: Estimate for Hierarchical Finite Element Discretizations of Elasticity Problems"*, Num.Meth. for Part. Diff. Equ., 15,4, 469-488,1999.

[7] A. Meyer, *"A Parallel Preconditioned Conjugate Gradient Method Using Do-main Decomposition and Inexact Solvers on Each Subdomain"*, Computing,45,217-234,1990.

[8] S. Nepomnyaschikh, *"Ficticious Space Method on Unstructured Meshes"*, East-West J. Num. Math.,3(1),71-79,1990.

[9] P.Oswald, *"Multilevel Finite Element Approximation: Theory and Applications"*, Teubner Skripten zur Numerik, B.G.Teubner Stuttgart 1994.

[10] J.Schöberl, *"Solving the Signorini problem on the basis of domain decomposition techniques"*, Computing, 60, 323-344,1993.

[11] A. Signorini, *"Sopra akune questioni di elastistatica"*, Attil della Societa Italiana per il Progresso delle Scienzie, 1993.

[12] R. Verfürth, *"A Review of a posteriori Error Estimation and Adaptive Mesh Refinement Techniques"*, Wiley–Teubner, Chichester, Stuttgart, 1996.

[13] J. Xu, *"Iterative Methods by Space Decomposition and Subspace Correction"*, SIAM Rev., 34,4, 163-184, 1992.

[14] H. Yserentant, *"Two Preconditioners Based on the Multilevel Splitting of Finite Element Spaces"*, Numer. Math., 58,163-184,1990.

©2002, Saxe-Coburg Publications, Stirling, Scotland
Engineering Computational Technology
B.H.V. Topping and Z. Bittnar, (Editors)
Saxe-Coburg Publications, Stirling, Scotland, 167-192.

Chapter 8

Combining SGBEM and FEM for Modeling 3D Cracks

G.P. Nikishkov† and S.N. Atluri‡
† Department of Computer Software
   The University of Aizu, Aizu-Wakamatsu City, Fukushima, Japan
‡ Center for Aerospace Research and Education
   University of California at Los Angeles, United States of America

Abstract

The SGBEM-FEM alternating method suitable for the solution of elastic and elastic-plastic three-dimensional fracture mechanics problems is presented. The crack is modeled by the symmetric Galerkin boundary element method (SGBEM), as a distribution of displacement discontinuities in an infinite medium. The finite element method (FEM) is used for stress analysis of the uncracked finite body. The solution for the structural component with the crack is obtained in an iterative procedure, which alternates between FEM solution for the uncracked body, and the SGBEM solution for the crack in an infinite body. Both elastic and elastic-plastic alternating procedure are developed. In the elastic alternating procedure residual forces on the surface of the finite element model are sought in order to balance tractions induced by presence of the crack. Elastic-plastic alternating procedure is based on volume residuals due to presence of the crack and elastic-plastic material behavior. Initial stress approach is used inside alternating iterative procedure for elastic-plastic problems. Computational procedure for fatigue crack growth of nonplanar cracks is presented. It is assumed that crack growth rate and the direction of crack growth are determined by the ΔJ-integral.

Keywords: Crack, FEM, SGBEM, elastic-plastic, fracture mechanics, fatigue crack growth.

1   Introduction

Numerical modeling of nonplanar cracks and their growth in structural components is important for many engineering fields. At present it is not widely used because of its complexity and lack of available numerical tools. The finite element method (FEM) is employed for modeling in many engineering areas including fracture mechanics. Various finite element techniques for fracture mechanics analysis have been developed. The use of energetic methods and in particular the equivalent domain integral method [1, 2] allows one to obtain fracture mechanics parameters for an arbitrary three-dimensional crack. A serious difficulty in applying the finite element method to the analysis of three-dimensional cracks is related to generation of appropriate meshes for non-planar cracks and especially to modifying finite element mesh during crack growth. Recently introduced the extended finite element method [3] allows to model cracks which surfaces are not necessary coincide with element surfaces. However the extended finite element method usually requires refined meshes.

In the boundary element method (BEM) for linear problems, the mesh should be created only for the boundary of the structure, and for the crack surface. Generation of a boundary element mesh for the surface is simpler than generation of a finite element mesh for the entire body with a crack. However, for the growing crack, both the crack-surface mesh and the mesh for the surface of the structure should be modified. Among disadvantages of the traditional BEM, it is possible to mention the non-symmetrical matrix of the equation system, and the hypersingular kernels contained in the integral relations.

The symmetric Galerkin boundary element method (SGBEM) [4] is based on satisfying the boundary integral equations of elasticity in a Galerkin weak form. The SGBEM equations are characterized by weakly singular kernels. Using special coordinate transformation, which removes the singularity from kernels, the boundary element matrices can be integrated with the use of usual Gaussian rules.

Employing the superposition principle it is possible to represent solution for a cracked body as a sum of two solutions: solution for a body without a crack and solution for a crack in an infinite medium. Some special fictitious loads should be added to both superimposed problems in order to account for mutual influence of the crack and the body surface. A method for combining SGBEM and FEM procedures in crack analysis is proposed in Reference [5]. Here we present further development of the combined SGBEM-FEM procedure. Fatigue crack growth of the surface crack is modeled. It is shown that the SGBEM-FEM procedure can be applied to elastic-plastic crack analysis.

2   Overview of the SGBEM-FEM procedure

Using jointly the symmetric Galerkin boundary element method for modeling an arbitrary non-planar crack in an infinite body, and the finite element method for an uncracked finite body, in fracture mechanics problems, allows us to employ advantages of both methods.

The finite element method is a robust method for elastic and elastic-plastic problems. It can easily incorporate various types of boundary conditions. The finite element method is widely used in industry. There are commercial preprocessor programs, which are capable of transforming any CAD model into a finite element model.

The boundary element method is most suitable for modeling cracks in infinite bod-

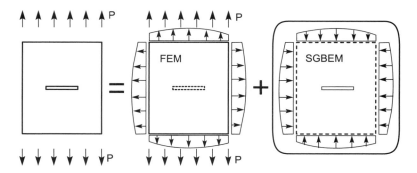

Figure 1: Superposition principle for a finite body with a crack.

ies. The displacement discontinuity approach provides for a simple modeling of the crack. Only one surface of the crack should be discretized. The independence of the crack model and the finite element model of the body allows to easily change the crack model in order to simulate crack growth under monotonic or cyclic loading.

According to the superposition principle, the solution for a finite body with a crack can be obtained as a superposition of two solutions:

1. finite element solution for a finite body under external loading, without a crack;

2. boundary element solution for an infinite body with a crack modeled.

Using the superposition principle, it is possible to employ a direct method for creating a system of equations for the fracture mechanics problem. The matrix of the combined equation system includes finite element and boundary element matrices and two interaction matrices. Such direct approach for combining the traditional hypersingular boundary element method, and the finite element method, is used by the authors of Reference [6]. An obvious disadvantage of the direct approach is the large size of the matrices that characterize the interaction between the finite element and the boundary element global matrices, and consequently the large computing time for the assembly and solution of the equation system.

Illustration of the superposition principle is presented in Figure 1. For a correct superposition corresponding to the solution for a finite body with a crack, fictitious forces on the boundary of the finite element model should be found in order to compensate for the stresses caused by the presence of a crack in an infinite body. While this can be done with a direct procedure, the alternating method [12] provides for a more efficient solution, without assembling the joint SGBEM-FEM matrix.

The SGBEM-FEM alternating method alternates between the finite element solution for an uncracked body and the boundary element solution for a crack in an infinite body.

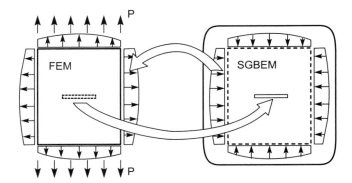

Figure 2: Transfer of stresses from the finite element model to the boundary element model and vice versa.

The basic steps of the elastic SGBEM-FEM alternating iterative procedure are as follows:

1. Using FEM, obtain the stresses at the location of the crack in a finite uncracked body subjected to given boundary conditions.

2. Using SGBEM, solve the problem of a crack, the faces of which subjected to tractions, as found from FEM analysis of the uncracked body.

3. Determine the residual forces at the outer boundaries of the finite body, from displacement discontinuities at the crack surface.

4. Using FEM, solve a problem for a finite uncracked body under residual forces from SGBEM analysis.

5. Obtain the stresses at the location of the crack corresponding to FEM solution.

6. Repeat Steps (2)-(5) until the residual load is small enough.

7. By summing all the appropriate contributions, compute the total solution for a finite body with the crack.

Transfer of stresses from the finite element model to the boundary element model and vice versa in the elastic alternating procedure is illustrated in Figure 2.

In the elastic-plastic alternating procedure the sources of residual forces are both unknown stress boundary conditions due to presence of the crack and nonlinear material behavior. Because of this the residuals in the elastic-plastic case are computed for the volume of the finite element model.

3 FEM procedure

3.1 Elastic FEM procedure

The finite element procedure for structural analysis is well known [7]. The global equation system of equilibrium for elastic problems in terms of displacements has the following appearance:

$$[K_{FEM}]\{u_{FEM}\} = \{P\}, \tag{1}$$

where $[K_{FEM}]$ is the global stiffness matrix, $\{u_{FEM}\}$ are displacements at nodes and $\{P\}$ is the load vector.

The global stiffness matrix $[K_{FEM}]$ is assembled of element stiffness matrices k_{ij}^{mn}:

$$
\begin{aligned}
k_{ii}^{mn} &= \int\limits_{V} \left[(\lambda + 2\mu)\frac{\partial N_m}{\partial x_i}\frac{\partial N_n}{\partial x_i} + \mu \left(\frac{\partial N_m}{\partial x_{i+1}}\frac{\partial N_m}{\partial x_{i+1}} + \frac{\partial N_m}{\partial x_{i+2}}\frac{\partial N_m}{\partial x_{i+2}} \right) \right] dV, \\
k_{ij}^{mn} &= \int\limits_{V} \left(\lambda\frac{\partial N_m}{\partial x_i}\frac{\partial N_n}{\partial x_j} + \mu\frac{\partial N_m}{\partial x_j}\frac{\partial N_n}{\partial x_i} \right) dV, \quad i \neq j.
\end{aligned}
\tag{2}
$$

Here λ and μ are the Lamé constants; indices i and j are related to coordinate axes x_1, x_2, x_3; m and n are local node numbers inside the finite element.

Elastic stresses are determined by the relation:

$$\{\sigma_{FEM}\} = [D][B]\{u_{FEM}\}. \tag{3}$$

Here $[D]$ is the elasticity matrix and $[B]$ is the displacement differentiation matrix.

3.2 Elastic-plastic FEM procedure

In elastic-plastic problems the stress increment can be related to the displacement increment in a complicated manner. For small increments the elastic-plastic stress-displacement relation is similar to (3):

$$\{\Delta\sigma_{FEM}\} = [D^{ep}][B]\{\Delta u_{FEM}\}, \tag{4}$$

where $[D^{ep}]$ is the elastic-plastic stress-strain matrix. In actual computer implementations the finite strain increment is divided into subincrements or special integration schemes are used [8]. A simple way of solving elastic-plastic problems is the initial stress iterative procedure:

$$\{u_{FEM}^{(0)}\} = \{0\}$$
$$\{\Psi^{(0)}\} = \{P\}$$

do iterations $i = 1,\ 2\ \dots$

$$\{\Delta u_{FEM}^{(i)}\} = [K_{FEM}]^{-1}\{\Psi^{(i-1)}\}$$
$$\{u_{FEM}^{(i)}\} = \{u_{FEM}^{(i-1)}\} + \{\Delta u_{FEM}^{(i)}\} \qquad (5)$$
$$\{\sigma^{(i)}\} = \{\sigma^{ep}(\{u_{FEM}^{(i)}\})\}$$
$$\{\Psi^{(i)}\} = \{P\} - \int\limits_V [B]^T\{\sigma^{(i)}\}dV$$

while $||\Psi^{(i)}||/||P|| > \varepsilon$

Here (i) is the iteration number. Iterations start from elastic solution with the applied load $\{P\}$. Then residual vector $\{\Psi^{(i)}\}$ due to unbalanced stresses is employed as an artificial volume load. The iteration loop is finished when residuals become small enough.

4 SGBEM procedure

4.1 Boundary integral equation

The following weakly-singular boundary integral equation is valid for a crack in an infinite medium [4, 9, 10]:

$$-\int\limits_S \int\limits_S D_\alpha u_i^*(z)C_{\alpha i\beta j}(\xi - z)D_\beta u_j(\xi)dS(\xi)dS(z) \;=\; \int\limits_S u_k^*(z)t_k dS(z) \qquad (6)$$

Here S is one of crack surfaces; u_i are displacement discontinuities for the crack surface; u_i^* are the components of a continuous test function; and t_k are crack face tractions. The two-point weakly singular kernel $C_{\alpha i\beta j}$ is given by the following expression:

$$C_{\alpha i\beta j}(\zeta) = \frac{\mu}{4\pi(1-\nu)r}\left((1-\nu)\delta_{i\alpha}\delta_{j\beta} + 2\nu\delta_{i\beta}\delta_{j\alpha} - \delta_{ij}\delta_{\alpha\beta} - \frac{\zeta_i\zeta_j}{r^2}\delta_{\alpha\beta}\right)$$
$$\zeta = \xi - z \qquad\qquad\qquad\qquad\qquad\qquad\qquad\qquad\qquad\qquad (7)$$
$$r^2(\zeta) = \zeta_i\zeta_i$$

where ν is Poisson's ratio and μ is the shear modulus. A tangential operator D_α is defined as follows:

$$D_\alpha = \frac{1}{J}\left(\frac{\partial}{\partial\eta_1}\frac{\partial x_\alpha}{\partial\eta_2} - \frac{\partial}{\partial\eta_2}\frac{\partial x_\alpha}{\partial\eta_1}\right)$$
$$J = |\mathbf{s} \times \mathbf{t}| \qquad\qquad\qquad\qquad\qquad\qquad\qquad (8)$$
$$\mathbf{s} = \partial\mathbf{x}/\partial\eta_1 , \quad \mathbf{t} = \partial\mathbf{x}/\partial\eta_2,$$

where η_1, η_2 are the surface coordinates on the crack surface and $\mathbf{s}$, $\mathbf{t}$ are vectors in the plane that is tangent to the crack surface.

4.2 Discretization of the boundary integral equation

The discretized SGBEM equilibrium equation system for a crack in an infinite medium can be written in the form similar to (1):

$$[K_{BEM}]\{u_{BEM}\} = \{T\}, \tag{9}$$

where $[K_{BEM}]$ is the SGBEM global matrix (symmetric), $\{u_{BEM}\}$ are SGBEM nodal displacement discontinuities at the crack surface and $\{T\}$ are the equivalent nodal forces from tractions at the crack surface. Using index notation the global boundary element equation system can be rewritten in the following form:

$$K_{iajb}u_{jb} = H_{aq}t_{iq} \tag{10}$$

The global matrix is composed of element matrices:

$$k_{iajb}^{mn} = -\int\limits_{S_m}\int\limits_{S_n} C_{\alpha i \beta j} D_\alpha N_a(z) D_\beta N_b(\xi) dS(\xi) dS(z) . \tag{11}$$

$$h_{aq}^m = \int\limits_{S_m} N_a N_q dS(z) \tag{12}$$

Here S_m and S_n are areas of the pair of boundary elements; i and j are indices related to coordinate axes x_1, x_2, x_3 ; a and b are local node numbers in elements m and n; N_a are shape functions.

4.3 Integration of element matrices

An integral of a "stiffness" matrix coefficient (11), for the combination of boundary elements n and m, can be presented as follows:

$$\begin{aligned} I_{nm} &= \int\limits_{S_n}\int\limits_{S_m} f(\mathbf{x}, \tilde{\mathbf{x}}) dS_x dS_{\tilde{x}} \\ &= \int\limits_0^1\int\limits_0^1\int\limits_0^1\int\limits_0^1 F(\mathbf{x}(\eta_1, \eta_2), \tilde{\mathbf{x}}(\tilde{\eta}_1, \tilde{\eta}_2)) d\eta_1 d\eta_2 d\tilde{\eta}_1 d\tilde{\eta}_2 \end{aligned} \tag{13}$$

Taking into account that area elements can be presented in the form:

$$\begin{aligned} dS(\mathbf{x}) &= J(\mathbf{x}) d\eta_1 d\eta_2 \\ dS(\tilde{\mathbf{x}}) &= J(\tilde{\mathbf{x}}) d\tilde{\eta}_1 d\tilde{\eta}_2 \end{aligned} \tag{14}$$

and that the determinant of Jacobi matrix is contained in the expression for the tangential operator D_α the integrand can be written in the following form:

$$F = -C_{\alpha i \beta j} \left(\frac{\partial N_a}{\partial \eta_1} \frac{\partial x_\alpha}{\partial \eta_2} - \frac{\partial N_a}{\partial \eta_2} \frac{\partial x_\alpha}{\partial \eta_1} \right) \left(\frac{\partial N_b}{\partial \tilde{\eta}_1} \frac{\partial \tilde{x}_\beta}{\partial \tilde{\eta}_2} - \frac{\partial N_b}{\partial \tilde{\eta}_2} \frac{\partial \tilde{x}_\beta}{\partial \tilde{\eta}_1} \right) \tag{15}$$

The integral I_{nm} can be estimated using the Gaussian integration rule:

$$I_{nm} = \sum_i \sum_j \sum_k \sum_l F(\eta_{1i}, \eta_{2j}, \tilde{\eta}_{1k}, \tilde{\eta}_{2l}) w_i w_j w_k w_l \tag{16}$$

where η_{1i}, η_{2j} are the abscissas of the Gaussian integration rule and $w_i w_j$ are the corresponding weights. Regular integration procedure is appropriate for a pair of boundary elements, which have no common points. In order to provide sufficient integration accuracy for elements, which are coincident or have one edge or one vertex in common special integration approach should be used.

Special integration approach [5, 11] for coincident elements and for elements with common edge or common vertex, is based on the division of the four-dimensional integration domain $0 \leq \eta_1, \eta_2, \tilde{\eta}_1, \tilde{\eta}_2 \leq 1$ into several integration subdomains. In each subdomain, a special coordinate transformation is introduced, which cancels the singularity. The integral I_{nm} for a special case of elements with common points is computed as a sum of subdomain integrals:

$$I_{nm} = \int_0^1 \int_0^1 \int_0^1 \int_0^1 \sum_{i=1}^s F(\mathbf{x}(\eta_1^i, \eta_2^i), \tilde{\mathbf{x}}(\tilde{\eta}_1^i, \tilde{\eta}_2^i)) J_i d\omega d\xi_1 d\xi_2 d\xi_3 \tag{17}$$

where s is the number of integration subdomains, $\eta_1^i, \eta_2^i, \tilde{\eta}_1^i, \tilde{\eta}_2^i$ are local coordinates in subdomain i, which are expressed through integration variables $0 \leq \omega$, ξ_1, ξ_2, $\xi_3 \leq 1$ and J_i is the subdomain transformation Jacobian. The number of subdomains for coincident elements, for elements with common edge and for elements with common vertex is equal to 8, 6 and 4 correspondingly. Table for expressing local subdomain coordinates $\eta_1^i, \eta_2^i, \tilde{\eta}_1^i, \tilde{\eta}_2^i$ through integration variables ω, ξ_1, ξ_2, ξ_3 and relations for subdomain transformation Jacobians J_i can be found in Reference [5].

4.4 Displacement and stresses

After determining crack surface discontinuities by solving global equation (9) displacements and stresses at any point are calculated by integration over the crack surface:

$$u_p(\mathbf{x}) = - \int_S n_i(\xi) S_{ij}^p(\xi - \mathbf{x}) u_j(\xi) dS(\xi) \tag{18}$$

$$\sigma_{kl}(\mathbf{x}) = -\int_S E_{klpq} e_{iqm} S_{ij}^p(\xi - \mathbf{x}) D_m u_j(\xi) dS(\xi) \tag{19}$$

Here E_{klpq} is the elasticity tensor; e_{iqm} is the permutation symbol and S_{ij}^p is the stress fundamental solution:

$$
\begin{aligned}
S_{ij}^p(\zeta) &= \frac{1}{8\pi(1-\nu)r^2}\left(\frac{(1-2\nu)}{r}\left(\zeta_p\delta_{ij} - \zeta_i\delta_{pj} - \zeta_j\delta_{pi}\right) - \frac{3\zeta_p\zeta_i\zeta_j}{r^3}\right) \\
\zeta &= \xi - z \\
r^2(\zeta) &= \zeta_i\zeta_i
\end{aligned}
\tag{20}
$$

5   Combined SGBEM-FEM procedure

For a correct superposition corresponding to the solution for a finite body with a crack, fictitious forces on the boundary of the finite element model should be found in order to compensate for the stresses caused by the presence of a crack in an infinite body. The alternating method [12, 5] provides an efficient solution of this problem, without assembling the joint SGBEM-FEM matrix. The SGBEM-FEM alternating method employs the finite element solution for an uncracked body and the boundary element solution for a crack in an infinite body. Using an iterative procedure, correct tractions at the crack surface, are sought. The alternating approach can be applied to the solution of both elastic and elastic-plastic crack problems.

5.1   Elastic alternating procedure

The finite element surface residuals $\{\Psi\}$, and the crack face boundary element tractions $[T]$, are estimated through a similar integration:

$$
\begin{aligned}
\{\Psi\} &= \int_S [N_{FEM}][n]\{\sigma_{BEM}\}dS, \\
\{T\} &= \int_S [N_{BEM}][n]\{\sigma_{FEM}\}dS,
\end{aligned}
\tag{21}
$$

where $[N_{FEM}]$, $[N_{BEM}]$ are the finite element and boundary element shape functions, and $[n]$ are normal vectors to the finite element surface or to the crack surface. The residuals $\{\Psi\}$ are computed by integration over the finite element surface. Crack face tractions $[T]$ are estimated through integration over the crack surface. Boundary element stresses are used to calculate the finite element residual vector, and vice versa, finite element stresses are involved in the calculation of crack surface tractions. The elastic SGBEM-FEM alternating procedure can be presented as follows:

$$\{u_{FEM}^{(0)}\} = \{0\}$$
$$\{\Psi^{(0)}\} = \{P\}$$
$$\{u_{BEM}^{(0)}\} = \{0\}$$

do iterations $i = 1, 2 \dots$

$$\{\Delta u_{FEM}^{(i)}\} = [K_{FEM}]^{-1}\{\Psi^{(i-1)}\}$$
$$\{u_{FEM}^{(i)}\} = \{u_{FEM}^{(i-1)}\} + \{\Delta u_{FEM}^{(i)}\}$$
$$\{\Delta \sigma_{FEM}^{(i)}\} = [D][B]\{\Delta u_{FEM}^{(i)}\}$$
$$\{\Delta T^{(i)}\} = \int_{S} [N_{BEM}][n]\{\Delta \sigma_{FEM}^{(i)}\}dS \tag{22}$$
$$\{\Delta u_{BEM}^{(i)}\} = [K_{BEM}]^{-1}\{\Delta T^{(i)}\}$$
$$\{u_{BEM}^{(i)}\} = \{u_{BEM}^{(i-1)}\} + \{\Delta u_{BEM}^{(i)}\}$$
$$\{\Delta \sigma_{BEM}^{(i)}\} = \{\Delta \sigma(\{\Delta u_{BEM}^{(i)}\})\}$$
$$\{\Psi^{(i)}\} = \int_{S} [N_{FEM}][n]\{\Delta \sigma_{BEM}^{(i)}\}dS$$

while $||\Psi^{(i)}||/||P|| > \varepsilon$

In the elastic alternating procedure the FEM is used for stress analysis of the un-cracked body and the SGBEM is employed for crack modeling. After termination of the iterative procedure correct tractions at the crack surface are determined thus making possible to compute correct values of the stress intensity factors at the crack front.

5.2   Elastic-plastic alternating procedure

Elastic-plastic alternating method based on combination of FEM solution for the un-cracked body and an analytical solution for the crack has been proposed in Reference [13]. Here this algorithm is generalized for the SGBEM-FEM elastic-plastic alternating method. The finite element method is employed for stress analysis of the uncracked body. The residual vector $\{\Psi\}$ at nodes of the finite element model is calculated as the volume integral using accumulated stresses:

$$\{\Psi\} = \{P\} - \int_{V} [B]^{T}\{\sigma\}dV \tag{23}$$

where $\{P\}$ is the current load level, $[B]$ is the displacement differentiation matrix of the finite element method, $\{\sigma\}$ are accumulated elastic-plastic stresses and V is the volume of the finite element model. The symmetric Galerkin boundary element method is used for elastic analysis of the crack in an infinite medium. Both methods can be combined in the following iterative procedure of the elastic-plastic SGBEM-FEM alternating method:

$$\{u_{FEM}^{(0)}\} = \{0\}$$
$$\{\Psi^{(0)}\} = \{P\}$$
$$\{u_{BEM}^{(0)}\} = \{0\}$$

do iterations $i = 1,\ 2\ ...$

$$\{\Delta u_{FEM}^{(i)}\} = [K_{FEM}]^{-1}\{\Psi^{(i-1)}\}$$
$$\{u_{FEM}^{(i)}\} = \{u_{FEM}^{(i-1)}\} + \{\Delta u_{FEM}^{(i)}\}$$
$$\{\Delta\sigma_{FEM}^{e\ (i)}\} = [D][B]\{\Delta u_{FEM}^{(i)}\}$$
$$\{\Delta T^{(i)}\} = \int_S [N_{BEM}][n]\{\Delta\sigma_{FEM}^{e\ (i)}\}dS \qquad (24)$$

$$\{\Delta u_{BEM}^{(i)}\} = [K_{BEM}]^{-1}\{\Delta T^{(i)}\}$$
$$\{u_{BEM}^{(i)}\} = \{u_{BEM}^{(i-1)}\} + \{\Delta u_{BEM}^{(i)}\}$$
$$\{\sigma^{(i)}\} = \{\sigma^{ep}(\{u_{FEM}^{(i)} + u_{BEM}^{(i)}\})\}$$
$$\{\Psi^{(i)}\} = \{P\} - \int_V [B]^T\{\sigma^{(i)}\}dV$$

while $||\Psi^{(i)}||/||P|| > \varepsilon$

The solution procedure alternates between the FEM solution for the uncracked body and the SGBEM solution for the crack. The stresses at the finite element integration points are determined with the use of elastic-plastic constitutive material equations. The iterative procedure is terminated when the residual vector becomes small enough in comparison to the applied load.

6   Fracture mechanics parameters and crack growth

6.1   Calculation of the stress intensity factors

The elastic fracture mechanics parameters (stress intensity factors K_I, K_{II} and K_{III}) are determined by using asymptotic formulae for displacements in the vicinity of the crack front:

$$K_I = \frac{E}{(1-\nu^2)}\frac{u_3}{4\sqrt{2r/\pi}}$$
$$K_{II} = \frac{E}{(1-\nu^2)}\frac{u_2}{4\sqrt{2r/\pi}} \qquad (25)$$
$$K_{III} = \frac{E}{(1+\nu)}\frac{u_1}{4\sqrt{2r/\pi}}$$

where E is the elasticity modulus; ν is the Poisson's ratio; r is the distance from the point to the crack front and u_1, u_2 and u_3 are components of the displacement discontinuities at points at the crack surface in a local crack front coordinate system. The axis x_3 of the crack front coordinate system is tangent to the crack front, and the axis x_2 is normal to the crack surface.

Boundary elements with proper modeling of square-root stress singularity should be used at the crack front, in order to obtain values of the stress intensity factors with

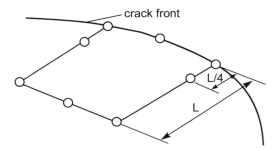

Figure 3: Singular boundary element with 8 nodes: two midside nodes are shifted by quater of the side length.

good precision. A convenient form of the boundary element with stress singularity is an 8-node element, with two midside nodes shifted towards the crack front by one quarter of the side length as shown in Figure 3. The values of the stress intensity factors K_{I}, K_{II} and K_{III} can be determined directly at the quarter-point nodes of the boundary elements or can be calculated at the corner and quarter-point nodes and extrapolated to the crack front.

In the elastic-plastic case the stress intensity factors are not valid fracture mechanics parameters. Some other fracture mechanics parameters such as J-integral components should be determined. However, in the case of applying elastic-plastic alternating method to fracture analysis the total solution is a sum of the SGBEM solution and the FEM solution. While elastic-plastic material relations are used to compute stress increment of the total strain, the SGBEM solution remains elastic at each iteration. The elastic asymptotic distribution dominates displacement and stress fields in the small vicinity of the crack front. Therefore, relations (25) can be used for the calculation of fracture mechanics parameters in the elastic-plastic case. Determined stress intensity factors K_I, K_{II} and K_{III} should be treated as elastic-plastic stress intensity factors, which correspond to the J-integral components through the following equivalence relations:

$$\begin{aligned} J_1 &= \tfrac{1-\nu^2}{E}(K_I^2 + K_{II}^2) + \tfrac{1+\nu}{E}K_{III}^2 \\ J_2 &= -2\tfrac{1-\nu^2}{E}K_I K_{II} \\ J &= \sqrt{J_1^2 + J_2^2} \end{aligned} \tag{26}$$

6.2 Fatigue crack growth

The SGBEM-FEM alternating method is very attractive for modeling of fatigue crack growth. Since the boundary element model and the finite element model are independent, only the boundary element model should be modified during crack growth simulation. To advance the front of a nonplanar crack it is necessary to know the direction of crack growth and the amount of crack growth.

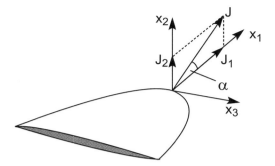

Figure 4: Local coordinate system and J-integral components for the crack front point.

The J-integral [14] is chosen here as a criterion for fatigue crack growth. According to the J-integral crack growth criterion: 1) crack grows in the direction of vector $\Delta\vec{J}$ as shown in Figure 4 ; 2) crack growth rate is determined by magnitude of ΔJ.

The $\Delta\vec{J}$ vector is normal to the crack front. Hence a point at the crack front moves in the plane normal to the crack front at the angle α from the plane which is tangential to the crack surface. The ranges of the J-integral ΔJ_1 and ΔJ_2 are expressed through ranges of the stress intensity factors ΔK_{I}, ΔK_{II} and ΔK_{III} using relations (26).

Typically, material fatigue crack growth models (such as Paris, Forman or NAS-GRO models) express the functional relationship for crack growth rate through the range of the effective stress intensity factor ΔK_{eff}:

$$\frac{da}{dN} = f(\Delta K_{eff}), \tag{27}$$

where da/dN is the crack growth per cycle. The range of the effective stress intensity factor ΔK_{eff} is related to the range ΔJ by the energy equivalence principle (26).

Modeling of fatigue crack growth is performed by finite increments. The crack-front advancement is performed by adding extra element layer to the existing crack model. At each increment the maximum crack advance is specified as Δa_{max}. The crack advance for a particular point at the crack front is calculated as follows:

$$\Delta a = \Delta a_{\mathrm{max}} \frac{(da/dN)}{(da/dN)_{\mathrm{max}}}. \tag{28}$$

7   Code development

7.1   Selection of programming language

During last decades a considerable experience has been accumulated in the development of finite element and boundary element codes. Fortran and C were mostly used as programming languages for FEM and BEM codes. Recently, developers of the engineering software started to use the object-oriented approach and C++ programming language. The object-oriented approach to the software development allows to develop and to debug software faster. The created codes are more reliable than codes developed with functional languages, these codes are easier to support, modify and reuse. A survey of object-oriented programming for structural mechanics is given in Reference [15].

Relatively new computer language Java inherited the best features of C++. However, Java has principal differences from C++. Some C++ features, which are considered potentially harmful (pointers, preprocessor, multiple inheritance etc.), do not exist in Java.

Java has some attractive features, which C++ does not possess. It is fully object-oriented language. Everything except primitive types for numbers should be an object in Java. Automatic garbage collection prevents Java codes from memory leaks. Java has APIs (application programming interfaces) for creating graphical user interfaces and for two-dimensional and three-dimensional visualization of geometry models and result fields. One of the most important features of Java is that Java codes are truly portable. A Java program is compiled into bytecode, which is run by a Java Virtual Machine (JVM). Most computer systems have JVMs. Thus once developed Java code can be executed on various platforms.

It is possible to conclude that Java offers a better software development environment for the development of engineering codes than Fortran or C++. Nevertheless, Java is not widely used in engineering computations. Often, performance of the Java code is considered insufficient for high-performance engineering computations. However, at present, Java Virtual Machines, which are used for the execution of Java codes, include Just-In-Time compiler and provide reasonable speed for typical finite element and boundary element computationally intensive routines. In Reference [16] it is shown that simple tuning of a finite element direct equation solver can help Java to provide roughly the same performance as the C language. Based on the above considerations we have chosen Java as a programming language for the development of SGBEM-FEM alternating code for crack analysis.

7.2   Object-oriented design of the code

From previous sections it is clear that the SGBEM-FEM alternating code should contain Java classes implementing the alternating procedure, SGBEM classes and FEM classes. Besides this, it is useful to have some other class groups. In Java, groups of classes and interfaces are placed in packages, which can be considered as tools for managing a large namespace and avoiding conflicts.

Our Java code consists of the following packages:

altern - alternating procedure;

sgbem - symmetric Galerkin boundary element method;

hsbem - hypersingular boundary element method;

fem - finite element method;

material - elastic and elastic-plastic material;

crackgen - generation of the crack mesh;

visual - visualization of models and results;

util - utility classes.

Alternating procedure package **altern** contains all classes with main methods and other classes necessary to organize interaction between the SGBEM and the FEM. Packages **sgbem** and **hsbem** are composed of classes implementing the symmetric Galerkin boundary element method and the hypersingular collocation boundary element method. After implementation of two different boundary element methods we have found that the SGBEM is more efficient for fracture mechanics analysis of non-planar cracks. Object-oriented approach to the code development allows easy implementation of two different packages for analyzing cracks without significant changes in other packages. Both packages interact with the same package **fem**, which include classes for the finite element method and additional classes data for BEM-FEM interaction.

The **sgbem** package consists of the following classes:

    class CrackModel - discrete model of the crack;
    class CrackNode - node of the crack model;
    class CrackElementData - data related to crack elements;
    class CrackElement - abstract boundary element;
        class CrackElement8N - quadrilateral boundary element with 8 nodes;
    class CrackSolver - abstract boundary element equation assembler/solver;
        class CrackSolverLDU - symmetric LDU equation assembler/solver;
    class CrackTractions - tractions at the crack surface;
    class FETractions - tractions at the surface of the finite element model;
    class StressIntensity - stress intensity factors at the crack front;
    class CrackResultFileWriter - writing results file.

Package **material** is composed of the following classes:

    class Material - abstract material class;
        class ElasticMaterial - implementation of elastic material;
        class ElasticPlasticMaterial - elastic-plastic material;
    class MidpointIntegration - mid-point integration of elastic-plastic increment;
    class FatigueModel - abstract fatigue material model;
        class ParisModel - Paris fatigue material model;
        class NasgroModel - NASGRO fatigue material model.

Constants and different utilities classes are collected in package **util**:

interface CNST - collection of constants;
class DATE - class for date and time;
class GaussRule - Gaussian integration rules;
class DirCosMatrix - matrix of direction cosines;
class DataFileReader - helper class for reading data files;
class Spline - spline approximation of the crack-front segments;
class Vector3D - operations on 3D vectors;
class XmlReader - helper class for reading files in XML format.

Object-oriented approach is suitable for creation reusable, extensible, and reliable components. It is worth noting that the extensive use of the object-oriented paradigm might not be always ideal for computationally intensive portions of the codes. Object creation and destruction in Java are expensive operations. The use of large amount of small objects can lead to considerable time and space overhead. Thus, we tried to employ useful features of the Java language in designing the SGBEM-FEM alternating code and to find a compromise between using objects and providing high efficiency for the computationally intensive parts of the code. A possible way to increase computing performance is reducing expenses for object creation in the code by using primitive types in place of objects. Expensive operations are calculation of the finite element and boundary element matrices and solution of the equation system. Only primitive type variables and one-dimensional arrays are used in computation of element matrices and in solution of the equation system thus excluding overhead associated with a full object-oriented programming.

Currently the Java compiler practically does not have actual means for powerful code optimization. Because of this attention should be devoted to a code tuning. It is necessary to identify code segments, which consume major computing time and to tune them manually. Tuning of the equation systems solution procedure is discussed in Reference [16]. Tuning efficiency can be illustrated by an example of tuning of the integration routine for a pair of boundary elements in the SGBEM method (see equations (11) and (17)). Unrolling of two inner loops and rearrangement of computations in outer loops produced a ten-times speedup on a Windows computer system. While tuning requires some additional efforts, we found that the use of Java leads to an overall development time reduction, because of easier programming and debugging in comparison to other languages.

7.3   Visualization

Visualization tools for finite element or boundary element models and analysis results are included in many commercial codes. A survey of these techniques can be found in Reference [17]. However commercial tools can not be directly used for visualization of results obtained with the SGBEM-FEM alternating code. The SGBEM-FEM alternating code produces results with high gradients near the crack front line using sparse meshes of boundary elements and finite elements.

Previously, when graphics libraries supported just simple graphical primitives, the development of visualization software was a complicated task. At present Java provides a rich graphics library, for three-dimensional visualization, which is called the Java3D. The Java 3D API contains Java classes for a sophisticated three-dimensional graphics rendering. The programmer works with high-level interface for creating and manipulating 3D geometric objects. The Java3D employs linear triangular and quadrilateral filled polygons for surface representation. Because of this the visualization of FEM/BEM models consisting of simplest elements is almost straightforward. However, for higher order elements the transformation of element surfaces into triangular polygons should be done carefully taking into account both geometry features and result field gradients.

The input data for the visualization consists of a set of nodes defined by spatial coordinates, a set of elements that is defined by nodal connectivities, and a set of result values. Primary results (displacements) are obtained at nodes of the finite element model and at nodes of the crack model. Total displacement field inside the three-dimensional body can be calculated as a sum of finite element displacements and of boundary element displacements in an infinite medium using Equation (18). Secondary finite element results, which are expressed through derivatives of the primary results, usually have the best precision at some points inside elements. For models composed of 20-node finite elements stresses have the most precise values at reduced Gaussian integration points $2 \times 2 \times 2$.

The visualization algorithm consists of the following main steps.

1. Obtain continuous field of finite element results by extrapolation from reduced integration points inside elements to element nodes with subsequent averaging. Add boundary element field.

2. Create the surface of the finite element model or create model section where results will be displayed.

3. Subdivide curved surfaces into flat triangles on the basis of surface curvature and gradient of results.

4. Create contour pictures by specifying coordinates of one-dimensional color pattern at triangle vertices.

In order to obtain continuous stress fields, stresses at reduced integration points are extrapolated to finite element nodes and are averaged with the use of contributions from adjacent finite elements. After this, nodal stresses can be interpolated inside elements using quadratic shape functions. Stresses induced by the presence of the crack can be computed at any point according to Equation (19).

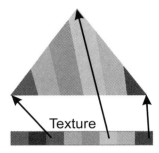

Figure 5: Using texture interpolation to produce color contours inside a triangle.

The surface of the finite element model is created from outer element surfaces. Outer surfaces are mentioned in the model connectivity array only once while inner surfaces are mentioned exactly two times. A fast approach to selecting outer element surfaces is to characterize each element surface by two connectivity numbers instead of eight. To avoid ambiguity we selected the following pair of connectivity numbers: the first connectivity number is the minimum global node number for the surface and the second connectivity number is the global number of node that is diagonal to the first node. If necessary to produce a body section and the section surface coincides with element surfaces then unnecessary elements are deleted from the model and the model surface is determined. In the case of arbitrary section a problem is more complicated since new elements should be created after model cut by the section surface.

Subdivision of quadratic element surfaces depends on two factors: curvature of the surface and range of result function over the surface. The subdivision into triangular elements is performed in the following steps: 1) determination of number of subdivisions along quadrilateral surface sides; 2) generating points inside quadrilateral surfaces and 3) Delaunay triangulation.

Geometry side subdivision depends on the side curvature. Curvature radius can be calculated using one-dimensional shape functions and coordinates of three nodes that define the side of a finite element surface or the side of a boundary element.

In order to obtain good quality of color contours the size parameter controlling triangulation should be selected such that each triangle contains the specified number of color intervals. This can be done by determining the range of result function over the surface and by computing the number of color bands along the surface side.

Java 3D provides tree-dimensional rendering of polygons with a possibility of texture interpolation. We use this possibility to create color contours inside triangles produced after subdivision of curved element surfaces. A one-dimensional texture containing desired number of color bands is generated. Values of functions at triangular vertices are transformed to texture coordinates using specified scale. Texture interpolation produces contours as shown in Figure 5. These texture coordinates are supplied to the Java 3D rendering engine, which generates a three-dimensional image.

Figure 6: Equivalent stress for a section of a cube with internal circular crack. Left: 16-color texture, center: 256-color texture, right: subdivision into triangles.

An example of contour images for a section of a tensile cube with an internal circular crack is presented in Figure 6. This example shows that the contour scale can be easily changed by generating new color texture with different number of color intervals. The texture with continuous color change produces a contour picture, which is usually created with a technique known as direct color interpolation.

8   Numerical examples

8.1   Inclined elliptical crack under tension

First we demonstrate the efficiency of the SGBEM for the determination of the stress intensity factors on the example of an inclined elliptical crack in an infinite medium under tension. An elliptical crack with semi-axes c and a is shown in Figure 7 where α is an angle of crack orientation in respect to direction of tension.

The elliptical crack inclined at 45 degrees with $a/c = 05$ is considered. It is characterized by the distribution of the stress intensity factors K_I, K_{II} and K_{III} along the crack front. A boundary element mesh used for crack modeling is shown in Figure 8. The mesh consists of 68 quadratic boundary elements and 229 nodes. Singular quarter-point elements are placed at the crack front. SGBEM results for the stress intensity factors K_I, K_{II} and K_{III} normalized as $K_i/(\sigma\sqrt{\pi a})$ are presented in Figure 9. Quite a satisfactory agreement of SGBEM results with theoretical solution [18] is observed.

8.2   Inclined semielliptical surface crack

The alternating method can be applied to surface crack modeling. Let us consider a semielliptical surface crack in a plate under mixed-mode loading conditions as shown in Figure 10,a. Geometric parameters are: major semi-axis ratio $a/c = 0.5$, orientation $\alpha = 45$ degrees, $a/t = 0.2$, $c/W = 0.2$. The crack is discretized by 34 quadratic

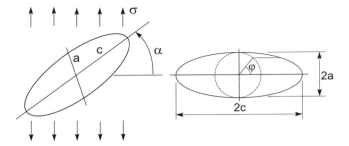

Figure 7: Inclined elliptical crack in an infinite medium under tensile loading.

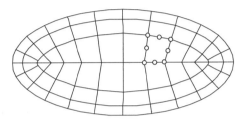

Figure 8: Boundary element mesh for modeling an inclined elliptical crack.

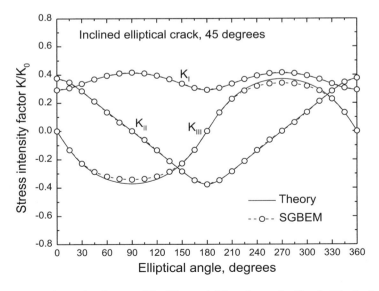

Figure 9: Stress intensity factors K_I, K_{II} and K_{III} for an inclined elliptical crack in an infinite medium under tensile loading.

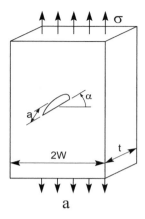 

a   b

Figure 10: Finite element and boundary element models for surface crack in a plate under mixed-mode loading.

boundary elements; and the finite element model consists of 320 brick-type 20-node elements (Figure 10,b). One layer of fictitious boundary elements outside the plate is added to the crack model in order to constrain rotations of boundary element nodes at the surface. Results for the stress intensity factors obtained by the SGBEM-FEM alternating method are compared to published solutions [19, 20] in Figure 11 where $K_0 = \sigma\sqrt{\pi a}$. One can see good agreement between SGBEM-FEM results and other numerical solutions.

8.3 Nonplanar fatigue crack growth

The fatigue growth of a surface crack under mixed-mode loading conditions is simulated starting from an inclined semielliptical precrack in a plate subject to a uniform tensile loading, which was modeled in a previous problem.

The Paris material fatigue model $da/dN = c(\Delta K_{eff})^m$ is chosen for fatigue crack growth with material parameters $c = 1.49 \cdot 0^{-8}$ and $m = 3.321$ (7075 Aluminum). According to the J-integral vector orientation and magnitude, the corner nodes at the crack front are advanced to new positions with scaling to the specified maximum crack advance $da_{\max}$. Locations of the midside nodes at the crack front are determined by spline interpolation. A new layer of elements is generated between old and new crack front lines. It is assumed that the movement of nodes located at the body surface are same as of the neighboring inside nodes. Then the new crack model is analyzed and *etc*. Five crack advancements with specified $da_{\max}/a = 0.3, 0.4, 0.5, 0.6$ and 0.7 were performed for the example problem. Stress intensity factors K_{I}, K_{II} and K_{III} after crack increments are given in Figure 12 with normalization $K_i/(\sigma\sqrt{\pi a})$. A three-dimensional view of the crack after five increments and its projection on the plane XY are presented in Figure 13.

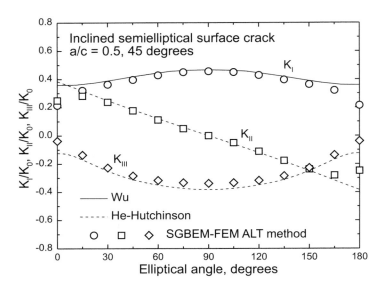

Figure 11: Stress intensity factors K_I, K_{II} and K_{III} for a semielliptical surface crack in a plate under mixed-mode loading.

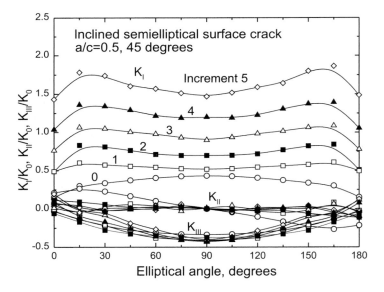

Figure 12: Stress intensity factors K_I, K_{II} and K_{III} after 5 crack growth increments for an inclined surface crack in a plate.

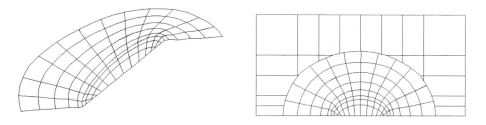

Figure 13: Three-dimensional view of the crack after five crack growth increments and its projection on the plane XY.

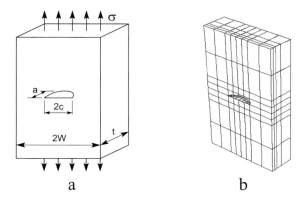

Figure 14: Semi-elliptical surface crack in a tensile plate and SGBEM-FEM model.

8.4 Elastic-plastic problem for a surface crack in a plate

The elastic-plastic algorithm of the SGBEM-FEM alternating method is used to determine the elastic-plastic stress intensity factor for a semi-elliptical surface crack in a tensile plate. Schematic of the problem is shown in Figure 14,a. The aspect ratio of the crack is $a/c = 2/3$, the relative crack depth is $a/t = 1/4$ and the thickness-width ratio is $t/W = 0.5$.

The crack is modeled by 40 quadratic boundary elements with 145 nodes. Singular boundary elements with quarter-point nodes are placed at the crack front. The finite element model of the plate contains 256 20-node elements and 1449 nodes (Figure 14,b).

Flow theory of plasticity with Mises yield surface is used. The material deformation curve with linear hardening and with the following parameters is selected:

$$\frac{\sigma}{\sigma_Y} = 1 + k\varepsilon^p, \qquad k = 0.1E/\sigma_Y, \qquad E/\sigma_Y = 00 ,$$

where σ_Y is the yield stress, k is the hardening coefficient, ε^p is the equivalent plastic

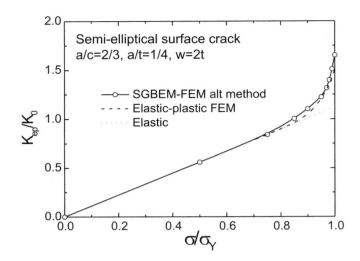

Figure 15: Elastic-plastic stress intensity factor for a surface semi-elliptical crack in a tensile plate.

strain and E is the elasticity modulus.

Tensile loading is applied in nine non-equal steps up to maximum load $\sigma/\sigma_Y = 1$. The elastic-plastic alternating iterative procedure is used at each load increment for determining displacement and stress fields. The iterative procedure is terminated when relative norm of the displacement increment reaches the specified tolerance ($\|\Delta u\|/\|u\| < 0.0\theta$). The number of iterations varied from 4 to 11 at different load steps. The elastic-plastic stress intensity factor K_{ep} is calculated using elastic asymptotic equations for displacements near the crack front. Values of K_{ep} determined by the SGBEM-FEM alternating procedure at the central point of the crack front are compared in Figure 15 to the elastic-plastic solution performed by the finite element method with explicit modeling of the crack. The elastic-plastic stress intensity factors are normalized as K_{ep}/K_0 where K_0 is equal:

$$K_0 = \sigma\sqrt{\pi a/Q}\,, \quad Q = 1 + 1464\,(a/c)^{1.65}.$$

Satisfactory agreement of both solutions can be observed.

9 Conclusion

The paper describes combination of the symmetric Galerkin boundary element method (SGBEM) and of the finite element method (FEM) for analyzing arbitrary three-dimensional cracks and their growth in structural components.

The SGBEM-FEM alternating method is very suitable for modeling of fatigue crack growth. Since the boundary element model and the finite element model are independent, only the boundary element model of the crack should be modified during crack growth modeling. The advancement of points at the front of a nonplanar crack is performed according to the J-integral fracture criterion.

The SGBEM-FEM alternating method is generalized for the solution of elastic-plastic crack problems. Flow theory of plasticity and initial stress method are used for elastic-plastic analysis. Both alternating procedure and initial stress procedure are placed inside the main iteration loop for the load step.

Results of several problems are given including simulation of fatigue crack growth of an inclined semi-elliptical crack and elastic-plastic analysis of a semi-elliptical surface crack.

References

[1] G.P. Nikishkov, S.N. Atluri, *"Calculation of fracture mechanics parameters for an arbitrary three-dimensional crack by the 'equivalent domain integral' method"*, Int. J. Numer Meth. Engng, 24, 851-867, 1987.

[2] K.N. Shivakumar, I.S. Raju, *"An equivalent domain integral method for three-dimensional mixed-mode fracture problems"*, Eng. Fract. Mech., 42, 935-959, 1992.

[3] N. Sukumar, N. Moes, B. Moran, T. Belytschko, *"Extended finite element method for three-dimensional crack modelling"*, Int. J. Numer. Meth. Engng, 48, 1549-1570, 2000.

[4] M. Bonnet, G. Maier, C. Polizzotto, *"Symmetric Galerkin boundary element methods"*, Appl. Mech. Rev., 51, 669-704, 1998.

[5] G.P. Nikishkov, J.H. Park, S.N. Atluri, *"SGBEM-FEM alternating method for analyzing 3D non-planar cracks and their growth in structural components"*, Computer Modeling in Engineering and Sciences, 2, 401-422, 2001.

[6] W.D. Keat, B.S. Annigeri, M.P. Cleary, *"Surface integral and finite element hybrid method for two- and three-dimensional fracture mechanics analysis"*, Int. J. Fracture, 36, 35-53, 1988.

[7] K.-J. Bathe, *"Finite Element Procedures"*, Prentice-Hall, Englewood Cliffs, NJ, USA, 1996.

[8] G.P. Nikishkov and S.N. Atluri, *"Implementation of a generalized midpoint algorithm for integration of elastoplastic constitutive relations for von Mises' hardening material"*, Computers and Structures, 49, 1037-1044, 1993.

[9] S. Li, M.E. Mear, *"Singularity-reduced integral equations for displacement discontinuities in three-dimensional linear elastic media"*, Int. J. Fract., 93, 87-114, 1998.

[10] S. Li, M.E. Mear, L. Xiao, *"Symmetric weak-form integral equation method for three-dimensional fracture analysis"*, Comput. Meth. Appl. Mech. Engng, 151, 435-459, 1998.

[11] A. Frangi, G. Novati, R. Springhetti, M. Rovizzi, *"Fracture mechanics in 3D by the symmetric Galerkin boundary element method"*, VIII Conf. On Numerical

Methods in Continuum Mechanics, 19-24 Sept. 2000, Liptovsky Jan, Slovak Republic.

[12] S.N. Atluri, *"Structural Integrity and Durability"*, Tech Science Press, Forsyth, 1997.

[13] G.P. Nikishkov, S.N. Atluri, *"An analytical-numerical alternating method for elastic-plastic analysis of cracks"*, Computational Mechanics, 13, 427-442, 1994.

[14] G.P. Cherepanov, *"Mechanics of Brittle Fracture"*, McGraw-Hill, New York, 1979.

[15] R.I.Mackie, *"Object oriented programming for structural mechanics: A review"*, in "Civil and Structural Engineering Computing: 2001", B.H.V. Topping (Editor), Saxe-Coburg Publications, Stirling, Scotland, 137-159, 2001.

[16] G.P. Nikishkov, Yu.G. Nikishkov, V.V. Savchenko, *"Comparison of C and Java performance in finite element computations"* (submitted to Computers and Structures), 2002.

[17] *"Computer Visualization. Graphics Techniques for Scientific and Engineering Analysis* (Ed. R.S.Gallagher). CRC Press, Boca Raton, 1994.

[18] H. Tada, P.C Paris, G.R. Irwin, *"The Stress Analysis of Cracks Handbook"*, ASME Press, 2000.

[19] X.R. Wu, *"Stress intensity factors for half-elliptical surface cracks subjected to complex crack face loading"*,Eng. Fract. Mech., 19, 387-405, 1984.

[20] M.Y. He, J.W. Hutchinson, *"Surface crack subject to mixed mode loading"*, Eng. Fract. Mech., 65, 1-14, 2000.

Chapter 9

©2002, Saxe-Coburg Publications, Stirling, Scotland
Engineering Computational Technology
B.H.V. Topping and Z. Bittnar, (Editors)
Saxe-Coburg Publications, Stirling, Scotland, 193-219.

Domain Decomposition Preconditioning for Parallel PDE Software

P.K. Jimack
School of Computing
University of Leeds, United Kingdom

Abstract

Domain decomposition methods have been applied to the solution of engineering problems for many years. Over the past two decades however the growth in the use of parallel computing platforms has ensured that interest in these methods, which offer the possibility of parallelism in a very natural manner, has become greater than ever. This interest has led to research that has yielded significant advances in both the theoretical understanding of the underlying mathematical structure behind domain decomposition methods and in the variety of domain decomposition algorithms that are available for use by the engineering community. In this paper we provide a brief overview of some of the main categories of domain decomposition algorithm and then focus on a particular variant of the overlapping Schwarz algorithm that is based upon the use of a hierarchy of finite element grids. Throughout the paper we consider domain decomposition methods as preconditioners for standard, Krylov subspace, iterative solvers however they may also be used directly as iterative methods in their own right. All of the theoretical results that are described apply equally in both cases.

Keywords: partial differential equations, domain decomposition, parallel computing.

1 Introduction

The majority of this paper considers the parallel finite element (FE) solution of the following two-dimensional variational problem, which is derived from a second order self-adjoint partial differential equation (PDE). All of the results and algorithms described can be generalized to three-dimensional problems however, and in section 4 practical extensions to a class of non-self-adjoint problems are also considered.

Problem 1.1 *Find $u \in H_E^1(\Omega)$ such that*

$$\mathcal{A}(u, v) = \mathcal{F}(v), \quad \forall v \in H_0^1(\Omega) , \tag{1}$$

where $\Omega \in \Re^2$ is the problem domain,

$$H_E^1(\Omega) = \{u \in H^1(\Omega) : u|_{\partial\Omega_E} = u_E(\underline{x})\} \tag{2}$$

and

$$H_0^1(\Omega) = \{u \in H^1(\Omega) : u|_{\partial\Omega_E} = 0\} . \tag{3}$$

Here $\partial\Omega_E$ is the (non-empty) part of the boundary, $\partial\Omega$, upon which essential (Dirichlet) boundary conditions are imposed and $\mathcal{A}(\cdot, \cdot)$ and $\mathcal{F}(\cdot)$ are the bilinear and linear forms

$$\mathcal{A}(u, v) = \int_\Omega (A(\underline{x})\underline{\nabla}u) \cdot \underline{\nabla}v \, d\underline{x} \quad \text{and} \quad \mathcal{F}(v) = \int_\Omega fv \, d\underline{x} + \int_{\partial\Omega_N} gv \, ds , \tag{4}$$

where $A(\underline{x})$ is symmetric and strictly positive-definite, and $\partial\Omega_N = \partial\Omega - \partial\Omega_E$ is the part of the boundary subject to Neumann boundary conditions: $\underline{n} \cdot (A(\underline{x})\underline{\nabla}u) = g(\underline{x})$.

In order to approximate this solution from a finite dimensional space of trial functions, $S^h(\Omega)$ say, it is necessary to solve the following discrete problem.

Problem 1.2 *Find $u^h \in S^h(\Omega) \cap H_E^1(\Omega)$ such that*

$$\mathcal{A}(u^h, v^h) = \mathcal{F}(v^h), \quad \forall v^h \in S^h(\Omega) \cap H_0^1(\Omega) . \tag{5}$$

This problem may in turn be expressed as the matrix equation

$$K\underline{u} = \underline{b} , \tag{6}$$

where K is the stiffness matrix, $\underline{b}$ is the load vector and $\underline{u}$ is a vector of nodal displacements which is to be determined. Note that the matrix K is strictly positive-definite and, for the usual choices of FE trial space and basis (e.g. [33, 49]), sparse. Hence an iterative solution method for (6), such as the conjugate gradient (CG) method, [23], is most appropriate.

In fact, for the FE method it is well known that when $S^h(\Omega)$ is a space of piecewise polynomial functions defined on a mesh of elements covering Ω with edge size h, $\mathcal{T}^h$ say, the condition number of K grows like $O(h^{-2})$ as $h \to 0$ (see [33] for example). For this reason it is necessary to apply a preconditioned version of the CG algorithm, or other iterative solver, for realistic mesh sizes h (again see [23]). The majority of this paper is concerned with this preconditioning step however for the rest of this section we consider the other major issues associated with solving (6) in parallel using an iterative method.

1.1 Mesh Partitioning and Parallel Finite Element Assembly

Let us assume that the problem domain Ω is a bounded polygonal region which is discretized into a non-overlapping set of triangles $\mathcal{T}$. Furthermore, consider the simplest FE trial space consisting of piecewise linear functions on $\mathcal{T}$. Hence u^h takes the form

$$u^h = \sum_{i=1}^{n} u_i N_i(\underline{x}) \; + \; \sum_{i=n+1}^{n+m} u_i N_i(\underline{x}) \,, \tag{7}$$

where $N_i(\underline{x})$ are the usual piecewise linear basis functions on $\mathcal{T}$ (which has n interior vertices and m on the Dirichlet boundary), and $u_{n+1}, ..., u_{n+m}$ are known to equal u_E from the Dirichlet boundary condition. The terms in the matrix equation (6) may therefore be written explicitly as:

$$K_{ji} \;=\; \int_{\Omega} A(\underline{x}) \underline{\nabla} N_j \cdot \underline{\nabla} N_i \, d\underline{x} \,, \tag{8}$$

$$b_j \;=\; \int_{\Omega} f N_j \, d\underline{x} + \int_{\partial\Omega_N} g N_j \, ds \;-\; \sum_{i=n+1}^{n+m} u_i \int_{\Omega} A(\underline{x}) \underline{\nabla} N_j \cdot \underline{\nabla} N_i \, d\underline{x} \,. \tag{9}$$

Before being able to solve this problem in parallel it is first necessary to form the system of equations (6) in parallel. This may be achieved by partitioning the triangulation $\mathcal{T}$ so that each processor works only with a subset of $\mathcal{T}$. In particular it is necessary to produce a set of sub-triangulations, $\{\mathcal{T}_1, ..., \mathcal{T}_p\}$ say, such that

$$\bigcup_{i=1}^{p} \mathcal{T}_i = \mathcal{T} \tag{10}$$

and

$$\mathcal{T}_i \cap \mathcal{T}_j = \phi \quad \text{when } i \neq j \,. \tag{11}$$

It is then possible to assemble the contributions to the matrix K and the vector $\underline{b}$ in (6) independently (and concurrently) on p different processors, $i = 1, ..., p$, with processor i working only on $\mathcal{T}_i$. There are many possible algorithms for producing a suitable partition of $\mathcal{T}$ however a description of these is beyond the scope of this paper. The main requirements of the partition are that $|\mathcal{T}_i| \approx |\mathcal{T}|/p$ (i.e. each processor deals with an approximately equal number of elements[1]) and the number of vertices lying on the partition boundary is as small as possible (this will minimize communication overheads). Further details of partitioning algorithms may be found in, for example, [16, 19, 22, 24, 25, 34, 44, 50, 52], and the papers cited therein.

Suppose that, once the triangulation $\mathcal{T}$ has been partitioned, the unknowns, $\underline{u}$, are ordered in the following manner. Each of the interior vertices in $\mathcal{T}_1$ ($\underline{u}_1$ say), followed by each of the interior vertices in $\mathcal{T}_2$ ($\underline{u}_2$ say), etc., up to each of the interior vertices

---

[1] We assume here that a homogeneous parallel computing system is being used for which each processor has the same performance characteristics.

in $\mathcal{T}_p$ ($\underline{u}_p$), followed by each of the vertices lying on the partition boundary ($\underline{u}_s$ say). It then follows that the system (6) may be written in block matrix form as

$$
\begin{bmatrix}
A_1 & & & & B_1 \\
& A_2 & & & B_2 \\
& & \ddots & & \vdots \\
& & & A_p & B_p \\
B_1^T & B_2^T & \cdots & B_p^T & A_s
\end{bmatrix}
\begin{bmatrix}
\underline{u}_1 \\
\underline{u}_2 \\
\vdots \\
\underline{u}_p \\
\underline{u}_s
\end{bmatrix}
=
\begin{bmatrix}
\underline{b}_1 \\
\underline{b}_2 \\
\vdots \\
\underline{b}_p \\
\underline{b}_s
\end{bmatrix}.
\tag{12}
$$

In (12) the block-arrowhead structure of the matrix K stems from the local support of the FE basis functions. Moreover, each of the blocks A_i, B_i (and therefore B_i^T) and $\underline{b}_i$ may be computed entirely by processor i (which works only with sub-mesh $\mathcal{T}_i$), for $i = 1, ..., p$. The blocks A_s and $\underline{b}_s$ both have contributions from elements in all of the sub-meshes however it is possible for each processor to assemble the contributions to each of these only from those elements on its own sub-mesh. On processor i we will refer to these contributions as $A_{s(i)}$ and $\underline{b}_{s(i)}$ respectively.

Once each processor has (concurrently) assembled each of the blocks A_i, B_i, $\underline{b}_i$, $A_{s(i)}$ and $\underline{b}_{s(i)}$, the system (12) is stored in a distributed manner. It is then ready to be solved. As indicated above the type of solver that we consider here is one based upon Krylov subspace iterations, such as the CG method. Apart from the preconditioning step, the main two computational tasks that must be undertaken within each iteration of such an algorithm are the formation of at least one matrix-vector product and at least one inner product of two vectors. The manner in which these may be undertaken in parallel is now described.

1.2   Parallel Matrix-Vector Products and Inner Products

Consider the calculation of the following matrix-vector product: $\underline{q} = K\underline{r}$, where K is the stiffness matrix that appears in (6). Such a product must be calculated at least once per iteration when using a Krylov subspace solver such as CG. This product may be written in the block matrix notation of (12), to give

$$
\begin{bmatrix}
\underline{q}_1 \\
\underline{q}_2 \\
\vdots \\
\underline{q}_p \\
\underline{q}_s
\end{bmatrix}
=
\begin{bmatrix}
A_1 & & & & B_1 \\
& A_2 & & & B_2 \\
& & \ddots & & \vdots \\
& & & A_p & B_p \\
B_1^T & B_2^T & \cdots & B_p^T & A_s
\end{bmatrix}
\begin{bmatrix}
\underline{r}_1 \\
\underline{r}_2 \\
\vdots \\
\underline{r}_p \\
\underline{r}_s
\end{bmatrix}.
\tag{13}
$$

From this it is clear that, for $i = 1, ..., p$,

$$
\underline{q}_i = A_i \underline{r}_i + B_i \underline{r}_s
\tag{14}
$$

and

$$
\begin{aligned}
\underline{q}_s &= \sum_{i=1}^{p} B_i^T \underline{r}_i + A_s \underline{r}_s \\
&= \sum_{i=1}^{p} \left(B_i^T \underline{r}_i + A_{s(i)} \underline{r}_s \right) .
\end{aligned}
\tag{15}
$$

Given that A_i, B_i and $A_{s(i)}$ are all stored on processor i, equations (14) and (15) show that, provided $\underline{r}_i$ is stored on processor i and $\underline{r}_s$ is stored on all processors, the matrix-vector product may be computed with only a single communication (corresponding to the summation in (15)). In fact, if we store $\underline{q}_s$ in a distributed pattern, where $\underline{q}_{s(i)} = B_i^T \underline{r}_i + A_{s(i)} \underline{r}_s$ is stored on processor i, then no inter-processor communication is required when forming a matrix-vector product.

In order to form the inner product of two vectors, $\underline{r}$ and $\underline{q}$ say, it is again helpful to use the block notation of (12). Thus

$$
\tau = \underline{r} \cdot \underline{q} =
\begin{bmatrix}
\underline{r}_1 \\ \underline{r}_2 \\ \vdots \\ \underline{r}_p \\ \underline{r}_s
\end{bmatrix}
\cdot
\begin{bmatrix}
\underline{q}_1 \\ \underline{q}_2 \\ \vdots \\ \underline{q}_p \\ \underline{q}_s
\end{bmatrix} .
\tag{16}
$$

If we again assume that $\underline{r}_i$ and $\underline{q}_i$ are stored on processor i, and if we further assume that $\underline{r}_s$ is stored on all processors but that $\underline{q}_{s(i)}$ is stored on processor i, then we may write (16) as:

$$
\tau = \sum_{i=1}^{p} \left(\underline{r}_i \cdot \underline{q}_i + \underline{r}_s \cdot \underline{q}_{s(i)} \right) .
\tag{17}
$$

It is clear from this that each inner product may be calculated with the need for just one inter-processor communication.

Full details of the parallel implementation of the multiplication operations considered in this subsection, along with further details on the parallel implementation of a CG solver and the FE assembly, may be found in [32]. This reference makes use of the library of parallel communication functions called MPI (Message Passing Interface), [38].

1.3 Schur Complement Methods

Before moving on to consider domain decomposition (DD) preconditioning for the iterative solution of (12) it is worth noting an alternative strategy for the solution of linear systems of this form. This strategy is based upon elimination of all of the interior unknowns $\underline{u}_1, ..., \underline{u}_p$ in (12). This is achieved by noting that

$$
A_i \underline{u}_i + B_i \underline{u}_s = \underline{b}_i \qquad \text{for } i = 1, ..., p
\tag{18}
$$

and

$$\sum_{i=1}^{p} B_i^T \underline{u}_i + A_s \underline{u}_s = \underline{b}_s \ . \tag{19}$$

Combining these two expressions gives

$$\left(A_s - \sum_{i=1}^{p} B_i^T A_i^{-1} B_i\right)\underline{u}_s = \underline{b}_s - \sum_{i=1}^{p} B_i^T A_i^{-1} \underline{b}_i \tag{20}$$

which may be written in the form

$$S\underline{u}_s = \underline{g} \ , \tag{21}$$

where

$$S = A_s - \sum_{i=1}^{p} B_i^T A_i^{-1} B_i \tag{22}$$

and the right-hand-side vector, $\underline{g}$, is given by

$$\underline{g} = \underline{b}_s - \sum_{i=1}^{p} B_i^T A_i^{-1} \underline{b}_i \ . \tag{23}$$

This system is known as the Schur complement, or Capacitance, system and the matrix S is known as the Schur complement, or Capacitance, matrix. Note that if one is able to obtain a solution to (21) for the interface unknowns $\underline{u}_s$, then (18) may be used to solve for the remaining unknowns $\underline{u}_i$ for $i = 1, ..., p$.

The solution of (21) is usually undertaken iteratively. This is because, although the matrix S is generally dense, it is computationally expensive to form S explicitly. If an iterative method, such as the CG algorithm is applied to solve (21) then all that is required is the product of S with a vector, $\underline{q}_s = S\underline{r}_s$ say. Hence the matrix S itself is not required explicitly. Furthermore, note that

$$S\underline{r}_s = \sum_{i=1}^{p} A_i \underline{r}_s - \sum_{i=1}^{p} B_i^T \left(A_i^{-1} \left(B_i \underline{r}_s\right)\right) \ . \tag{24}$$

Hence each product may be obtained using only local matrix-vector products and interior subdomain solves independently on each subdomain (followed by a single communication step). If an efficient sequential solver is available on each processor then this technique can be highly competitive. For very large problems however it is usually necessary to solve the subdomain problems iteratively and so the the Schur complement approach becomes less attractive. We do not consider it further in this paper but refer the reader to works such as [1, 2, 18, 21, 28, 35, 45, 46] for further details.

2 Preconditioning

There are many possible ways in which the system (6) can be preconditioned. Some of these are purely algebraic, such as incomplete Cholesky factorization [6, 37] or using sparse approximate inverses [7, 15], whilst others make use of the underlying FE derivation of the system, such as element-by-element preconditioning [27, 51] or domain decomposition preconditioners, which are the subject of this paper. The essential idea behind any preconditioning strategy for (6) is to find a positive-definite matrix, M say, that has two properties.

1. The matrix $M^{-1}K$ should have a small condition number.

2. The system $M\underline{s} = \underline{r}$ should be computationally cheap to solve.

(In fact the above properties refer to what is known as left preconditioning, where the system (6) is expressed as

$$(M^{-1}K)\underline{u} = M^{-1}\underline{b} . \tag{25}$$

This is the form of preconditioning that is considered in this paper, however all of the issues discussed apply equally to symmetric or right preconditioning.) When seeking to solve the system (6) on a parallel computer there is an additional requirement.

3. The system $M\underline{s} = \underline{r}$ should be easy to solve in parallel.

These properties are required because the rate of convergence of the preconditioned conjugate gradient (PCG) algorithm is dependent upon the condition number of the preconditioned matrix $M^{-1}K$, and the major computational step in this algorithm at each iteration is the solution of a system of the form $M\underline{s} = \underline{r}$ ([23]). As we will see, domain decomposition preconditioning can satisfy all three of these requirements.

In this section we provide an overview of the two main classes of DD preconditioner: iterative substructuring methods and Schwarz methods. There are other variants in the literature too, perhaps most notably the FETI approach of Farhat et al, [21], which is a variant of the substructuring approach based upon the use of Lagrange multipliers. Details of such variants however are beyond the scope of this paper.

2.1 Iterative Substructuring

These techniques are based upon a partition of the FE triangulation $\mathcal{T}$ into a set of non-overlapping sub-triangulations $\{\mathcal{T}_1, ..., \mathcal{T}_p\}$, such that not only do (10) and (11) both hold, but also the interface between subdomains forms a valid (coarse) triangulation of the domain Ω. Figure 1 illustrates an allowable decomposition of an example FE triangulation $\mathcal{T}$.

Given an allowable decomposition of the mesh it is possible to write the algebraic system (6) in the block form (12). Now suppose that the nodes on the interface, whose solution values are identified by the vector $\underline{u}_s$, are broken down into two types: those

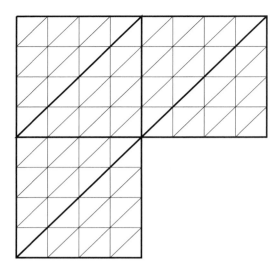

Figure 1: An allowable decomposition of a finite element mesh.

lying on edges of the coarse grid (E) and those forming vertices of the coarse grid (V). Figure 2 illustrates this for the mesh partitioned in Figure 1, showing the nodes on substructure edges as circles and the nodes at substructure vertices as squares: solid shapes indicate unknowns whilst unfilled shapes indicate that the nodal values are prescribed (due to Dirichlet boundary conditions for example). Hence the vectors $\underline{u}_s$ and $\underline{b}_s$ and the matrix block A_s in (12) may themselves be broken down accordingly:

$$\underline{u}_s = \begin{bmatrix} \underline{u}_E \\ \underline{u}_V \end{bmatrix}, \quad \underline{b}_s = \begin{bmatrix} \underline{b}_E \\ \underline{b}_V \end{bmatrix} \quad \text{and} \quad A_s = \begin{bmatrix} A_{EE} & A_{EV} \\ A_{EV}^T & A_{VV} \end{bmatrix}. \tag{26}$$

With this particular ordering of the unknowns the following expression defines a possible preconditioner for the system (12):

$$M = \begin{bmatrix} A_1 & & & & & \\ & A_2 & & & & \\ & & \ddots & & & \\ & & & A_p & & \\ & & & & \text{diag}(A_{EE}) & \\ & & & & & \tilde{K} \end{bmatrix}. \tag{27}$$

Here $\text{diag}(A_{EE})$ is the diagonal of the matrix block A_{EE} and $\tilde{K}$ is the stiffness matrix which arises when solving the original PDE problem (1) on the coarse triangulation formed by the subdomains.

This idea may be extended by replacing the diagonal submatrix $\text{diag}(A_{EE})$ by a corresponding block diagonal, $\text{block}(A_{EE})$ say, which has a block, $A_{EE_{(j)}}$, for each

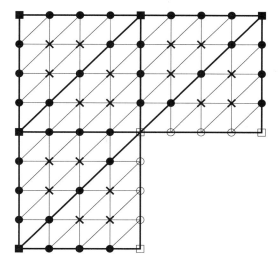

× **Node in subdomain interior**

● ■ **Node on substructure boundary**

Figure 2: The different types of unknown in a typical mesh.

of the substructure edges. That is,

$$
\text{block}(A_{EE}) = \begin{bmatrix} A_{EE_{(1)}} & & & \\ & A_{EE_{(2)}} & & \\ & & \ddots & \\ & & & A_{EE_{(N)}} \end{bmatrix}, \tag{28}
$$

where N is the number of edges in the coarse substructure not lying on the Dirichlet part of the boundary. In [9] Bramble, Pasciak and Schatz analyze their own preconditioner which is based upon this idea of separating the internal, the substructure edge and the substructure vertex unknowns. This is the first of a series of papers analyzing the construction of preconditioners by substructuring [9, 10, 11, 12], and in it they prove that the condition number of their preconditioned problem grows in proportion to $(1 + \log (H/h))^2$, where H represents the size of elements in the coarse mesh (substructure).

To see how such a preconditioner may be constructed recall that the preconditioning step in the PCG algorithm ([23]) requires the solution of the system $M\underline{s} = \underline{r}$ at each iteration. Taking

$$
M = \begin{bmatrix}
A_1 & & & & & \\
& A_2 & & & & \\
& & \ddots & & & \\
& & & A_p & & \\
& & & & \mathrm{block}(A_{EE}) & \\
& & & & & \tilde{K}
\end{bmatrix}, \tag{29}
$$

with $\mathrm{block}(A_{EE})$ given by (28), implies that this preconditioning step may be written as

$$
\underline{s} = M^{-1}\underline{r} = \sum_{i=1}^{p} R_{I_{(i)}}^{T} A_{II_{(i)}}^{-1} R_{I_{(i)}}\underline{r} + \sum_{j=1}^{N} R_{E_{(j)}}^{T} A_{EE_{(j)}}^{-1} R_{E_{(j)}}\underline{r} + R_{V}^{T}\tilde{K}^{-1}R_{V}\underline{r} . \tag{30}
$$

Here $R_{I_{(i)}}$ denotes the restriction operator which, given a vector of length n (the dimension of $\underline{r}$), extracts only those components associated with the interior of subdomain i. Similarly, $R_{E_{(j)}}$ extracts only those components associated with edge j of the substructure and R_V extracts the components associated with the substructure vertices. Further details, along with a discussion of a practical parallel implementation, may be found in [29]. Other works in which this type of approach is considered, and parallel implementation issues are addressed, include [14, 20, 28, 26, 35, 39, 46]

2.2   Schwarz Methods

Unlike the iterative substructuring approach described in the previous subsection, Schwarz methods are DD solvers that make use of overlapping, as opposed to non-overlapping (see (11)), subdomains. The original ideas behind these techniques may be traced back to the work of Schwarz, who formulated what has become known as the Schwarz alternating method as long ago as 1869 [43]. The idea behind this approach is to divide Ω into a number of overlapping subdomains (we use two for this illustration), Ω_1 and Ω_2 say. Now let the part of the boundary of Ω_1 that lies inside Ω_2 be labeled Γ_1, and the part of the boundary of Ω_2 that lies inside Ω_1 be labeled Γ_2. The Schwarz alternating algorithm is then:

- Guess values of solution unknowns on Γ_1.

- Repeat steps 1 and 2 until convergence:

    1. Use latest values on Γ_1 to solve problem in Ω_1,
    2. Use latest values on Γ_2 to solve problem in Ω_2.

This idea may easily be incorporated into the PCG algorithm by simply applying steps 1 and 2 above as the preconditioner at each iteration. As it stands however the approach is inherently sequential. If there is a larger number of subdomains than processors then parallelism may be introduced by colouring the subdomains so that any two which overlap are coloured differently. This then allows solves to be performed in all of the subdomains of the same colour in parallel (see [26, 39, 47] for details). This is known as multiplicative additive Schwarz preconditioning.

An alternative way of introducing parallelism into the above preconditioner is to decouple the subproblems so that all of the subdomain solves may be performed simultaneously. This may be viewed as an overlapping block Jacobi preconditioner as opposed to an overlapping block Gauss-Seidel preconditioner ([39]). In [17] however, Dryja and Widlund describe this decoupling more algebraically. Following [36], they define the overlapping Schwarz method in terms of the product of a number of projections onto subspaces (corresponding to the subregion problems). They then define the decoupled approach to preconditioning in terms of sums of these projection operators, referring to such techniques as additive (as opposed to multiplicative) Schwarz methods (see below).

With both additive and multiplicative Schwarz methods the number of iterations required for convergence grows significantly with the number of subdomains. Hence, in [18], Dryja and Widlund introduce an extra step to the additive preconditioner (i.e. in addition to the decoupled overlapping subdomain solves). This step requires the additional solution of a problem on a much coarser triangulation than $\mathcal{T}$, whose elements are of size H say. The solution of this coarse grid problem is rather like the substructure solves given by $\tilde{K}^{-1}$ in the previous subsection, however the elements of coarse grid do not need to correspond to the subdomains in this case (generally each subdomain will be made up of a number of coarse grid elements). Although this modification places a restriction on the way in which the domain Ω may be decomposed into subdomains it has the advantage that, provided the overlap distance, δ, is bounded below by a fixed fraction of the coarse grid size, H, the condition number of the preconditioned system is independent of both H and h. These arguments may be formalized by the following theoretical results.

Suppose that we have a coarse grid $\mathcal{T}^0$ and the fine grid $\mathcal{T}$ may be obtained through a refinement of this grid[2]. Then the FE space $\mathcal{V}_0$, made up of all possible piecewise linear (say) polynomials on $\mathcal{T}^0$ is a subspace of $\mathcal{V}$, the FE space of all possible piecewise linear functions on $\mathcal{T}$. Furthermore, suppose that $\mathcal{T}_i$ is the subset of $\mathcal{T}$ that covers the subdomain Ω_i, for $i = 1, ..., p$ (where the subdomains, and therefore the triangulations $\mathcal{T}_i$, may overlap). Then each $\mathcal{V}_i$, the FE space of all possible piecewise linear functions on $\mathcal{T}_i$, is also a subspace of $\mathcal{V}$. Given the usual FE basis for each of the spaces $\mathcal{V}, \mathcal{V}_0, \mathcal{V}_1, ..., \mathcal{V}_p$ it is possible to define rectangular matrices, R_i say, corresponding to the L_2 projections from $\mathcal{V}$ to $\mathcal{V}_i$ for $i = 0, 1, ..., p$. The additive Schwarz (AS) preconditioner is then given by the following algebraic expression,

$$M^{-1} = \sum_{i=0}^{p} R_i^T K_i^{-1} R_i , \qquad (31)$$

---

[2]This restriction that the fine grid should be a refinement of the coarse grid is not strictly necessary but it does simplify the description that follows sufficiently to justify being made an assumption throughout this paper. See, for example [47], for a discussion of the use of non-nested grids.

where $K_i = R_i K R_i^T$ for $i = 0, 1, ..., p$. An alternative way of viewing K_i is that it is the FE stiffness matrix derived from $\mathcal{A}_i$, the restriction of $\mathcal{A}$ (from (1) and (4)) to $\mathcal{V}_i \times \mathcal{V}_i$ given by:

$$\mathcal{A}_i(u_i, v_i) = \mathcal{A}(u_i, v_i), \quad \forall u_i, v_i \in \mathcal{V}_i . \tag{32}$$

The following theorem which is proved in [53] for example (or see [47] for a slightly more general form), provides the main theoretical justification for this pre-conditioner.

Theorem 2.1 *The matrix M defined by (31) is symmetric and positive-definite. Furthermore, if we assume that there is some constant $C > 0$ such that: for all $v \in \mathcal{V}$ there are $v_i \in \mathcal{V}_i$ such that $v = \sum_{i=0}^{p} v_i$ and*

$$\sum_{i=0}^{p} \mathcal{A}_i(v_i, v_i) \leq C\mathcal{A}(v, v) , \tag{33}$$

then the spectral condition number of $M^{-1}K$ is given by

$$\kappa(M^{-1}K) \leq n_c C , \tag{34}$$

where n_c is the minimum number of colours required to colour the subdomains Ω_i in such a way that no neighbours are the same colour.

Furthermore the following result, which is also proved in [47] and [53], demonstrates that the condition (33) is satisfied for a constant C that is independent of h, H and p provided that the overlap between the subdomains Ω_i is always proportional to the size of elements in the coarse triangulation. This is often referred to as a "generous" overlap, and the preconditioner is said to be "optimal" in this case.

Theorem 2.2 *Provided the overlap between the subdomains Ω_i is of size $O(H)$, where H represents the mesh size of $\mathcal{T}^0$, then there exists $C > 0$, which is independent of h, H and p, such that for any $v \in \mathcal{V}$ there are $v_i \in \mathcal{V}_i$ such that $v = \sum_{i=0}^{p} v_i$ and*

$$\sum_{i=0}^{p} \mathcal{A}_i(v_i, v_i) \leq C\mathcal{A}(v, v) . \tag{35}$$

The requirement for a generous overlap between subdomains in order to obtain an optimal preconditioner is extremely demanding. In two dimensions, as $\mathcal{T}$ is refined (assuming uniform global refinement for simplicity), the number of elements of $\mathcal{T}$ in the overlap regions is $O(h^{-2})$, and in three dimensions it is $O(h^{-3})$. In practice therefore this requirement is usually dropped in one of two different ways. The simplest option is to stick to a fixed number of layers of overlapping elements of the triangulation $\mathcal{T}$ as it is refined. This means that the overlap region decreases as $h \to 0$ and so the condition number of the preconditioned system grows. Fortunately however this growth is typically quite slow and so the additional iterations required by the

PCG solver are justified by the reduced cost of applying the preconditioner at each iteration (since there are less unknowns associated with each subdomain problem in comparison to the optimal preconditioner with an $O(H)$ overlap), [47]. Nevertheless, the preconditioner is no longer optimal.

The other common approach taken to avoid the need for a generous overlap is to make use of a sequence of nested grids (as opposed to just a coarse grid and a fine grid). The coarse grid problem in (31) is then replaced by the solution of a problem on a grid which is only one level of refinement less than the fine grid problem. This problem is then solved using a two level approximation based upon the grid that is two levels of refinement less than the fine grid problem. Repetition of this two level approach at repeatedly coarser levels leads to a "multilevel" Schwarz algorithm that is rather more complex than the two level algorithm given by (31) but which yields an optimal preconditioner with only a minimal amount of overlap. Again see, for example, [47] for further details.

The main drawback with the multilevel AS algorithm is the need for a series of projections from one grid to the next at each preconditioning step. On a regular sequence of grids obtained through uniform refinement this is relatively straightforward to manage in parallel, however on a sequence of grids generated with local, rather than global, mesh refinement this can become a complex programming task. Throughout the rest of this paper therefore we describe an alternative to the multilevel approach that is a two level AS algorithm based upon the use of a hierarchical sequence of grids. As such, this may be considered to be a cross between the classical two level AS algorithm with a generous overlap and the classical multilevel AS algorithm.

3   A Hierarchical Two Level Schwarz Algorithm

This section introduces an alternative two level AS algorithm based upon the use of a "weakly overlapping" hierarchy of nested grids, as described in [5]. An overview of the parallel implementation is also included.

3.1   A Weakly Overlapping Additive Schwarz Preconditioner

As before, let $\mathcal{T}^0$ be a coarse triangulation of Ω and suppose that $\mathcal{T}^0$ consists of N_0 triangular elements, $\tau_j^{(0)}$, each of size $O(H)$. Now assume that $\mathcal{T}^0$ is partitioned into p *non-overlapping* subdomains Ω_i such that:

$$\overline{\Omega} = \bigcup_{i=1}^{p} \overline{\Omega}_i, \tag{36}$$

$$\Omega_i \cap \Omega_j = \phi \quad (i \neq j), \tag{37}$$

$$\overline{\Omega}_i = \bigcup_{j \in I_i} \tau_j^{(0)} \quad \text{where } I_i \subset \{1, ..., N_0\} \quad (I_i \neq \phi). \tag{38}$$

Here the overbar is used to denote the closure of a set of points.

Now permit $\mathcal{T}^0$ to be refined several times, to produce a family of triangulations, $\mathcal{T}^0, ..., \mathcal{T}^J$, where each triangulation, $\mathcal{T}^k$, consists of N_k elements, $\tau_j^{(k)}$, such that

$$\overline{\Omega} = \bigcup_{j=1}^{N_k} \tau_j^{(k)} \quad \text{and} \quad \mathcal{T}^k = \{\tau_j^{(k)}\}_{j=1}^{N_k} . \tag{39}$$

The successive mesh refinements that define this sequence of triangulations need not be global and may be non-conforming, however they must satisfy a number of conditions, as in [8] for example:

1. $\tau \in \mathcal{T}^{k+1}$ implies that either

    (a) $\tau \in \mathcal{T}^k$, or

    (b) τ has been generated as a refinement of an element of $\mathcal{T}^k$ into four similar children,

2. the level of any triangles which share a common point can differ by at most one,

3. only triangles at level k may be refined in the transition from $\mathcal{T}^k$ to $\mathcal{T}^{k+1}$.

(Here the level of a triangle is defined to be the least value of k for which that triangle is an element of $\mathcal{T}^k$.) In addition to the above it is also necessary that:

4. in the final mesh, $\mathcal{T}^J$, all pairs of triangles on either side of the boundary of each subdomain Ω_i have the same level as each other.

Having defined a decomposition of Ω into subdomains and a nested sequence of triangulations of Ω, [5] next defines the restrictions of each of these triangulations onto each subdomain by

$$\Omega_{i,k} = \{\tau_j^{(k)} : \tau_j^{(k)} \subset \overline{\Omega}_i\} , \tag{40}$$

and, in order to introduce a certain amount of overlap between neighbouring subdomains,

$$\tilde{\Omega}_{i,k} = \{\tau_j^{(k)} : \tau_j^{(k)} \text{ has a common point with } \overline{\Omega}_i\} . \tag{41}$$

Following this, finite element spaces associated with these local triangulations are introduced. Let G be some triangulation and denote by $\mathcal{S}(G)$ the space of continuous piecewise linear functions on G. Then the following definitions can now be made:

$$\mathcal{V} = \mathcal{S}(\mathcal{T}^J) \tag{42}$$
$$\mathcal{V}_0 = \mathcal{S}(\mathcal{T}^0) \tag{43}$$
$$\mathcal{V}_{i,k} = \mathcal{S}(\Omega_{i,k}) \tag{44}$$
$$\tilde{\mathcal{V}}_{i,k} = \mathcal{S}(\tilde{\Omega}_{i,k}) \tag{45}$$
$$\mathcal{V}_i = \tilde{\mathcal{V}}_{i,0} + ... + \tilde{\mathcal{V}}_{i,J} . \tag{46}$$

Note that the spaces $\mathcal{V}$ and $\mathcal{V}_0$ are the same as those defined in the previous section but that the spaces $\mathcal{V}_i$, for $i = 1, ..., p$ are different (since they now correspond to an overlap of just one element at each level of the mesh hierarchy). Nevertheless, it is still evident that

$$\mathcal{V} = \mathcal{V}_0 + \mathcal{V}_1 + \dots + \mathcal{V}_p \ . \tag{47}$$

This means that for all $v \in \mathcal{V}$ there are $v_i \in \mathcal{V}_i$ such that $v = \sum_{i=0}^{p} v_i$, and this is the decomposition of $\mathcal{V}$ that [5] proposes for the alternative two level AS preconditioner of the form (31).

In [5] the following theorem is proved for problems in both two and three dimensions. When combined with Theorem 2.1, this shows that the preconditioner (31), with the rectangular matrices now representing projections from $\mathcal{V}$ to the new spaces $\mathcal{V}_i$, is optimal.

Theorem 3.1 *Let $\mathcal{V}$, $\mathcal{V}_0$ and $\mathcal{V}_i$ (for $i = 1, \dots, p$) be defined by (42), (43) and (46) respectively. Then there exists $C > 0$, which is independent of h, H and p, such that for any $v \in \mathcal{V}$ there are $v_i \in \mathcal{V}_i$ such that $v = \sum_{i=0}^{p} v_i$ and*

$$\sum_{i=0}^{p} \mathcal{A}_i(v_i, v_i) \leq C\mathcal{A}(v, v) \ . \tag{48}$$

3.2 Implementation

In order to implement the above AS preconditioner in parallel, [5] combines the coarse grid solve associated with K_0^{-1} in (31) with each of the subdomain solves. This is achieved by assigning a copy of the entire coarse mesh, $\mathcal{T}^0$, to each processor but only allowing processor i to refine this mesh inside $\tilde{\Omega}_{i,k-1}$ at step k of the refinement process (i.e. from $\mathcal{T}^{k-1}$ to $\mathcal{T}^k$). The continuous piecewise linear FE spaces on the resulting meshes are then given by

$$\tilde{\mathcal{V}}_i = \mathcal{V}_0 \cup \mathcal{V}_i \tag{49}$$

for $i = 1, \dots, p$. Corresponding meshes (with $p = 3$) are illustrated for a simple 2-d example in Figure 3. Note that in this figure there are a number of "slave" nodes in each processor's mesh which cause these meshes to be non-conforming. The solution values at these nodes are not free: they are determined by the nodal values at the ends of the edges on which the slave nodes lie. For a practical implementation it turns out to be simpler to allow the solution values at these nodes to be free by performing an interior refinement of those elements on the unrefined sides of the edges that have "hanging" nodes on them. In the 2-d example of Figure 3 this simply involves bisecting all triangular elements containing a hanging node and for 3-d problems a similar intermediate refinement strategy may be used (see [48] for example).

Having obtained a mesh on each processor it is possible for processor i to assemble the stiffness matrix, K_i, for its own mesh independently of the other processors. These

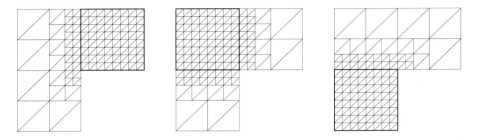

Figure 3: An example of the meshes produced when $p = 3$ and $J = 2$ for a coarse mesh containing 24 elements.

	$p = 2$	$p = 4$	$p = 8$	$p = 16$
$J = 2$	6	9	12	14
$J = 3$	6	8	12	13
$J = 4$	6	8	12	13
$J = 5$	6	7	11	12
$J = 6$	6	7	11	12

Table 1: The number of PCG iterations required to reduce the residual by a factor of 10^6 when solving Poisson's equation on a coarse mesh of 256 elements (taken from [5]).

are the matrices that are used in a modified version of the preconditioner (31) which takes into account the merging of $\mathcal{V}_0$ with $\mathcal{V}_i$:

$$M^{-1} = \sum_{i=1}^{p} \tilde{R}_i^T K_i^{-1} \tilde{R}_i \ . \tag{50}$$

In (50) each $\tilde{R}_i$ is the rectangular matrix that represents the projection from $\mathcal{V}$ to $\tilde{\mathcal{V}}_i$. Note that the application of these projections (and the corresponding prolongations $\tilde{R}_i^T$) requires interprocessor communications but that the subdomain solves (corresponding to K_i^{-1}) may be undertaken independently and concurrently. A number of results are presented in [5] however Table 1 shows a sample of these for solving Poisson's equation on a square domain in two dimensions on between 2 and 16 subdomains, using between 2 and 6 levels of uniform refinement. As predicted by the theory, the number of iterations taken appears to be bounded independently of h and p for a fixed choice of H.

4   Extensions

In this section two possible extensions of the work of [5], outlined in the previous section, are considered. The first of these is based upon an observation made by Cai and Sarkis in [13] concerning traditional AS preconditioners of the form (31). The second extension is to a larger class of problem than those given by Problem 1.1. In addition to the straightforward generalization to three dimensional problems (which is included in [5]), it is possible to apply the preconditioning techniques reviewed in this paper to non-self-adjoint problems. This is described for convection-diffusion equations, based upon the work appearing in [4, 30, 31, 40, 41].

4.1   A Restricted Version of the Preconditioner

In [13] an AS preconditioner of the form (31) is considered. It is noted that for each subdomain i, for $i = 1, ..., p$, the operations R_i and R_i^T each require a certain amount of computational work in order to restrict vectors between $\mathcal{T}$ and $\mathcal{T}_i$, and prolongate vectors from $\mathcal{T}_i$ to $\mathcal{T}$, respectively. Furthermore, in a parallel implementation on p processors, each of these operations requires interprocessor communication between neighbouring processors. In [13] therefore, a restricted AS preconditioner is proposed in which R_i^T is replaced by a different prolongation matrix, $\hat{R}_i^T$ say, for $i = 1, ..., p$. This matrix represents a simpler prolongation to $\mathcal{T}$ from those nodes of $\mathcal{T}_i$ in the region of Ω_i that does not overlap with neighbours, with overlapping nodal values set to zero. Hence no interprocessor communication is required for the prolongation steps and the total communication cost of each preconditioning step is halved. Interestingly, [13] reports that, with this restricted AS preconditioner, iteration counts are actually slightly lower than those obtained using the full AS preconditioner (31). It should be noted however that this new preconditioner is not symmetric positive-definite (SPD) and so the PCG algorithm can no longer be used.

In [4] and [41] results are presented (in two and three dimensions respectively) using a GMRES solver (see, for example, [3, 23, 42]) with a preconditioner based upon a restricted version of the new preconditioner (50):

$$M^{-1} = \sum_{i=1}^{p} \hat{R}_i^T K_i^{-1} \tilde{R}_i \,. \tag{51}$$

Here $\tilde{R}_i$ and K_i are as in (50) and $\hat{R}_i^T$ is a rectangular prolongation matrix that maps $\underline{\zeta} \in \Re^{\tilde{n}_i}$ to $\underline{z} \in \Re^n$ in the following manner (where the index j is used to enumerate the n nodes in the fine mesh $\mathcal{T}$).

$$
\begin{aligned}
z_j &= \zeta_k &&\text{when node } j \text{ is in the interior of} \\
&&&\text{the subregion owned by processor } i \\
&&&\text{(and is numbered } k \text{ on processor } i\text{),} \\[1em]
z_j &= \zeta_k / \nu_k &&\text{when node } j \text{ is on the boundary of} \\
&&&\text{the subregion owned by processor } i, \\
&&&\text{(and is numbered } k \text{ on processor } i\text{),} \\[1em]
z_j &= 0 &&\text{otherwise.}
\end{aligned}
$$

	AS preconditioner (50)				Reduced AS preconditioner (51)			
	$p = 2$	$p = 4$	$p = 8$	$p = 16$	$p = 2$	$p = 4$	$p = 8$	$p = 16$
$J = 1$	6	11	16	19	3	3	4	4
$J = 2$	7	12	17	19	3	5	6	6
$J = 3$	8	13	19	20	4	6	7	8
$J = 4$	8	13	19	20	4	8	9	9

Table 2: The number of preconditioned GMRES iterations required to reduce the residual by a factor of 10^5 when solving Poisson's equation on a coarse mesh of 384 tetrahedral elements (taken from [41]).

Here ν_k represents the total number of processors for which node k of mesh i lies on their subregion boundary and $\tilde{n}_i$ denotes the dimension of $\tilde{\mathcal{V}}_i$ as defined by (49).

As with the restricted version of (31) reported in [13], the preconditioner (51) typically performs better than its symmetric counterpart. For example, Table 2 illustrates results included in [41] for the solution of Poisson's equation on a cube-like domain in three dimensions using between 2 and 16 subdomains and 1 to 4 levels of uniform refinement. It is clear that for the same partitions into 2, 4, 8 and 16 weakly overlapping subdomains the restricted preconditioner performs considerably better. The original preconditioner is optimal but the upper bound on the number of iterations appears to be much higher than for the restricted preconditioner. In the latter case there is, as yet, no theoretical proof of the optimality of the preconditioner however numerical evidence, such as that presented in Table 2 and [4, 40, 41], suggests that it is the better choice in practice.

4.2 Generalization to Convection-Diffusion Problems

So far this paper has only considered the application of DD techniques to self-adjoint PDEs that may be expressed in the form given by Problem 1.1. When discretized this type of equation naturally leads to a matrix system (6) which is SPD. In practice however, many practical problems are not self-adjoint and so it is desirable to be able to apply DD techniques to a wider class of equation. In this subsection we follow [4, 30, 31, 40, 41] in considering the extension of the restricted weakly overlapping AS preconditioner (51) to convection-diffusion problems of the following form.

Problem 4.1 *Find $u \in H_E^1(\Omega)$ such that*

$$\mathcal{A}(u, v) + \mathcal{C}(u, v) = \mathcal{F}(v), \quad \forall v \in H_0^1(\Omega) , \tag{52}$$

where $\Omega \in \Re^2$ or $\Re^3$ and $H_E^1(\Omega)$ and $H_0^1(\Omega)$ are as in (2) and (3) respectively.

Here $\mathcal{A}(u, v)$ and $\mathcal{F}(v)$ are as in (4) and

$$\mathcal{C}(u, v) = \int_\Omega \underline{b} \cdot \underline{\nabla} uv \, d\underline{x} \tag{53}$$

	$p = 2$	$p = 4$	$p = 8$	$p = 16$	$p = 32$
$J = 2$	3	4	4	4	5
$J = 3$	3	4	5	5	5
$J = 4$	3	4	5	5	5
$J = 5$	3	4	5	5	5
$J = 6$	3	4	5	5	5
$J = 7$	3	4	5	5	5

Table 3: The number of preconditioned GMRES iterations required to reduce the residual by a factor of 10^6 when solving (56) on a coarse mesh of 64 elements (taken from ([4]).

for some $\underline{b} \neq \underline{0}$. The standard Galerkin FE discretization of this problem leads to a matrix system (6) which is no longer SPD. In fact, each entry of the stiffness matrix K is now given by

$$K_{ji} = \int_{\Omega} A(\underline{x}) \underline{\nabla} N_j \cdot \underline{\nabla} N_i \, d\underline{x} + \int_{\Omega} N_j \underline{b} \cdot \underline{\nabla} N_i \, d\underline{x} , \tag{54}$$

with entries of the load vector (9) similarly modified (from the Dirichlet boundary terms). Following the block matrix notation of (12) the matrix system may now be written as

$$\begin{bmatrix} A_1 & & & & B_1 \\ & A_2 & & & B_2 \\ & & \ddots & & \vdots \\ & & & A_p & B_p \\ C_1 & C_2 & \cdots & C_p & A_s \end{bmatrix} \begin{bmatrix} \underline{u}_1 \\ \underline{u}_2 \\ \vdots \\ \underline{u}_p \\ \underline{u}_s \end{bmatrix} = \begin{bmatrix} \underline{b}_1 \\ \underline{b}_2 \\ \vdots \\ \underline{b}_p \\ \underline{b}_s \end{bmatrix} , \tag{55}$$

and a Krylov subspace solver, such as GMRES, applied in parallel. All of the observations made in Subsections 1.1 and 1.2 concerning parallel assembly and parallel matrix-vector products still apply to this new system but with B_i^T replaced by C_i for $i = 1, ..., p$.

When it comes to preconditioning the system (55) it is natural to use the restricted preconditioner (51) rather than (50) since (55) is already non-symmetric. Results, taken from [4], are presented in Table 3 for the solution of the PDE

$$-\underline{\nabla}(\underline{\nabla} u) + \begin{bmatrix} 1 \\ 1 \end{bmatrix} \cdot \underline{\nabla} u = f \tag{56}$$

on a square domain in two dimensions. As before we see that, for a given coarse grid $\mathcal{T}^0$, the number of preconditioned GMRES iterations appears to be bounded independently of h (equivalently J) and p.

In [30, 31, 40, 41] Problem 4.1 is considered in the case where the convection term dominates the diffusion term, i.e. $\|A(\underline{x})\| \ll \|\underline{b}\|$ (with $A(\underline{x})$ and $\underline{b}$ as in (4) and (53) respectively). In this situation the Galerkin method is known to be unstable unless the

	$\varepsilon = 10^{-2}$				$\varepsilon = 10^{-3}$			
	$p = 2$	$p = 4$	$p = 8$	$p = 16$	$p = 2$	$p = 4$	$p = 8$	$p = 16$
$J = 1$	2	3	3	3	2	3	3	3
$J = 2$	3	3	4	4	3	4	4	4
$J = 3$	3	4	4	5	3	4	4	4
$J = 4$	4	5	6	6	3	4	4	4

Table 4: The number of preconditioned GMRES iterations required to reduce the residual by a factor of 10^5 when solving Problem 4.1 with $A(\underline{x}) = \varepsilon I$, $\underline{b} = (1, 0, 0)^T$, and a coarse mesh of 384 tetrahedral elements (taken from [41]).

mesh Peclet number (essentially the ratio of h to $\|A(\underline{x})\|/\|\underline{b}\|$) is sufficiently small (see [33] for example). In order to be able to solve such problems on realistic meshes therefore, a more stable FE discretization is required and so, as in [30, 31, 40, 41], the streamline-diffusion FE method may be considered. This again alters the definition of the matrix K in (6) so that, assuming piecewise linear FE elements are used for simplicity,

$$K_{ji} = \int_{\Omega} A(\underline{x}) \underline{\nabla} N_j \cdot \underline{\nabla} N_i \, d\underline{x} + \int_{\Omega} N_j \underline{b} \cdot \underline{\nabla} N_i \, d\underline{x} + \alpha \int_{\Omega} (\underline{b} \cdot \underline{\nabla} N_j)(\underline{b} \cdot \underline{\nabla} N_i) \, d\underline{x} . \quad (57)$$

Here, α is a streamline-diffusion parameter which also appears in a similarly modified form of the load vector $\underline{b}$.

It is apparent that the AS techniques described in this paper, and in particular the restricted weakly overlapping AS preconditioner (51), may be trivially adapted for the case where the stiffness matrix K is given by (57) as opposed to (8) or (54). Results are presented in [30, 31, 40, 41] showing that this again leads to an apparently optimal preconditioner, although no formal proof of this is offered. Table 4 shows a typical set of iteration counts, taken from [30], where $A(\underline{x}) = \varepsilon I$, $\underline{b} = (1, 0, 0)^T$, $\Omega = (0, 2) \times (0, 1) \times (0, 1)$ and the coarse grid $\mathcal{T}^0$ contains 768 tetrahedral elements.

4.3 Some Parallel Performance Results

We conclude this section with a small number of sample parallel results taken from [4] and [40]. It should be noted that there are a number of important factors that affect the quality of these results and so they should be regarded as illustrative only. Further details may be found in [4, 40].

The main factors that affect the parallel performance of the preconditioner (51) may be summarized as follows.

- The way in which Ω is decomposed into p subdomains. In the two examples below a simple recursive coordinate bisection algorithm, [44], has been used on $\mathcal{T}^0$, but for more complex problems a more sophisticated strategy should be used. Furthermore, for convection-dominated problems there may be some advantage to be had in aligning the subdomains with the convection direction where this is possible.

- The accuracy of the subproblem solves (corresponding to K^{-1} in (51)) on each processor. If these are solved too accurately then unnecessary computational effort is expended, however if they are not solved sufficiently accurately the number of iterations taken by the preconditioned GMRES solver may grow. In practice a relative residual reduction of between 10^1 and 10^2 appears to be best.

- The efficiency of the parallel implementation. Where possible all communication should be overlapped with computation for example. Idle processor time should be kept to a minimum through a good load-balancing strategy: see the note on the decomposition of Ω above.

- The quality of the DD solver when compared to the best available sequential solver. In the tables below all timings are contrasted against those of a fast sequential algebraic multilevel ILU preconditioned solver, see [6] for example.

The timings given in Tables 5 and 6 are taken from [4] and [40] respectively. In each case the "Speedup" row compares the parallel solution time (on a SGI Origin 2000 computer) with the best sequential solution time, whereas the "Parallel speedup" row compares the parallel solution time on p processors with the sequential solution time for the p subdomain algorithm.

Table 5 shows timings for the solution of the two-dimensional problem (56) using the Galerkin FE method on a square domain for which $\mathcal{T}^0$ has 256 elements and a uniform refinement level of $J = 6$ (hence $\mathcal{T}^J = \mathcal{T}^6$ and contains 1048576 elements). Table 6 shows timings for the solution of the three-dimensional problem

$$
-\frac{1}{1000}\nabla(\nabla u) + \begin{bmatrix} 1 \\ 1 \\ 1 \end{bmatrix} \cdot \nabla u = f \tag{58}
$$

using the streamline-diffusion FE method. Here $\Omega = (0,2) \times (0,1) \times (0,1)$, $\mathcal{T}^0$ contains 768 elements and a uniform refinement level of $J = 4$ is used (hence $\mathcal{T}^J = \mathcal{T}^4$ and contains 3145728 elements). It can be see that in both cases excellent parallel speedups and very useful speedups are achieved.

5   Summary

The aim of this paper has been to introduce the reader to some of the main aspects of domain decomposition preconditioning. This includes a motivation for DD methods through their applicability to the solution of PDEs using parallel computing architectures. Given that this is the main motivation for using these methods the paper also provides a short overview of the parallel assembly of finite element systems of

	$p=1$	$p=2$	$p=4$	$p=8$	$p=16$
Sequential time	234.4	351.0	301.2	271.0	262.6
Parallel time	—	191.0	83.1	42.9	20.6
Speedup	—	1.2	2.8	5.5	11.4
Parallel speedup	—	1.8	3.6	6.3	12.7

Table 5: Solution times (in seconds) and speedups for the restricted weakly overlapping algorithm on the two-dimensional problem (56) (taken from [4]).

	$p=1$	$p=2$	$p=4$	$p=8$	$p=16$
Sequential time	559.7	634.1	760.6	868.3	941.6
Parallel time	—	323.4	196.6	114.2	65.8
Speedup	—	1.7	2.8	4.9	8.5
Parallel speedup	—	2.0	3.9	7.6	14.3

Table 6: Solution times (in seconds) and speedups for the restricted weakly overlapping algorithm on the three-dimensional problem (58) (taken from [40]).

equations based upon a geometric decomposition of the problem. This decomposition requires that the FE grid be partitioned and that this partition be mapped onto the processor network in some way.

Before discussing preconditioning algorithms the introductory material in Section 1 also describes the main computational steps that are required by a typical iterative solution algorithm such as CG or GMRES. The assumption is made that the FE equations have been assembled in parallel and that a parallel solution is required. This requires the ability to undertake distributed matrix-vector multiplications in parallel and to compute distributed inner products in parallel. Both of these operations are considered. The final part of the introductory section is included for completeness and is not built upon in the rest of the paper. This describes the Schur complement approach to the solution of block-arrowhead systems of the form (13). Although not discussed here, it should be noted that there are many similarities between the domain decomposition methods that may be used to precondition the Schur complement system (21) and the full system (13): see, for example, [35] for further details.

Section 2 of the paper introduces the notion of preconditioning for the iterative solution of linear systems of equations. The motivation for the need to precondition comes from the earlier observation that the condition number of the stiffness matrix K grows like $O(h^{-2})$ as $h \to 0$. The main properties required for a preconditioning matrix are discussed and then two possible classes of DD preconditioner are considered in turn.

A simple example of an iterative substructuring technique is described both in terms of the construction of a preconditioning matrix and the action of the multiplication of a vector by the inverse of this matrix (i.e. the solution of a linear system). The main feature of this type of preconditioner is that the subdomains themselves act as a very coarse grid upon which a solution is required as part of the preconditioning process. In a parallel implementation this tends to lead to there being more than one subdomain per processor in order to obtain a substructure that is not too coarse. Since this method also requires frequent subdomain solves (one per iteration of the PCG solver) having many subdomains per processor means that these subdomain problems are smaller, although there are more of them.

The rest of the paper is concerned with Schwarz preconditioners. The distinction between additive and multiplicative Schwarz methods is made and a simple additive Schwarz preconditioner, (31), is introduced. The potential use of subdomain colouring in order to implement a multiplicative algorithm in parallel is commented upon but otherwise the focus is on the properties and parallel implementation of (31), and similar AS preconditioners, with one subdomain per processor. In particular two fundamental theorems are presented concerning the quality of (31). The first of these, Theorem 2.1, states that, provided the finite element subspaces that are associated with each subdomain form a stable splitting (as defined by (33)) of the global finite element space, then the condition number of the preconditioned system is bounded. The second result, Theorem 2.2, then shows that if there is a generous overlap between the subdomains then the subdomain spaces do indeed form a stable splitting of the global space (provided a coarse grid solve is included as part of the preconditioner).

Unfortunately the computational cost associated with maintaining an overlap that is independent of h as the finest mesh is refined is not justified in terms of practical performance, and so the use of a smaller overlap and the use of multilevel methods are both discussed. Instead of these approaches a new alternative is then proposed which combines many of the features of the optimal two level algorithm with those of optimal multilevel algorithms. This is based upon a two level preconditioner of the form (31) but with a different splitting into subdomain problems which makes use of a nested hierarchy of finite element grids. In this splitting each subdomain problem has an overlap layer of precisely one element at each level of the refinement. This has many fewer elements in the overlap layer than the conventional generous overlap splitting however, as demonstrated by Theorem 3.1, it still yields an optimal preconditioner.

The practical implementation of this weakly overlapping preconditioner is then discussed in Subsection 3.2. This involves maintaining a copy of the entire coarse grid $\mathcal{T}^0$ on each subdomain and then building the subdomain problems on top of this on each processor by only refining the grids in, or immediately next to, the subdomain owned by that processor. The combination of the coarse grid problem with each subdomain problem is described algebraically by (49) and the modified form of the AS preconditioner is given by (50). Results, taken from [5], are provided to illustrate the practical behaviour of the algorithm on a simple two-dimensional test problem. The immediate extension to three dimensions is also noted.

The final main section of the paper discusses two extensions to the weakly overlapping AS preconditioner. The first of these is to introduce a restricted version of this preconditioner, (51), following the work of [13] for conventional AS preconditioners.

This is shown not only to require less communication when implemented in parallel but also to significantly reduce the number of iterations required over the preconditioner (50). The only potential drawback of the restricted approach is that symmetry is lost and so the PCG algorithm has to be replaced by a more general iterative method such as GMRES. It is clear however that the small additional cost per iteration of applying GMRES is more than offset by the reduced number of iterations taken.

The second extension considered takes the loss of symmetry one step further by applying the restricted weakly overlapping technique to the solution of problems that are themselves non-symmetric. In particular convection-diffusion problems are considered. Two discretization methods are discussed: standard Galerkin FE and streamline-diffusion FE (the latter being necessary for the stability of the solution of convection-dominated problems). In both cases, and in both two and three dimensions, the observed numerical results suggest that the preconditioner is still optimal (or close to optimal) even though no theoretical proof of this is available.

Section 4 concludes by presenting some typical parallel performance results, as reported in [4] and [40]. These show that the DD technique parallelizes very efficiently but that the speed of the sequential version of the p subdomain algorithm is generally less than that of the best available sequential algorithm. This imposes a restriction on the overall efficiency of the algorithm when compared to the best available sequential algorithm. Nevertheless, the parallel timings are extremely encouraging and demonstrate that it is possible to obtain practical results that are in line with theoretical expectations.

References

[1] M. Ainsworth, *"A Preconditioner Based on Domain Decomposition for h-p Finite Element Approximation on Quasi-Uniform Meshes"*, SIAM J. on Numerical Analysis, 33, 1358–1376, 1996.

[2] M. Ainsworth, *"A Hierarchical Domain Decomposition Preconditioner for h-p Finite Element Approximation on Locally Refined Meshes"*, SIAM J. on Scientific Computing, 17, 1395–1413, 1996.

[3] S.F. Ashby, T.A. Manteuffel and P.E. Taylor, *"A Taxonomy for Conjugate Gradient Methods"*, SIAM J. on Numerical Analysis, 27, 1542–1568, 1990.

[4] R.E. Bank and P.K. Jimack, *"A New Parallel Domain Decomposition Method for the Adaptive Finite Element Solution of Elliptic Partial Differential Equations"*, Concurrency and Computation: Practice and Experience, 13, 327–350, 2001.

[5] R.E. Bank, P.K. Jimack, S.A. Nadeem and S.V. Nepomnyaschikh, *"A Weakly Overlapping Domain Decomposition Preconditioner for the Finite Element Solution of Elliptic Partial Differential Equations"*, SIAM J. on Scientific Computing, 23, 1818-1842, 2002.

[6] R.E. Bank and R.K. Smith, *"An Incomplete Factorization Multigraph Algorithm"*, SIAM J. on Scientific Computing, 20, 1349–1364, 1999.

[7] M. Benzi, J.K. Cullum and M. Tuma, *"Parallel Implementation and Practical use of Sparse Approximate Inverse Preconditioners with A Priori Sparsity Patterns"*, SIAM J. on Scientific Computing, 22, 1318–1332, 2000.

[8] V. Bornemann and H. Yserentant, *"A Basic Norm Equivalence for the Theory of Multilevel Methods*, Numerische Mathematik, 64), 455–476, 1993.

[9] J. Bramble, J. Pasciak and A.H. Schatz, *"The Construction of Preconditioners for Elliptic Problems by Substructuring, I"*, Mathematics of Computation, 47, 103–134, 1986.

[10] J. Bramble, J. Pasciak and A.H. Schatz, *"The Construction of Preconditioners for Elliptic Problems by Substructuring, II"*, Mathematics of Computation, 49, 1–16, 1987.

[11] J. Bramble, J. Pasciak and A.H. Schatz, *"The Construction of Preconditioners for Elliptic Problems by Substructuring, III"*, Mathematics of Computation, 51, 415–430, 1988.

[12] J. Bramble, J. Pasciak and A.H. Schatz, *"The Construction of Preconditioners for Elliptic Problems by Substructuring, IV"*, Mathematics of Computation, 53, 1–24, 1989.

[13] X.-C. Cai and M. Sarkis, *"An Restricted Additive Schwarz Preconditioner for General Sparse Linear Systems"*, SIAM J. on Scientific Computing, 21, 792–797, 1999.

[14] T.F. Chan and J.P. Shao, *"Parallel Complexity of Domain Decomposition Methods and Optimal Coarse Grid Size"*, Parallel Computing, 21, 1033–1049, 1995.

[15] E. Chow, *"A Priori Sparsity Patterns for Parallel Sparse Approximate Inverse Preconditioners"*, SIAM J. on Scientific Computing, 21, 1804–1822, 2000.

[16] G. Cybenko, *"Dynamic Load Balancing for Distributed Memory Multiprocessors"*, J. of Parallel and Distributed Computing, 7, 279–301, 1989.

[17] M. Dryja and O.B. Widlund, *"Towards a Unified Theory of Domain Decomposition Algorithms for Elliptic Problems"*, in Third International Symposium on Domain Decomposition Methods (T.F. Chan *et al*, eds.), SIAM publications, Philadelphia, 1990.

[18] M. Dryja and O.B. Widlund, *"Some Domain Decomposition Algorithms for Elliptic Problems"*, in Iterative Methods for Large Linear Systems, Academic Press, 1990.

[19] C. Farhat, *"A Simple and Efficient Automatic FEM Domain Decomposer"*, Computers and Structures, 28, 579–602, 1988.

[20] C. Farhat, P.-S. Chen and P. Stern, *"Towards the Ultimate Iterative Substructuring Method: Combined Numerical and Parallel Scalability, and Multiple Load Cases"*, Computing Systems in Engineering, 5, 337-350, 1994.

[21] C. Farhat, J. Mandel and F.X. Roux, *"Optimal Convergence Properties of the FETI Domain Decomposition Method"*, Computer Methods for Applied Mechanics and Engineering, 115, 365–385, 1994.

[22] G. Globisch, *"PARMESH: A Parallel Mesh Generator"*, Parallel Computing, 21, 509–524, 1995.

[23] G.H. Golub and C.F. Van Loan, *"Matrix Computations"*, John Hopkins Press,

2nd edition, 1989.

[24] B. Hendrickson and R. Leland, *"An Improved Spectral Graph Partitioning Algorithm for Mapping Parallel Computations"*, SIAM J. on Scientific Computing, 16, 452–469, 1995.

[25] D.C. Hodgson and P.K. Jimack, *"Efficient Mesh Partitioning for Parallel Elliptic Differential Equation Solvers"*, Computing Systems in Engineering, 6, 1–12, 1995.

[26] K.H. Hoffman and J. Zou, *"Parallel Efficiency of Domain Decomposition Methods"*, Parallel Computing, 19, 1375–1391, 1993.

[27] T.J.R. Hughes, I. Levit and J. Winget, *"An Element-by-Element Solution Algorithm for Problems of Structural and Solid Mechanics"*, Computer Methods for Applied Mechanics and Engineering, 36, 241–254, 1983.

[28] D.C. Hodgson and P.K. Jimack, *"A Domain Decomposition Preconditioner for a Parallel Finite Element Solver on Distributed Unstructured Grids"*, Parallel Computing, 23, 1157–1181, 1997.

[29] P.K. Jimack and D.C. Hodgson, *"Parallel Preconditioners Based Upon Domain Decomposition"*, in Parallel and Distributed Processing for Computational Mechanics: Systems and Tools, ed. B.H.V. Topping (Saxe-Coburg Publications), 207-223, 1999.

[30] P.K. Jimack and S.A. Nadeem, *"Parallel Application of a Novel Domain Decomposition Preconditioner for the Stable Finite Element Solution of Three-Dimensional Convection-Dominated PDEs"*, in Euro-Par 2001 Parallel Processing: 7th International Euro-Par Conference Manchester, UK, August 2001 Proceedings, ed. R. Sakellariou et al. (Lecture Notes in Computer Science 2150, Springer), 592–601, 2001.

[31] P.K. Jimack and S.A. Nadeem, *"A Parallel Domain Decomposition Algorithm for the Adaptive Finite Element Solution of 3-D Convection-Diffusion Problems"*, in Computational Science – ICCS 2002: International Conference Amsterdam, The Netherlands, April 2002 Proceedings Part II, ed. P.M.A. Sloot et al. (Lecture Notes in Computer Science 2330, Springer), 797–805, 2002.

[32] P.K. Jimack and N. Touheed, *"Developing Parallel Finite Element Software Using MPI"*, in High Performance Computing for Computational Mechanics, ed. B.H.V. Topping and L. Lammer (Saxe-Coburg Publications), 15–38, 2000.

[33] C. Johnson *"Numerical Solution of Partial Differential Equations by the Finite Element Method"*, Cambridge University Press, 1987.

[34] G. Karypis and V. Kumar, *"Parallel Multilevel k-way Partition Scheme for Irregular Graphs"*, SIAM Review, 41, 278–300, 1999.

[35] D.E. Keyes and W.D. Gropp, *"A Comparison of Domain Decomposition Techniques for Elliptic PDEs and their Parallel Implementation"*, SIAM J. on Scientific and Statistical Computing, 8, 166–202, 1987.

[36] P.L. Lions, *"Interprétation Stochastique de la Méthode Alternée de Schwarz"*, C. R. Acad. Sci. Paris, 268, 325–328, 1978.

[37] T.A. Mantueffel, *"Shifted Incomplete Cholesky Factorization"*, in Sparse Matrix Proceedings (I.S. Duff and G.W. Stewart eds.), SIAM Publications, Philadelphia,

1978.

[38] Message Passing Interface Forum, *"MPI: A Message-Passing Interface Standard"*, International Journal of Supercomputer Applications, 8, No. 3/4, 1994.

[39] G. Meurant, *"Domain Decomposition Methods for PDEs on Parallel Computers"*, Int. J. Supercomputer Applications, 2, 5–12, 1988.

[40] S.A. Nadeem *"Parallel Domain Decomposition Preconditioning for the Adaptive Finite Element Solution of Elliptic Problems in Three Dimensions"*, Ph.D. thesis, University of Leeds, 2001.

[41] S.A. Nadeem and P.K. Jimack, *"Parallel Implementation of an Optimal Two Level Additive Schwarz Preconditioner for the 3-D Finite Element Solution of Elliptic Partial Differential Equations"*, to appear in Int. J. Num. Meth. Fluids, 2002.

[42] Y. Saad and M. Schultz, *"GMRES: A Generalized Minimal Residual Algorithm for Solving Nonsymmetric Linear Systems"*, SIAM J. on Scientific Computing, 7, 856–869, 1986.

[43] H.A. Schwarz, *"Uber Einige Abbildungsaufgaben"*, J. fur Die Reine und Angewandte Mathematik, 70, 105–120, 1869.

[44] H.D. Simon, *"Partitioning of Unstructured Problems for Parallel Processing"*, Computing Systems in Engineering, 2, 135–148, 1991.

[45] B.F. Smith, *"An Optimal Domain Decomposition Preconditioner for the Finite Element Solution of Linear Elasticity Problems"*, SIAM J. on Scientific Computing, 13, 364–378, 1992.

[46] B.F. Smith, *"A Parallel Implementation of an Iterative Substructuring Algorithm for Problems in Three Dimensions"*, SIAM J. on Scientific Computing, 14, 406–423, 1993.

[47] B. Smith, P. Bjorstad and W. Gropp, *"Domain Decomposition: Parallel Multilevel Methods for Elliptic Partial Differential Equations"*, Cambridge University Press, 1996.

[48] W. Speares and M. Berzins, *"A 3-d Unstructured Mesh Adaptation Algorithm for Time-Dependent Shock Dominated Problems"*, International Journal for Numerical Methods in Fluids, 25, 81–104, 1997.

[49] G. Strang and G.J. Fix, *"An Analysis of the Finite Element Method"*, Prentice-Hall, Englewood-Cliffs, 1973.

[50] C. Walshaw and M. Cross, *"Parallel Optimisation Algorithms for Multilevel Mesh Partitioning"*, Parallel Computing, 26, 1635–1660, 2000.

[51] A.J. Wathen, *"An Analysis of some Element-by-Element Techniques"*, Computer Methods for Applied Mechanics and Engineering, 74, 271–287, 1989.

[52] R.D. Williams, *"Performance of Dynamic Load Balancing for Unstructured Mesh Calculations"*, Concurrency: Practice and Experience, 3, 457–481, 1991.

[53] J. Xu, *"Iterative Methods by Space Decomposition and Subspace Correction"*, SIAM Review, 34, 581–613, 1992.

Parallel and Distributed
Finite Element Analysis of Structures

E.D. Sotelino and Y. Dere
School of Civil Engineering, Purdue University
West Lafayette, Indiana, United States of America

Abstract

Parallel processing has been perceived as a means of achieving the necessary computational power for the realistic simulation of structural engineering applications. The finite element method has played a prominent role in structural engineering. This paper describes several approaches that have been used to parallelise the finite element analysis of structures. Particular emphasis is placed in parallel finite element methods for structural dynamics. A number of concurrent algorithms based on domain decomposition and substructuring techniques are described. Issues related to how workload balancing among processors in a parallel or distributed computer environment affects parallel performance are discussed. This discussion ranges from automatic domain partitioning algorithms to dynamic load balancing techniques. Finally, an object-oriented environment that has been developed to facilitate and promote the reuse, rapid prototyping and portability of parallel finite element software is described.

Keywords: parallel processing, distributed computing, finite element analysis, domain decomposition, substructuring, load balancing.

1   Introduction

Several sophisticated commercial finite element software packages are available. However, the realistic simulations of structural engineering applications is, in general, computationally intensive and often cannot be achieved using traditional computing facilities. This is especially true for three-dimensional nonlinear dynamic simulations of structural behaviour. Parallel processing and distributed computing provide possible venues to achieve the needed computational power. Since the late 1980's much research has been carried out to take advantage of this technology in many engineering areas, and, in particular, in structural engineering.

The finite element method is the most popular numerical method used to solve structural engineering problems. Various schemes have been proposed to parallelise this method. These attempts date back to the late 1970's, when, for example, the Finite Element Machine [1, 2, 3] was developed. In this approach, the nodal structure of the finite element mesh is mapped onto the hardware, i.e., each microprocessor in the array of processors is mapped to a node in the structure being analysed. Other researchers focused in the research of methods that optimise the solution of systems of linear algebraic equations, which in many finite element applications is the most costly phase. Finally, concurrent algorithms based on domain decomposition and substructuring techniques have also been the focus of much work. Both techniques are based on the partition of the physical structure into subdomains or substructures. This paper concentrates on the two latter approaches.

The performance of a parallel algorithm based on domain partitioning is highly dependent on how the structure is partitioned. Ideally, a perfect balance of the workload in each processor in a parallel or distributed computer is desired. Thus, many researchers have also studied the partitioning problem. The various approaches that have been used, ranging from automatic domain partitioning algorithms to dynamic load balancing techniques, are also discussed here.

The implementation of the various parallel methods into a single platform poses a great challenge. Research has been carried out on the development of software tools that facilitate and promote the reuse, rapid prototyping and portability of parallel structural engineering software. In particular, one such a tool is the SECSDE (Structural Engineering Concurrent Software Development Environment), which has been developed in the School of Civil Engineering at Purdue University. A brief description of the architecture of the SECSDE is also presented in this paper.

This article is organised as follows. First, background information on issues related to parallel and distributed computing is provided in Section 2. Section 3 gives an overview of integration methods used in time-dependent finite element analysis of structures. The two main styles that have been adopted for introducing concurrency in these applications, i.e., Domain Decomposition and Domain Splitting methods, are discussed in Sections 4 and 5, respectively. Section 6 describes the issue of load balancing between processors and gives some of the existing automatic domain decomposition algorithms, and a discussion on dynamic load balancing. Section 7 describes the SECSDE environment and Section 8 gives some concluding remarks.

2   Parallel and Distributed Processing

2.1   Parallel Architectures

Computer architectures can be categorised into four main groups according to their method of handling the data and instruction streams [4]. SISD (Single Instruction stream, Single Data stream), SIMD (Single Instruction stream, Multiple Data stream), MISD (Multiple Instruction stream, Single Data stream), and MIMD

(Multiple Instruction stream, Multiple Data stream). SISD machines (e.g. Intel Pentium series, AMD Athlon series) are single processor machines, which execute a single stream of instruction on a data set. In SIMD machines, each processor simultaneously executes the same single instruction on different data sets. Example of SIMD machines include array and vector processors (e.g. Cray-1). MISD machines involve pipelined architectures and are similar in design to SISD systems. MIMD machines can simultaneously execute different instructions upon different data sets. A cluster of workstations or PC's can be programmed to form a MIMD machine.

In the MIMD architecture, each processor has its own program to execute. This can be described as Multiple Program Multiple Data (MPMD) structure. Some of the programs may be copies of the same program. In the particular case in which a single program is written and processors execute their own individual copies, the paradigm is referred to as Single Program Multiple Data (SPMD). Although the same code is executed in each processor, the program may be written so that different each processor may process different portions of the code.

Parallel computers can also be classified according to the way data is stored. A multiprocessor system that has a main memory, which is shared by all processors, is called Shared Memory Multiprocessor System. In a shared memory system, processors do not have local memory. From a programming viewpoint, sharing the data among processors is convenient and desirable, however a bottleneck occurs when many processors try to access to the shared memory at a given time. In distributed memory multiprocessors, or multicomputers, each processor has its own local memory and can access only to its own memory. Therefore, the memory is distributed among the processors. The processors in a multicomputer system share data by sending messages to each other through an interconnecting network. Since shared memory systems are more attractive to programmers, but not very efficient, the concept of distributed shared memory systems in which processors have access to the whole memory using a single memory address was introduced. In distributed shared memory systems, although the memory is distributed, automated message passing occurs when processors try to access memory locations other than their local memory. The KSR1 multiprocessor, which has only cache memory local to each processor, is an example of a distributed shared memory system.

Parallel systems are characterized according to their processor granularity. This characteristic is defined based on the number of processors and the executed work units. If a small number of powerful processors execute large work units, this is called coarse-grain parallelism. On the other hand, the form of parallelism when a large number of simple processors execute small program statements is referred to as fine-grain parallelism. This paper focuses on the SPMD programming structure on message passing distributed memory computing environments. Furthermore, coarse-grain parallelism is considered.

2.2   Distributed Memory Systems: Interprocess Communication

As mentioned above, in distributed memory computers, such as a cluster of workstations or personal computers, data is shared via an interconnecting network. This is achieved through message passing between the processors. A number of high-level tools have been developed to aid parallel program developers. In particular, a group of academics and industrial partners developed a standard for message-passing systems, which is referred to as the Message Passing Interface (MPI) [5]. MPI has become a standard with its wide use and portability characteristics. A number implementations of the MPI standard are available, such as CHIMP from Edinburgh Parallel Computing Center, LAM from the Ohio Supercomputing Center, MPICH from Argonne International Laboratories and Mississippi State University. MPI provides the necessary library routines for message passing and related operations in a parallel-distributed environment. Communication domains, called communicators allow a set of processors to communicate between themselves. MPI applications require a message-passing architecture in which the processors have its own memory and are connected to each other. Network of workstations, IBM SP, Intel Paragon XP/S, etc. are examples of message passing architectures.

   While MPI is designed using the object-oriented philosophy, it currently provides only function-oriented interfaces through ANSI C and FORTRAN 77 language bindings. In [6] an object-oriented message-passing class library in C++, called Parallel Portability Interface (**PPI++**), was developed for portable parallel programming. **PPI++** is a programming interface, which makes the client parallel codes more understandable and easier to use by hiding implementation details of the Message Passing Interface. It uses MPICH libraries, as they were available in the Sun workstations and IBM SP multicomputer, which were used in the research. **PPI++** is a part of the research effort on the development of the Structural Engineering Concurrent Software Development Environment (SECSDE), which is discussed in more detail in Section 7 of this paper.

2.3   Performance Measurements

The performance of a parallel system is measured by comparing the performance obtained with single processor with the performance obtained from a multiprocessor system. Two major performance parameters are often used: speed-up and efficiency. The speed-up is the relative performance of a multiprocessor system and it is defined as

$$\text{Speed} - \text{up} = \frac{t_s}{t_p} \tag{1}$$

where,
t_s: Execution time using one processor

t_p: Execution time using multiprocessors

The maximum speed-up that can be achieved with N_p processors is N_p. This ideal speed-up is called "linear speed-up". When the speed-up with N_p processors is higher than N_p, it is called "super-linear speed-up". Super-linear speed-ups may occur when the sequential algorithm is not as optimised as the parallel algorithm. Another reason for super-linear speed-up may be due to some exceptional features that improve the performance of the parallel algorithm. Speed-up is usually lower than linear speed-up due to the inter-processor communication.

The efficiency parameter demonstrates the fraction of time that the processors are being used in the computation. The efficiency is given by the following expression:

$$\text{Efficiency} = \frac{\text{Speed} - \text{up}}{N_p} \text{x100\%} \tag{2}$$

For example, when the efficiency is 100%, this means that the processors are being fully utilized during the computations.

3   Time-Dependent Finite Element Analysis

The general equation of equilibrium that governs the dynamic response of a finite element model is given by

$$\mathbf{M\ddot{u}}(t) + \mathbf{f}_{int}(\mathbf{u}, \mathbf{\dot{u}}) = \mathbf{f}_{ext}(\mathbf{u}, t)$$

and in the linear case,

$$\mathbf{M\ddot{u}}(t) + \mathbf{C\dot{u}}(t) + \mathbf{Ku}(t) = \mathbf{f}_{ext}(t) \tag{3}$$

where $\mathbf{M}$, $\mathbf{C}$ and $\mathbf{K}$ are the mass, viscous damping and stiffness matrix; $\mathbf{u}$, $\mathbf{\dot{u}}$ and $\mathbf{\ddot{u}}$ are the displacement, velocity and acceleration vectors, $\mathbf{f}_{int}$ is the internal resisting force, and $\mathbf{f}_{ext}$ is the vector of time dependent externally applied loads. The equation of motion given by Equation 3 is a second order system of ordinary differential equations. In finite element analysis, the equation of motion can be solved by either direct integration or mode superposition methods.

In direct integration methods, the equation of motion is integrated numerically using step-by-step methods, in which, in general, the time duration t is divided into n equal time intervals h. The differential equations given by Equation 3 are satisfied at each time step. Finite difference methods are utilized to discretise the time derivatives in the equation of motion.

In the mode superposition method, an eigenvalue analysis of the matrices $\mathbf{M}$, $\mathbf{C}$ and $\mathbf{K}$ is required. The coupled degrees of freedom are decoupled using modal synthesis and the resulting system is integrated for each mode using a time-stepping algorithm. Modal superposition is then used to get the solution.

For linear structural dynamics, modal superposition is an effective method. However, for large nonlinear systems where frequent update of the stiffness matrix is required, they become impractical [7]. In these cases, direct time integration methods are preferred. Thus, time integration methods are the focus of this paper.

In a time integration algorithm, the solution can be obtained explicitly or implicitly. Explicit methods formulate the dynamic equilibrium at time the present time (t_n), whereas implicit solution algorithms formulate the structural dynamic equilibrium at a future time (for example at $t_{n+1} = t_n+h$). Finite difference relationships are introduced that relate displacements, velocities and accelerations at this future time to their known values at previous times.

In implicit methods, a system of simultaneous linear algebraic equations must be solved for each time step, while for explicit methods this is not necessary. Thus, explicit methods tend to be less computationally intensive than implicit methods. However, implicit methods are considered more effective for structural dynamics, since they are unconditionally stable, thus, permitting the use of large time step sizes. Explicit methods are conditionally stable and smaller time step sizes are required for stability reasons. For long duration applications, the small time-step sizes required by explicit methods can be prohibitive.

3.1   Newmark Method

The Newmark family of methods [8] is widely used for solving structural dynamic problems. Both explicit and implicit methods can be obtained from the general formulation, depending on the selection of the parameters γ and β (see below). The following finite difference formulas define the general form of the Newmark family of methods:

$$\dot{u}^{n+1} = \dot{u}^n + h(1-\gamma)\ddot{u}^n + h\gamma\ddot{u}^{n+1} \tag{4}$$

$$u^{n+1} = u^n + h\dot{u}^n + \frac{h^2}{2}(1-2\beta)\ddot{u}^n + h^2\beta\ddot{u}^{n+1} \tag{5}$$

where h is the time step size, γ and β are the free parameters related to integration accuracy and stability. As mentioned above, different methods are generated for different values of γ and β. Table 1 gives some of the most popular members of the Newmark family of methods.

Of these popular methods, the Linear Acceleration Method and the Average/Constant Acceleration methods are implicit schemes and unconditionally stable, while the Central Difference method is an explicit scheme and conditionally stable. In general, the stability characteristics of the Newmark's methods depend on the values of the free parameters γ and β. In particular, the unconditional stability requirement is given by

$$2\beta \geq \gamma \geq \frac{1}{2} \tag{6}$$

Method	γ	β
Linear Acceleration Method	1/2	1/6
Average/Constant Acceleration Method	1/2	1/4
Central Difference Method	1/2	0

Table 1: Members of the Family of Newmark methods

An incremental procedure, which uses implicit finite difference approximations to the displacements, velocities, and accelerations, is presented in [9]. In this procedure, displacements, velocities and accelerations are not independent values. Therefore, a residual vector, which is a function of displacements, can be written as (in the linear case)

$$\mathbf{r} = \mathbf{f}_{ext} - (\mathbf{M\ddot{u}} + \mathbf{C\dot{u}} + \mathbf{Ku}) = 0 \tag{7}$$

In the equilibrium configuration at time t_{n+1}, this expression can be rewritten as

$$\mathbf{r}^{n+1} = \mathbf{f}_{ext}^{n+1} - (\mathbf{M\ddot{u}}^{n+1} + \mathbf{C\dot{u}}^{n+1} + \mathbf{Ku}^{n+1}) = 0 \tag{8}$$

In the equation above, the externally applied force vector is assumed to be independent of the displacements. For the nonlinear case, the damping and stiffness matrices are replaced by their corresponding tangential matrices. The solution of Equation 3 can be obtained by requiring that the residual vector $\mathbf{r}$ be zero. If the structure is nonlinear, the solution for the displacements $\mathbf{u}^{n+1}$ usually requires an iterative procedure. The linearised residual vector for the next iteration can be written by assuming that the displacements at iteration $k+1$, $\mathbf{u}_{k+1}^{n+1}$, may be approximated by $\mathbf{u}_k^{n+1}$, using some type of linearisation, as

$$\mathbf{r}_k^{n+1} = \mathbf{r}_{k-1}^{n+1} + \mathbf{\hat{K}}_{k-1}^{n+1} \Delta \mathbf{u}_k^{n+1} \quad k = 1,2,\ldots \tag{9}$$

where $\mathbf{\hat{K}}$ is the effective stiffness matrix defined as

$$\mathbf{\hat{K}}_{k-1}^{n+1} = \frac{\partial \mathbf{r}_{k-1}^{n+1}}{\partial \mathbf{u}_{k-1}^{n+1}} \tag{10}$$

and the displacement increment is given by

$$\Delta \mathbf{u}_k^{n+1} = \mathbf{u}_k^{n+1} - \mathbf{u}_{k-1}^{n+1} \tag{11}$$

At the first iteration, the effective stiffness matrix becomes

$$\hat{\mathbf{K}} = \frac{\partial \mathbf{r}^{n+1}}{\partial \mathbf{u}^{n+1}} = \mathbf{M}\frac{\partial \ddot{\mathbf{u}}^{n+1}}{\partial \mathbf{u}^{n+1}} + \mathbf{C}\frac{\partial \dot{\mathbf{u}}^{n+1}}{\partial \mathbf{u}^{n+1}} + \mathbf{K} \tag{12}$$

For the Newmark family of methods, the effective stiffness matrix can be written as

$$\hat{\mathbf{K}} = \mathbf{K} + \frac{\gamma}{\beta h}\mathbf{C} + \frac{1}{\beta h^2}\mathbf{M} \tag{13}$$

The general step-by-step solution scheme can be written as follows:

1. Estimate the displacements, velocities and accelerations using the Newmark's finite difference approximations,

$$\ddot{\mathbf{u}}_0^{n+1} = \ddot{\mathbf{u}}^n \tag{14}$$

$$\dot{\mathbf{u}}_0^{n+1} = \dot{\mathbf{u}}^n + h(1-\gamma)\ddot{\mathbf{u}}^n + h\gamma\ddot{\mathbf{u}}^{n+1} \tag{15}$$

$$\mathbf{u}_0^{n+1} = \mathbf{u}^n + h\dot{\mathbf{u}}^n + \frac{h^2}{2}(1-2\beta)\ddot{\mathbf{u}}^n + h^2\beta\ddot{\mathbf{u}}^{n+1} \tag{16}$$

2. Compute the displacement increment as

$$\Delta\mathbf{u}_k^{n+1} = \left(\hat{\mathbf{K}}_{k-1}^{n+1}\right)^{-1}\mathbf{r}_{k-1}^{n+1} \tag{17}$$

3. Update the displacements, velocities and accelerations using the obtained corrections as

$$\mathbf{u}_k^{n+1} = \mathbf{u}_{k-1}^{n+1} + \Delta\mathbf{u}_k^{n+1} \tag{18}$$

$$\dot{\mathbf{u}}_k^{n+1} = \dot{\mathbf{u}}_{k-1}^{n+1} + \frac{\gamma}{\beta h}\Delta\mathbf{u}_k^{n+1} \tag{19}$$

$$\ddot{\mathbf{u}}_k^{n+1} = \ddot{\mathbf{u}}_{k-1}^{n+1} + \frac{1}{\beta h^2}\Delta\mathbf{u}_k^{n+1} \tag{20}$$

4. Increment time: n=n+1, t=t+h and go to the second step if not converged.

3.2   Central Difference Method

In the Central Difference method, the displacement solution for time t_{n+1} is found by using equilibrium equations written at time t_n. The equation of motion for linear analysis at time t_n is given by

$$\mathbf{M\ddot{u}}^n + \mathbf{C\dot{u}}^n + \mathbf{Ku}^n = \mathbf{f}^n_{ext} \tag{21}$$

This equation can be rewritten as

$$\mathbf{\ddot{u}}^n = \mathbf{M}^{-1}(\mathbf{f}^n_{ext} - \mathbf{C\dot{u}}^n - \mathbf{Ku}^n) \tag{22}$$

which provides an expression for the acceleration at the current time step. The displacements at the next time step are calculated using finite difference relations. For constant time steps, this relations is given by

$$\mathbf{u}^{n+1} = \mathbf{u}^n + \mathbf{\dot{u}}^n h + \frac{1}{2}h^2\,\mathbf{\ddot{u}}^n \tag{23}$$

From the displacements, the velocity and acceleration finite divide differences are obtained as

$$\mathbf{\dot{u}}^n = \frac{\mathbf{u}^{n+1} - \mathbf{u}^{n-1}}{2h} \tag{24}$$

$$\mathbf{\ddot{u}}^n = \frac{\mathbf{u}^{n+1} - 2\mathbf{u}^n + \mathbf{u}^{n-1}}{h^2} \tag{25}$$

After substituting $\mathbf{\dot{u}}^n$ and $\mathbf{\ddot{u}}^n$ into Equation 23, and rearranging, one obtains the following

$$\left(\frac{1}{h^2}\mathbf{M} + \frac{1}{2h}\mathbf{C}\right)\mathbf{u}^{n+1} = \mathbf{f}^n_{ext} - \left(\mathbf{K} - \frac{2}{h^2}\mathbf{M}\right)\mathbf{u}^n - \left(\frac{1}{h^2}\mathbf{M} - \frac{1}{2h}\mathbf{C}\right)\mathbf{u}^{n-1} \tag{26}$$

As it can be seen from Equation 26, the solution for the displacement $\mathbf{u}^{n+1}$ depends only on the values of displacements at previous times, i.e., time t_n and t_{n-1}. Thus, as mentioned above, the central difference method is explicit. If the mass matrix in Equation 26 is diagonal, and the damping matrix is proportional to the mass matrix, the solution can be found without having to solve a global system of equations. The central difference method is effective for nonlinear analysis, since the stiffness matrix does not need to be factorised as it is seen in Equation 26.

The central difference method requires small time step size due to stability requirements, since it is conditionally stable. For stability, the time step size must be chosen smaller than the critical time step size, which is given by

$$\Delta t < \Delta t_{critical} = \frac{T_{min}}{\pi} \qquad (27)$$

where T_{min} is the minimum period of vibration of the structure.

For problems with rapid load variations where the duration of the loading is of the order of the fundamental period of the structure, or smaller, the stability limit of the time step size does not cause extra analysis time. However, for long duration loading (e.g. 10 to 100 times the fundamental period of the structure), the time step size required by the stability limit is much lower than the time step size required for the accurate modelling of the system. Therefore the central difference method has been found to be unsuitable for this type of problems.

4   Substructuring-Based Methods

In parallel structural analysis, there are two main Domain Decomposition (DD) solution approaches: explicit DD approach in which the structural model is geometrically divided into a number of subdomains/substructures and implicit DD approach in which the system of equations for the entire structure is assembled and solved without physically partitioning the structure. This work is concerned with the former approach, and thus, the term "domain decomposition methods" is used here to refer to "explicit domain decomposition methods".

DD methods divide a large-scale problem into a number of smaller sub-problems and therefore they may be identified as "divide-and-conquer" algorithms. The resulting subdomains are then solved independently in each processor and the solution for the whole problem domain is obtained by exchanging interface data between subdomains periodically.

In Domain Decomposition (DD) or Parallel Substructuring methods the structure is divided into a number of substructures. The interior DOF are condensed out so that the size of the condensed system reduces to the number of interface DOF. The solution for the interior DOF is then recovered from the interface DOF. The condensation procedure can be performed in parallel by each processor independently. The reduced global system of equations for the interface DOF can also be solved in parallel using iterative or direct equation solvers. However, communication is needed in this case, since the effective stiffness matrix and the load vector have overlapping coefficients. When direct solvers are used, the overlapping coefficients must be gathered into one processor through message passing. If the number of interface DOF is small, then direct methods are often preferred. After solving the interface DOF in parallel, the solution for interior DOF is recovered.

In Reference [10] a parallel DD method was developed for the solution of finite element systems. In this work, the finite element mesh is partitioned using an automatic domain decomposition algorithm. The interior nodes are numbered first and the interface nodes last. With this numbering scheme, the form of the resulting global system of equations is such that the submatrices in the diagonal are uncoupled, and thus can be factorised concurrently.

Reference [11] presents an approach for parallel finite element analysis using a small number of processors. In their method, each substructure is analysed concurrently by communicating the condensed stiffness matrices and interface forces. The algorithm is given for the two-substructure case, however the authors point out that the method can be adapted for analyses with multi-substructures. According to the studied case study, around ~75-80% efficiency is obtained using two substructures. However, it was concluded that the flow of information and efficiency of the algorithm for multi-substructure case requires further research.

In Reference [12] a domain decomposition algorithm using the unconditionally stable Newmark-β constant average acceleration method for nonlinear transient analysis of three-dimensional framed structures is presented. In this approach, the structure is decomposed into substructures and each substructure is assembled and condensed in parallel. The preconditioned conjugate gradient method, which is an iterative solver, is used to solve the condensed system of equations. It was observed that domain division highly affects the number of conjugate gradient iterations. The single-process analysis, which is used as a reference to determine the efficiency of their parallel implementation, is performed using a direct solver. More specifically a Cholesky decomposition of the effective stiffness matrix is carried out. The method was found to be effective for the transient nonlinear analysis large frame problems.

When the substructuring concept is applied to dynamic analysis, it is called 'reduction'. In this approach the full set of dynamic equilibrium is reduced to the interface equilibrium set. In Reference [14] it is found that the application of the static condensation process directly to structural dynamic problems introduces errors. This is because possible internal movements of a substructure cannot be captured. In this work, the Component Mode Synthesis (CMS) technique is used in order for an accurate modelling of substructures. This technique can produce possible movements of a substructure by making use of component displacement modes. Reference [15] presents a review of available literature on CMS technique and some applications of CMS can be found in [16].

5   Domain-Splitting Methods

Similarly to domain decomposition methods, domain-splitting methods divide the physical problem into a number of smaller sub-problems and therefore they may also be identified as "divide-and-conquer" algorithms. The main difference between these methods and the previously described methods is that they do not use substructuring in their formulation. Instead, they adopt other approaches to restore uniqueness of the solution at the interface degrees of freedom.

5.1   Parallel Central Difference Scheme

Reference [13] the implementation of a parallel version of the central difference method for 3D nonlinear transient dynamic analysis of frames is presented. The central difference method is suitable for parallel processing since only neighbour communication is needed between processors per time step. In order to aid the parallel implementation, the structural nodes and elements were classified as shown in Figure 1. In their work, the structure is partitioned into subdomains and each subdomain is assigned to a processor. Coarse-grained parallelism is assumed, in which the number of finite elements in the structure is much larger than the number of subdomains.

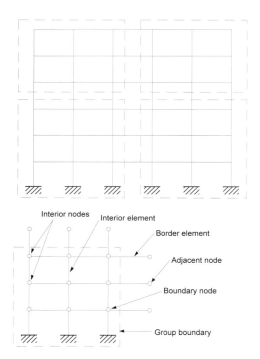

Figure1: The division of structural data for explicit analysis [13]

In this method, at each time step, each subdomain calculates the responses of the interior nodes independently and the displacements of the boundary nodes are sent to adjacent subdomains. The solution proceeds to the next time step after each subdomain receives the displacements from the adjacent subdomains. It should be noted that shared border elements belong to adjacent subdomains. Therefore, the force recovery and nonlinear updating for border elements are duplicated at each time step. Since the number of interior elements is significantly larger than the number of border elements (coarse-grain parallelism), the duplicate computations are found to be essentially negligible. Reference [13] reports ~80%, ~75% and ~70% efficiencies using 2, 3 and 4 subdomains from an example of a three-dimensional 3-story frame structure.

5.2   Finite Element Tearing and Interconnecting Algorithm

Reference [17] presents the Finite Element Tearing and Interconnecting (FETI) algorithm.   This algorithm is a domain-splitting method that uses Lagrange multipliers for the parallel solution of the finite element equilibrium equations. In the FETI method, the structure's domain Ω is partitioned into N_s subdomains and the compatibility of the displacements at the interface nodes is satisfied using Lagrange multipliers λ.

The general equilibrium equation for a structural system is given by,

$$\mathbf{Ku} = \mathbf{f} \tag{28}$$

where $\mathbf{K}$ is a symmetric positive-definite stiffness matrix, $\mathbf{u}$ is the displacement vector, and $\mathbf{f}$ is the external forces vector. This equation can be transformed into the following equivalent set of equations for each subdomain Ω^s:

$$\mathbf{K}^s \mathbf{u}^s = \mathbf{f}^s - \mathbf{B}^{s^T} \lambda, \ s = 1, \ldots, N_s \tag{29}$$

$$\sum_{s=1}^{N_s} \mathbf{B}^s \mathbf{u}^s = 0 \tag{30}$$

where $\mathbf{B}^s$ is the subdomain connectivity matrix. Equation 29 states the local equilibrium equation for subdomain Ω^s and Equation 30 is the compatibility equation for the subdomain interface. The Lagrange multipliers λ, represent interface interaction forces. The connectivity matrix for a subdomain is given as,

$$\mathbf{B}^s_{nxN} = [\mathbf{O}^s_{nxm} \pm \mathbf{I}^s_{nxn}] \tag{31}$$

where n is the number of interface DOF, m is the number of interior DOF, N is the total number of DOF, $\mathbf{O}$ is the null matrix and $\mathbf{I}$ is the identity matrix. Reorganizing Equation 29 one obtains

$$\mathbf{u}^s = \mathbf{K}^{s^{-1}} (\mathbf{f}^s - \mathbf{B}^{s^T} \lambda) \tag{32}$$

Pre-multiplying Equation 32 by $\mathbf{B}^s$ one gets

$$\mathbf{B}^s \mathbf{u}^s = \mathbf{B}^s \mathbf{K}^{s^{-1}} (\mathbf{f}^s - \mathbf{B}^{s^T} \lambda) \tag{33}$$

Computing the global sum of Equation 33 for the whole domain, and remembering the result from Equation 30, gives the following

$$\sum_{s=1}^{N_s} \mathbf{B}^s \mathbf{u}^s = \sum_{s=1}^{N_s} \mathbf{B}^s \mathbf{K}^{s^{-1}} (\mathbf{f}^s - \mathbf{B}^{s^T} \lambda) = 0 \tag{34}$$

By re-arranging the terms in Equation 34, the following equation is obtained

$$\left(\sum_{s=1}^{N_s} \mathbf{B}^s \mathbf{K}^{s^{-1}} \mathbf{B}^{s^T} \right) \lambda = \sum_{s=1}^{N_s} \mathbf{B}^s \mathbf{K}^{s^{-1}} \mathbf{f}^s \tag{35}$$

Once the Lagrange multipliers are obtained by solving Equation 35, the subdomain displacements can be determined using Equation 32.

It should be noted that when the domain is partitioned into subdomains, this might generate unconstrained (floating) or insufficiently constrained subdomains. In this case, local singularities may occur in those subdomains, and the subdomain stiffness matrix $\mathbf{K}^s$ becomes semi-definite. To overcome this problem, Equation 32 is replaced with the following equation

$$\mathbf{u}^s = \mathbf{K}^{s^+} (\mathbf{f}^s - \mathbf{B}^{s^T} \lambda) + \mathbf{R}^s \alpha^s \tag{36}$$

where $\mathbf{K}^{s^+}$ is a pseudo-inverse matrix, which satisfies $\mathbf{K}^s \mathbf{K}^{s^+} \mathbf{K}^s = \mathbf{K}^s$, $\mathbf{R}^s$ is a rectangular matrix representing rigid body (zero energy) modes of the subdomain, and α^s is the set of amplitudes that specifies the contribution of $\mathbf{R}^s$ to the solution $\mathbf{u}^s$. The coefficients λ and α^s can be determined by requiring that each subdomain problem be solvable mathematically, i.e., that each floating subdomain be self-equilibrating. This can be expressed as,

$$\mathbf{R}^{s^T} (\mathbf{f}^s - \mathbf{B}^{s^T} \lambda) = 0 \tag{37}$$

Substituting Equation 37 into Equation 36, Equation 30, which represents the interface compatibility, is replaced by the following equation,

$$\begin{bmatrix} \mathbf{F}_I & -\mathbf{G}_I \\ -\mathbf{G}_I^T & 0 \end{bmatrix} \begin{Bmatrix} \lambda \\ \alpha \end{Bmatrix} = \begin{Bmatrix} \mathbf{d} \\ \mathbf{e} \end{Bmatrix} \tag{38}$$

where

$$\mathbf{F}_I = \sum_{s=1}^{N_s} \mathbf{B}^s \mathbf{K}^{s^+} \mathbf{B}^{s^T}$$

$$\mathbf{d} = \sum_{s=1}^{N_s} \mathbf{B}^s \mathbf{K}^{s^+} \mathbf{f}^s$$

$$\mathbf{G}_I = [\mathbf{G}_I^1 \quad \dots \quad \mathbf{G}_I^{N_f}] = [\mathbf{B}^1 \mathbf{R}^1 \quad \dots \quad \mathbf{B}^{N_f} \mathbf{R}^{N_f}]$$

$$\boldsymbol{\alpha}_I = [\boldsymbol{\alpha}^{1^T} \quad \dots \quad \boldsymbol{\alpha}^{N_f^T}]^T$$

$$\mathbf{e} = [\mathbf{f}^{1^T} \mathbf{R}^1 \quad \dots \quad \mathbf{f}^{N_f^T} \mathbf{R}^{N_f}]^T$$

where N_f is the number of floating subdomains. After calculating λ and α from Equation 38, the subdomain displacements $\mathbf{u}^s$ can be determined using Equation 36.

In Reference [17], numerical examples are provided to demonstrate the performance of the FETI algorithm. The speed-ups obtained for a 3-D example problem are 15.4 and 28.8 using 16 and 32 processors, respectively. According to the authors, the FETI algorithm outperformed the substructuring method by exhibiting 20% higher speed-up for their numerical example. They claim that one of the reasons for this improvement in performance is that the interprocessor communication time per iteration for the substructuring is almost 3.3 times higher than that for the FETI algorithm. The FETI algorithm has a different interprocessor communication pattern than substructuring. In the FETI algorithm, the maximum number of neighbouring subdomains for interprocessor communication is reduced from 8 (i.e., 4 corners x 2 neighbours) to 4 (i.e. 4 edge neighbours) for a 2D mesh and 26 (i.e. 8 corners x 3 neighbours) to 6 (i.e. 6 face neighbours) for a 3D mesh.

5.3   Group Implicit Algorithm

The Group Implicit (GI) algorithm was first proposed in Reference [18] to solve heat conduction problems and it was later adapted for the transient structural dynamics problems [19, 20, 21, 22].

In the GI algorithm, the finite element mesh is first decomposed into a group of elements called subdomains. The unknown displacements, velocities, and accelerations in the equations of motion are solved in each subdomain independently by using any fully implicit analysis algorithm, such as the Average/Constant Acceleration Method method, over a time step.

The subdomain interface nodes are allowed to displace freely during the subdomain solution phase. Therefore, multiple solutions are obtained for the interface nodes. Although the results obtained at the subdomain interfaces are in equilibrium, they do not satisfy the compatibility condition. In order to restore the compatibility at the interface region, the different results obtained for the same interface nodes are averaged according to the mass matrices of the neighbouring subdomains. In Reference [18] it is shown that the mass averaging rule is the only choice, which results in consistency with the equations of motion for the heat conduction case. Later on, in Reference [21], it is shown that this is also true for structural dynamic applications. Furthermore, in this work it is shown that using the acceleration as the primary unknown provides the best results. The Mass Averaging Rule (MAR) for the interface nodes is given by,

$$\ddot{\mathbf{u}}_{n+1} = \mathbf{M}^{-1} \sum_{s=1}^{Nsub} \mathbf{M}^s \tilde{\tilde{\mathbf{u}}}_{n+1}^s \tag{39}$$

where $\ddot{\mathbf{u}}_{n+1}$ is the averaged interface acceleration vector, $\mathbf{M}$ is the mass matrix for the whole structure, $\mathbf{M}^s$ is the mass matrix of an individual subdomain s, $\tilde{\tilde{\mathbf{u}}}_{n+1}^s$ is the interface acceleration vector of the subdomain s, and Nsub is the number of subdomains. Figure 2 gives the steps of the GI algorithm for linear problems using a predictor-corrector formulation [21]. In Figure 3, the GI solution procedure is illustrated for a 1D mesh partitioned into two subdomains.

1	Predictor phase:
	$\tilde{\mathbf{u}}_{n+1} = \mathbf{u}_n + h\,\dot{\mathbf{u}}_n + \dfrac{h^2}{2}(1 - 2\beta)\ddot{\mathbf{u}}_n$
	$\tilde{\dot{\mathbf{u}}}_{n+1} = \dot{\mathbf{u}}_n + h(1 - \gamma)\ddot{\mathbf{u}}_n$
2	Equation solving phase:
	$\ddot{\mathbf{u}}_{n+1} = 0$
	a. For each subdomain s,
	$\qquad \tilde{\tilde{\mathbf{u}}}_{n+1}^s = -(\mathbf{M}^s + \gamma h\,\mathbf{C}^s + \beta h^2\,\mathbf{K}^s)^{-1}\,\mathbf{K}^s\,\mathbf{u}_{n+1}^s$
	$\qquad \ddot{\mathbf{u}}_{n+1} = \ddot{\mathbf{u}}_{n+1} + \mathbf{M}^s \tilde{\tilde{\mathbf{u}}}_{n+1}^s$
	$\ddot{\mathbf{u}}_{n+1} = \mathbf{M}^{-1} \ddot{\mathbf{u}}_{n+1}$
3	Corrector phase:
	$\mathbf{u}_{n+1} = \tilde{\mathbf{u}}_{n+1} + \beta h^2\,\ddot{\mathbf{u}}_{n+1}$
	$\dot{\mathbf{u}}_{n+1} = \tilde{\dot{\mathbf{u}}}_{n+1} + \gamma h\,\ddot{\mathbf{u}}_{n+1}$

Figure 2: Steps of the GI algorithm

The GI algorithm is highly parallelisable with very little communication overhead. Only an interface vector is exchanged through interprocessor communication. The other advantage of using GI comes from the decomposition of the domain into smaller and more computationally manageable subdomains. As a consequence, GI can speed-up the computations even in a single machine.

In Reference [22], the stability of the GI algorithm was investigated. In this work, it was shown that the GI algorithm has the same stability characteristics as the Newmark method. Thus, unconditional stability can be achieved if

$$\gamma \geq \frac{1}{2} \quad \text{and} \quad \beta \geq \frac{\gamma}{2} \tag{40}$$

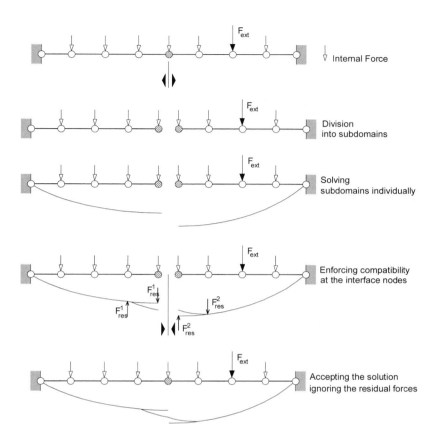

Figure 3. Illustration of the GI solution procedure

In Reference [19] the accuracy characteristics of GI algorithm was investigated. In this work, they established a Courant type limit for the time step size based on the characteristic dimension of subdomain, L_s and wave speed, c such that,

$$\Delta t \leq \frac{L_s}{c} \qquad (41)$$

The time step size restriction criterion is a requirement that controls the deterioration in the accuracy of the algorithm. It was shown, in the same work, that this criterion increases the number of time steps by a factor of $s^{1/2}$.

The accuracy of the GI algorithm was further investigated in Reference [23]. It was found that the above criterion is too stringent for the dynamic analysis of frames, since the characteristic dimension for frames is of the order of a few beam or column lengths, which limits the time-step size to that of 2 to 3 times the explicit time-step size. In this reference, three main reasons were given as the causes of the inaccuracies in the GI solution. First, the domain interfaces have approximate solutions due to the averaging of the computed results at each time step. Second, the domains with no structural supports are free to undergo rigid body motion since only inertia forces restrain them. Third, the method used to divide the lumped masses and applied loading among subdomains affects the final solution. They concluded that this algorithm has the potential to be one of the most efficient parallel methods for the solution of structural dynamics equations if the deterioration in accuracy could be avoided.

The flow of information of the GI algorithm is very similar to that of an implicit algorithm, except that interface nodes have multiple solutions. After mass averaging the compatibility at the interface is satisfied, the information from the interior adjacent nodes is kept the same as before the mass averaging. Therefore, some information is lost jeopardizing the accuracy of the method. As the number of subdomains increases, this information flow between subdomains is further reduced and a phase error is also introduced.

5.4   Iterative Group Implicit Algorithm

While the GI algorithm provides an efficient way of solving structural dynamics problems in parallel, its accuracy characteristics were found to too restrictive for some practical structural dynamics problems. Reference [24] proposes the Iterative Group Implicit (IGI), which is intended to eliminate the known accuracy problems of the GI algorithm. The IGI algorithm seeks to restore equilibrium at the subdomain interface degrees of freedom (DOF) iteratively. Similarly to the GI algorithm, the compatibility at the interface DOF is restored by means of a mass-averaging rule. However, in the IGI algorithm, both equilibrium and compatibility are satisfied within each subdomain.

In the IGI algorithm, equilibrating interface forces are applied iteratively, in the same magnitude and in opposite direction, during the process of restoring compatibility. Hence, both equilibrium and compatibility are satisfied everywhere once the method converges. The main steps of the IGI algorithm are provided in Figure 4 [24]. Figure 5 illustrates the IGI solution procedure for a 1D mesh partitioned into two subdomains.

The IGI algorithm eliminates the accuracy problems of the GI algorithm by enforcing equilibrium in addition to compatibility at the interface. Therefore the limitation on the time step size for the GI is not necessary. In the IGI algorithm, the flow of information between the subdomains is not constrained. In Reference [24], numerical studies were carried out that showed that significant speed-ups with high level of accuracy were obtained with the IGI algorithm.

(1) Estimate displacement, velocity, and acceleration of the interior and interface degrees of freedom. Compute the interface forces.
(2) Compute residual forces considering external forces, interface forces, and the current response.
(3) Solve iteratively until convergence:
   (a) Solve each subdomain for the computed residual forces. (*Note that compatibility is not satisfied at the interface*).
   (b) Average interface DOF using a mass-averaging rule. (*Note that equilibrium is no longer satisfied at both interface and adjacent nodes*).
   (c) Re-evaluate the interface reactive forces from either stress recovery of border elements, i.e., elements that contain interface DOF, or by matrix-vector multiplication between the effective stiffness matrix and the effective response vector.
   (d) Assemble the global interface reactive force vector and externally applied forces at the interface. If the norm of the assembled interface reactive forces is negligible, convergence is achieved. Stop iterating and go to the next time-step (Step 1).
   (e) If convergence is not achieved, i.e., interface nodes are still not in equilibrium, distribute the interface residual forces among the DOF in proportion to the subdomains effective stiffness.
   (f) Update the interface force vector, and go to Step 3a.

Figure 4. Steps of the IGI algorithm

6   Load Balancing and Automatic Domain Partitioning

When using algorithms that are based on the partition of the physical domain, the finite element mesh is partitioned/decomposed into a number of subdomains or substructures in order to be distributed and processed among the different processors in the parallel/distributed computing environment. A good partitioning algorithm ideally should be able to balance the workload and minimize communication between processors so that significant speed-up can be achieved in a parallel analysis. Domain partitioning is a Non-deterministic Polynomial time (NP) complete problem [25], thus, optimal solutions cannot be obtained. Heuristic algorithms such as the Simulated Annealing method are used to obtain near-optimal solutions.

Achievement of good partitioning of a large-scale, 3D and/or irregular shaped, finite element mesh based on visual inspection can be very cumbersome and time consuming. Therefore, automatic domain-partitioning algorithms have been proposed by a number of researchers. Some of these methods are: the GReedy (GR) algorithm [26], the Al-Nasra and Nguyen's Partitioning (ANP) algorithm [27], the Reduced Bandwidth Decomposition (RBD) algorithm [28, 29], and the Recursive Spectral Two-way (RST) algorithm [30].

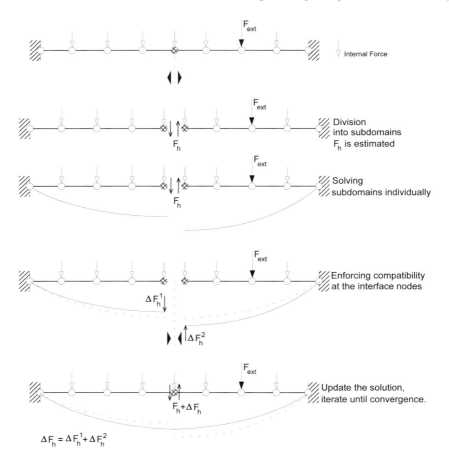

Figure 5: Illustration of the IGI solution procedure

Domain partitioning strategies can be classified into two main groups: element-based partitioning and node-based partitioning. In element based partitioning, the elements in the finite element mesh are partitioned, whereas in node-based partitioning the nodes are partitioned. The selection of the strategy depends on the parallel solution algorithm employed. As an example, implicit transient analysis methods generally requires element-based partitioning, while explicit methods may use either one of these strategies.

While the above-mentioned research concentrates on the quality of the initial partitioning of the domain, it does not address the issue of Dynamic Load Balancing (DLB). In some structural applications, especially those were localised nonlinearities occur, a single initial partitioning of the domain may not ensure that the workload will be well balance throughout the analysis. Many researchers operating systems research area have studied the issue of dynamic load balancing for more than two decades and various methods have been proposed for Dynamic Load Balancing (DLB) in distributed and parallel computing. In parallel finite element analysis of structures, only a handful of methods have been proposed to date.

According to Reference [31], most of the DLB methods lack a global view of loads across processors. They suggest that a systematic way of measuring and balancing all processor loads is needed for a method to be applicable to a variety of realistic applications. Although the main purpose of DLB is the same for all research areas, which is to optimise processor utilization while minimizing inter-processor communication, there have been different approaches to manage the DLB problem since the causes of load imbalance and the tasks (i.e. processes) considered in different areas may require special handling. Some of the research that has been carried out on DLB related the parallel finite element analysis is presented in the next few paragraphs.

Reference [32] addresses the DLB issue for Computational Fluid Dynamics (CFD) problems where adaptive type of mesh refinement is inevitable. In their work, a processor is assigned as a decision maker, i.e., a dynamic load balancer called *Jove*. This processor monitors the computational behaviour of the other processors at every predetermined number of iterations, while the other processors execute the CFD computations. Jove is responsible for the evaluation of the workload and to make the decision of moving data between processors. The workload in each processor is estimated by adding together the "weights" of the elements that form its subgroup. Initial values of these weights are assigned to the elements in the pre-processing stage. They are then updated throughout the analysis. The Jove processor measures the computational load based on the element weights and redistributes the workload when necessary. The details of the Jove framework are given in Reference [31]. In this work, Jove is implemented on an IBM SP2 and MPI is used to facilitate portability. According to their experimental results, mesh adaptation with load balancing gives more than six-fold improvement over one without load balancing and Jove gives a 24-fold speed-up on 64 processors compared to sequential execution.

Reference [33] presents a finite element program for coupled fluid-structure problems that uses a *Master/Slave*-based DLB approach. In their work, the domain is partitioned into many subdomains with the number of subdomains being much greater than the number of processors. The subdomain operations represent the tasks in the task queue. The subdomain data such as elements, nodes and loads are assigned to a master processor. Slave processors ask the master processor for another subdomain after the completion of their task. The main processor iterates over the subdomains and assigns a new subdomain to the next available slave processor. The master processor performs the domain repartitioning when necessary. In this method a great deal of communication occurs when the subdomain data is sent from master processor to slave processors and when the results data are sent from subdomains to the master processor. There is also a bottleneck in the communication since the slaves may be waiting in the queue if they have completed their job synchronously.

In Reference [34], an object-oriented DLB framework is described in which the domain partitioning is modified by migration of elements between subdomains. In this approach, the number of processors is assumed to be the same as the number of subdomains. The developed framework was tested on a few examples. Although a 2.5-fold speed-up is achieved when using 3 subdomains, a 1.6-fold slow-down occurred when using 6 subdomains. The authors attribute this slow-down to the fact that the DLB produced a partitioning which required more computations than the original partitioning.

In Reference [35], a software library for parallel DLB of finite element analysis code is developed under an ESPRIT (the European Union information technologies program) project called DRAMA, Dynamic Re-Allocation of Meshes for parallel finite element Applications. The computational cost is divided into element-based and node-based portions. The communication cost however consists of element-element, node-node and element-node based contributions. The total workload for a processor is computed by summing the computational cost and communication cost. The computational cost for element calculations are based on user-specified operation counts since the finite element mesh may contain various types of elements. The input to DRAMA is the distributed mesh and cost monitoring parameters such as the number of elements, the number of operations required for elements, etc. This information is converted to a weighted graph representation where vertex weights correspond to computation costs and edge weights correspond to potential communication costs. Then DRAMA computes a new partitioning either by direct mesh migration or by parallel graph repartitioning by interfacing a partitioning program. After computation of the new partitioning, DRAMA returns the calculated partitioning along with the relation between the old and the new partitioning of the mesh. This relational information may be helpful to the application program in deciding whether or not to migrate the data.

7   SECSDE Research

Research on the Structural Engineering Concurrent Software Development Environment (SECSDE) has been carried out for the past ten years in the School of Civil Engineering at Purdue University [36]. The goal of this research is to develop the necessary test bed to eliminate problems related to the reuse, rapid prototyping and portability of sequential and parallel structural engineering software and to promote efficient development, utilization and maintenance of parallel software using the object-oriented philosophy.

The SECSDE is an object-oriented environment for parallel finite element analysis of structures. It is implemented using the C++ programming language. The implementation of the overall skeleton of the SECSDE has been completed. Presently, the previously developed objects and frameworks are being maintained and improved by additions and modifications to the current skeleton.

The overall architecture of the SECSDE is shown in Figure 6. A brief description of the main components of the SECSDE is provided in the next few paragraphs.

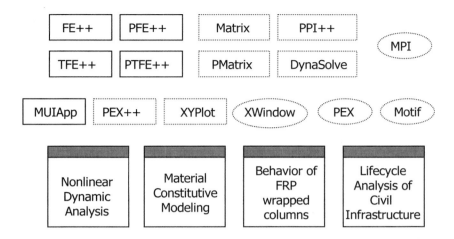

Figure 6: Overall architecture of the SECSDE

- *Object-Oriented Matrix Computation Libraries*: One of the first tasks of the SECSDE research involved the design and implementation of a matrix class library, referred to as **Matrix**, for sequential computing [37]. The resulting class library supports a variety of matrices required for finite element computations, such as symmetric, profile, banded, diagonal, and lower/upper triangular matrices. In order to facilitate parallel programming of matrix computations and to provide support for other "high-level" classes and components, a consistent and coherent hierarchy of parallel matrix classes for different types of matrix sparsity and distribution of data among processes was created [38]. A collection of software abstractions was developed to model parallel matrices in distributed computing environments, capturing the characteristics of the data distribution among the processes and taking into account matrix sparsity. This library, called **PMatrix**, was developed integrally with the parallel portability interface library PPI++, discussed below, for the encapsulation of both the data management as well as the message passing required by the parallel matrix algorithms.
- *Parallel Portability Interface*: To aid the portability of parallel codes in distributed memory concurrent systems, several packages are currently available. The message-passing standard called MPI [5] has been proposed to promote better support for development of parallel libraries. Although the MPI specification is designed using an object-oriented approach, it currently provides only function-oriented interfaces through ANSI C and FORTRAN 77 language bindings. As a consequence, MPI does not take full advantage of the many capabilities provided by object-oriented programming languages such as C++. An object-oriented parallel portability library called **PPI++** was created [6] as part of the SECSDE research. **PPI++** is built on top of MPI. The major goals of **PPI++** are to provide a stable (unchanging) interface between the client parallel program

and the rapidly evolving field of parallel code portability, and to support a consistent and easy-to-use interface for programming of message-passing using object-oriented methods.

- *Sequential and Parallel Finite Element Analysis Framework*: A framework for sequential finite element computing was been developed as part of the SECSDE research [39, 40]. This framework provides all the abstractions and mechanisms to perform a static displacement-based finite element analysis using sequential processing. A Parallel Finite Element Analysis Framework called **PFE++** [41] for static analysis was created on top of **FE++**. **PFE++** implements the general operations and data structures required for the decomposition of an analysis model into various domain partitions. It encapsulates the necessary data management and message passing associated with the substructuring problem in the analysis of solids and structures. **PFE++** supports the *concurrent server model*. That is, all the participating processes are concurrent servers. After initialisation, they read the data and perform the same task whenever the exchange of information is required. The parallel matrix library, [38], facilitates the management of matrix data in participating processes and carry out the matrix operations in a distributed environment. The message passing across the processes is done using the **PPI++** classes [6].

- *Sequential and Parallel Transient Finite Element Analysis Frameworks*: A framework for sequential and parallel finite element of structural dynamic problems was designed and implemented into SECSDE. These components are the Transient Finite Element framework (TFE++) and the Parallel Transient Finite Element framework, **PTFE++** [42]. Both the GI and IGI algorithms have been implemented using these frameworks. The class hierarchy diagram of **TFE++** and **PTFE++** is given in Figure 7 following Booch's notation [43].

- *Graphical User Interface Programming Framework*: An object-oriented GUI Framework was developed for the SECSDE [44]. This framework, titled **MUIApp** (*Motif User Interface Application Framework*) is layered on top of the X Window System and Motif [45, 46]. It provides high-level abstractions for creating user-interface components based on extended Motif widget sets as the basic building blocks. The abstractions in **MUIApp** are designed to capture the graphical user interface characteristics and a large portion of the control structure common to Motif graphics applications. **MUIApp** supports the development of graphical user interface components including menu bars with canvas areas for display of physical structural elements and/or data representations, and popup dialogs with various forms of command input and valuators.

- *Automatic mesh partitioning*: The Mesh Partitioning Environment, **MPE++** [47], was designed and implemented to facilitate the domain decomposition tasks involved in parallel applications. **MPE++** is based on the RST algorithm [30]. As part of this development, an interactive graphical component, **MPView** was also developed to aid in the evaluation and fine-tuning of the partitioning results. In Figure 6, the partitioning of a sample framed structure and the nomenclature utilized for the subdomain nodes and elements are shown. In this example, the frame is partitioned into three subdomains. In this case an element-based partitioning strategy is illustrated.

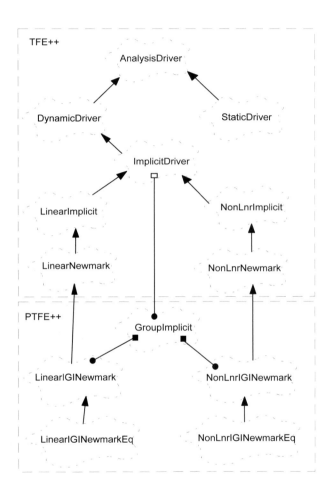

Figure 7: The class hierarchy of TFE++ and PTFE++ frameworks (Modak 1997).

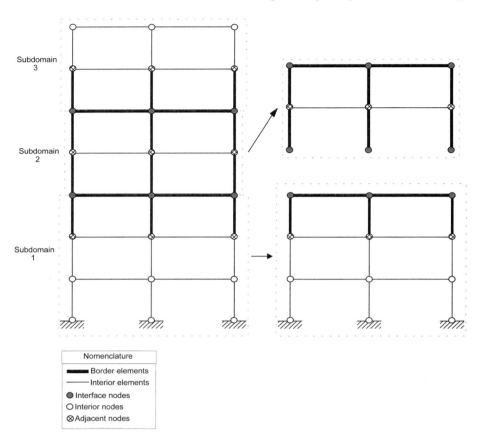

Figure 8. Partitioning of a frame structure into three subdomains using MPE++

8   Conclusion

In this paper, a number of research topics related to the parallel finite element analysis for structural dynamic applications are discussed. This is an exciting research area and much work is current underway. It should be noted that this work is not intended as a state-of-the-art review of the field. Instead, its goal is to provide some background information in parallel and distributed computing as it applies to the finite element analysis of structures and to discuss some of the research activities that have been carried out in this area. Much research in this area exists and a complete literature review would require a different forum.

References

[1] Jordan, H., *"A Special Purpose Architecture for Finite Element Analysis,"* Proc. 1978 Conf. on Parallel Processing, Wayne State University, Detroit, Michigan, 263-266, 1978.

[2] Jordan, H., *"The Finite Element Machine - Programmer's Reference Manual,"* CS DG-79-2, University of Colorado, Boulder, Colorado, USA, 1979.

[3] Jordan, H. F. and Sawyer, P. L., *"A Multi-Microprocessor System for Finite Element Structural Analysis"*, Computers & Structures, 10, 21-29, (1979).

[4] Flynn M. J. (1966), *"Very High Speed Computing Systems"*, Proceedings of 1996 IEEE, 12, 1901-1909, 1966.

[5] Message Passing Interface Forum, *"MPI: A Message-Passing Interface Standard"*, Supercomputer Applications and High Performance Computing, 8(3/4), 159-416, 1994.

[6] Hsieh, S.H., Sotelino, E.D., "*A Message-Passing Class Library in C++ for Portable Parallel Programming*", Engineering with Computers, 13, 20-34, 1997.

[7] Bathe, K. J., *"Finite Element Procedures"*, Prentice Hall, Englewood Cliffs, New Jersey, USA, 1996.

[8] Newmark, N.M. "*A Method of Computation for Structural Dynamics*," Journal of the Engineering Mechanics Division, Proceedings ASCE **85**(EM5), 67-90, 1959.

[9] Geradin M., Hogge M., and Idelsohn, S., *"Implicit Finite Element Methods"*, in T. Belytschko and T.J.R. Hughes (Editors): Computational Methods for Transient Analysis, 417-470, Elsevier Science Publishers B.V., Amsterdam, The Netherlands, 1983.

[10] Farhat, C. and Wilson, E., *"A New Finite Element Concurrent Computer Program Architecture"*, International Journal for Numerical Methods in Engineering, 24, 1771-1792, 1987.

[11] El-Sayed, M. E. M and Hsiung, C.K., *"Parallel Finite Element Computation with Separate Substructures"*, Computers & Structures, Vol. 36, No. 2, 261-265, 1990.

[12] Hajjar, J. F. and Abel, J. F., *"Parallel Processing for Transient Nonlinear Structural Dynamics of three-dimensional Framed Structures Using Domain Decomposition"*, Computers & Structures, 30(6), 1237-1254, 1988.

[13] Hajjar, J. F. and Abel, J. F., *"Parallel Processing of Central Difference Transient Analysis for three-dimensional Nonlinear Framed Structures,"* Communications in Applied Numerical Methods, 5, 39-46, 1989.

[14] Chen, H. M., "Distributed Object-Oriented Nonlinear Finite Element Analysis", PhD Thesis, Purdue University, West Lafayette, Indiana, USA, 2002.

[15] Seshu, P., *"Substructuring and Component Mode Synthesis,"* Shock and Vibration, 4 (3), 199-210, 1997.

[16] Archer, G. C., *"A Technique for the Reduction of Dynamic Degrees of Freedom,"* Earthquake Engineering and Structural Dynamics, 30, 127-145, 2001.

[17] Farhat, C. and Roux, F. X., "*A Method of Finite Element Tearing and Interconnecting and Its Parallel Solution Algorithm*," International Journal for Numerical Methods in Engineering, 32, 1205-1227, 1991.

[18] Ortiz, M., Nour-Omid, B., "*Unconditionally Stable Concurrent Procedures for Transient Finite Element Analysis*", Computer Methods in Applied Mechanics and Engineering, 58 (2), 151-174, 1986.

[19] Ortiz, M., Nour-Omid, B., Sotelino, E.D., "*Accuracy of A Class Of Concurrent Algorithms for Transient Finite Element Analysis*", International Journal for Numerical Methods in Engineering, 26(2), 379-391, 1988.

[20] Ortiz, M., Sotelino, E.D., Nour-Omid, B., "*Efficiency of Group Implicit Concurrent Algorithms*", International Journal for Numerical Methods in Engineering, 28, 2761-2776, 1989.

[21] Sotelino, E.D., "A Class of Concurrent Algorithms for Transient Finite Element Analysis", PhD dissertation, Brown University, Providence, Rhode Island, USA, 1990.

[22] Sotelino, E.D., "*Stability of the GI Concurrent Algorithms for Transient Structural Analysis*", International Journal for Engineering Analysis and Design, 1, 1-12, 1994.

[23] Hajjar, J.F., Abel, J.F., "*On the Accuracy of Some Domain-by-Domain Algorithms for Parallel Processing of Dynamics*", International Journal for Numerical Methods in Engineering, 28, 1855-1874, 1989.

[24] Modak, S., Sotelino, E.D., "*The Iterative Group Implicit Algorithm for Nonlinear Structural Analysis*", International Journal for Numerical Methods in Engineering 47(4), 869-885, 2000.

[25] Garey M.R. and Johnson D.S., "*Computers and Intractability: A Guide to the Theory and NP-Completeness*", W.H. Freeman and Company, New York, USA, 1979.

[26] Farhat, C., "*Multiprocessors in Computational Mechanics*", Ph.D. dissertation, University of California, Berkeley, California, USA, 1987.

[27] Al-Nasra, M. and Nguyen, D.T., "*An Algorithm for Domain Decomposition in Finite Element Analysis*", Computers & Structures, 39, 277-289, 1991.

[28] Malone, J.G. (1988), "*Automatic Mesh Decomposition and Concurrent Finite Element Analysis for Hypercube Multiprocessor Computers*", Comput. Methods Appl. Mech. Eng., 70, 27-58, 1988.

[29] Malone, J.G. (1990), "*Parallel Nonlinear Dynamic Finite Element Analysis of Three-Dimensional Shell Structures*", Computers & Structures, 35, 523-539, 1990.

[30] Hsieh, S.H., Paulino, G.H., Abel, J.F., "*Recursive Spectral Algorithms for Automatic Domain Partitioning in Parallel Finite Element Analysis*", Computer Methods in Applied Mechanics and Engineering, 121(1-4), 137-162, 1995.

[31] Sohn A., Biswas, R. and Simon, H.D., "*A Dynamic Load Balancing Framework for Unstructured Adaptive Computations on Distributed-Memory Multiprocessors*", Proceedings of the 8th Annual ACM Symposium on Parallel Algorithms and Architectures, Padua, Italy, 1996.

[32] Pramono, E., Simon, H.D. and Sohn, A., *"Dynamic Load Balancing for Finite Element Calculations on Parallel Computers"* in Bailey, D.H., Bjorstad, P.E., Gilbert, J.R., Mascagni, M., Schreiber, R.S., Simon, H.D., Torczon, V.J. and Watson L.T. (Editors): Proceedings of the Seventh SIAM Conference on Parallel Processing for Scientific Computing, 559-604, San Francisco, California, USA, February 15-17, 1995.

[33] Santiago, E.D. and Law, K.H., *"A Distributed Implementation of an Adaptive Finite Element Method for Fluid Problems"*, Computers and Structures, 74, 97-119, 1996.

[34] McKenna, F.T., "Object-Oriented Finite Element Programming: Frameworks for Analysis, Algorithms and Parallel Computing", PhD Thesis, University of California, Berkeley, California, USA, 1997.

[35] Maerten, B., Roose, D., Basermann, A., Fingberg, J., and Lonsdale, G., *"DRAMA: A Library for Parallel Dynamic Load Balancing of Finite Element Applications"*, ESPRIT LRT Project, Long Term Research Project No. 24953, Katholieke Universiteit, Leuven, Belgium, 1999.

[36] Sotelino, E.D., White, D.W., and Chen, W.F., *"Domain-Specific Object-Oriented Environment for Parallel Computing"*, Journal of Singapore Structures Steel Society, 3(1), 47-60, 1992.

[37] Lu, J., White, D. W., Chen, W. F., and Dunsmore, H. E. *"A Matrix Library in C++ for Structural Engineering Computing"*, Computers & Structures, 55(1), 95-111, 1995.

[38] Modak, S., Hsieh, S.H. and Sotelino, E.D. *"A Parallel Matrix Class Library in C++ for Computational Mechanics Applications"*, International Journal of Microcomputers in Civil Engineering, 12(2), 83-99, 1997.

[39] Lu, J., White, D. W., and Chen, W. F., *"Applying Object-Oriented Design to Finite Element Programming"*, Proceedings of the 1993 ACM/SIGAPP Symposium on Applied Computing, ACM, Indianapolis, IN, USA, 424-429, 1993.

[40] Lu, J., White, D. W., Chen, W. F. and Dunsmore, H. E. (1994), *"FE++: an Object-Oriented Application Framework for Finite Element Programming"*, Proceedings of the Second Annual Object-Oriented Numerics Conference, Sunriver, Oregon, USA, 438-447.

[41] Mukunda, G. R., Sotelino, E. D., and Hsieh, S. H. *"Distributed Finite Element Computations using Object-Oriented Techniques,"* Engineering with Computers, 14, 59-72, 1997.

[42] Modak, S., and Sotelino, E. D., *"An Object-Oriented Parallel Programming Framework for Linear and Nonlinear Transient Analysis of Structures"*, Computers and Structures, 80, 77-84, 2002.

[43] Booch, G., "Object-Oriented Design with Applications", 2nd. Edition, Benjamin/Cummings, Redwood City, California, USA, 1994.

[44] Mukunda, G. R., Hsieh, S. H., Sotelino, E. D., and White, D. W., *"MUIApp: An Object-Oriented Graphical User Interface Application Framework,"* Engineering Computations, 14(3), 256-280, 1987.

[45] Young, D. L., *"Object-Oriented Programming with C++ and OSF/Motif"*, Prentice Hall Inc., Reading, Massachusetts, USA, 1992

[46] Heller, D., "*Motif Programming Manual*", 1st Ed. O'Reilly and Associates, Inc., USA, 1991.

[47] Hsieh, S.H., Yang, Y.S., Cheng, W.C., Lu, M.D., Sotelino, E.D., "*MPE++: An Object-Oriented Mesh Partitioning Environment in C++*", Proceedings of the Sixth East Asia-Pacific Conference on Structural Engineering and Construction, Taipei, Taiwan, January 14-16, 313-318, 1998.

©2002, Saxe-Coburg Publications, Stirling, Scotland
Engineering Computational Technology
B.H.V. Topping and Z. Bittnar, (Editors)
Saxe-Coburg Publications, Stirling, Scotland, 251-269.

Chapter 11

Equilibrium Euler-Euler Modelling of Pulverized Coal Combustion

A.C. Benim
Department of Mechanical and Process Engineering
Düsseldorf University of Applied Sciences, Germany

Abstract

This paper demonstrates the application of an equilibrium Euler-Euler model of two-phase flow to model pulverized coal combustion. The state-of-the art two-phase flow models, which are being used in conjunction with the pulverized coal flames are majorly based on an Euler-Lagrange description. An alternative approach is the Euler-Euler description of the two phase flow, additionally assuming a fluid dynamical and thermodynamical equilibrium between the phases. The latter model offers the advantage of being potentially more economical due to the reduced number of momentum and energy transport equations and the missing problem of inter-phase coupling. The present paper shows that the equilibrium Euler-Euler model is able to deliver results with a similar degree of accuracy compared to the Euler-Lagrange model for the investigated configurations. This designates the equilibrium Euler-Euler model as a potentially attractive alternative especially for large-scale industrial applications.

Keywords: pulverized coal combustion, two-phase flow, equilibrium Euler-Euler vs. Euler-Lagrange modelling.

1 Introduction

State-of-the-art pulverized coal combustion models, are generally based on an Euler-Lagrange [1] description of the two-phase-flow. An alternative modelling approach is the Euler-Euler modelling of the two-phase-flow, by assuming fluid dynamical and thermodynamical equilibrium between the phases [2]. In this formulation, the treatment of the particle phase becomes analogous to that of a species in a single-phase mixture, and the modelling results in a single set of transport equations for both phases, i.e. for the "mixture". Therefore, this type of Euler-Euler modelling is less expensive compared to the Euler-Lagrange modelling, due to the fewer number of momentum and energy equations to be solved. The reduction in the computer time is additionally enhanced, since the equation system becomes more robust and shows better convergence properties due to the missing problem of inter-phase coupling.

The equilibrium assumption between the phases introduces an error in the modelling, of course. However, under the conditions of pulverized coal combustion, where the particulate loading is rather low in most parts of the combustion chamber, this assumption is not expected to be a substantial drawback.

Although the equilibrium Euler-Euler model was already applied by different authors [3-6] in modelling the pulverized coal flames, its comparison with the alternative Euler-Lagrange modelling approach has not sufficiently been demonstrated. The purpose of the present work is the application of the equilibrium Euler-Euler modelling to pulverised coal combustion, with emphasis on its assessment by comparing the results with experiments [7] and the predictions obtained by the Euler-Lagrange modelling [8].

2   Modelling

2.1   General

We assume that gas and solid phases behave as interpenetrating continua and can both be described using an Eulerian frame of reference. The additional assumption we introduce here is that the phases are in fluid dynamical and thermodynamical equilibrium. This assumption, i.e. the assumption of local equality of the mean velocities and temperatures of the two phases simplifies the problem in such a way that only a single set of momentum and energy transport equations needs to be solved, where the particle phase appears as a "component" of the "mixture". What remains is a convenient description of the relevant transport equations of the mixture, which is outlined in the following.

In the present analysis, we assume a statistically steady, high Reynolds number turbulent flow. Within this context, all equations presented below are to be understood as time averaged equations [9], adopting a Reynolds averaging [9] for the density and a Favre averaging [9] for the remaining dependent variables. The overbars and tildes, which are normally used to indicate Reynolds and Favre averaged variables are neglected for simplicity.

2.2   Transport Equations

The system of governing differential equations is shown in Table 1. In the momentum equations, the gravity is neglected. The turbulence is modelled by the turbulent viscosity approach [9], adopting a high Reynolds number k-ε model [10], for modelling the the turbulent viscosity, amended by a standard wall-functions approach [10] for modelling the near-wall turbulence. In the energy equation, written for the mixture static enthalpy (h) , the kinetic energy, pressure work and

Continuity equation	$$\frac{\partial}{\partial x_j}\left(\rho u_j\right)=0$$	(1)
Momentum equations	$$\frac{\partial}{\partial x_j}\left(\rho u_j u_i\right)=\frac{\partial}{\partial x_j}\left[\mu_e\left(\frac{\partial u_j}{\partial x_i}+\frac{\partial u_i}{\partial x_j}\right)-\frac{2}{3}\left(\mu_e\frac{\partial u_l}{\partial x_l}+\rho k\right)\delta_{ij}\right]-\frac{\partial p}{\partial x_i}$$	(2)
Transport equation of k	$$\frac{\partial}{\partial x_j}\left(\rho u_j k\right)=\frac{\partial}{\partial x_j}\left(\frac{\mu_e}{\sigma_k}\frac{\partial k}{\partial x_j}\right)+G_k-\rho\varepsilon$$	(3)
Transport equation of ε	$$\frac{\partial}{\partial x_j}\left(\rho u_j\varepsilon\right)=\frac{\partial}{\partial x_j}\left(\frac{\mu_e}{\sigma_\varepsilon}\frac{\partial\varepsilon}{\partial x_j}\right)+\frac{\varepsilon}{k}(C_1 G_k-C_2\rho\varepsilon)$$	(4)
Energy equation	$$\frac{\partial}{\partial x_j}\left(\rho u_j h\right)=\frac{\partial}{\partial x_j}\left(\frac{\mu_e}{\sigma_h}\frac{\partial h}{\partial x_j}\right)+4\pi K_a\left[I_o-\frac{\sigma}{\pi}T^4\right]$$	(5)
Radiation intensity transp. eq.	$$\frac{\partial}{\partial x_j}\left(\frac{1}{K_a}\frac{\partial I_o}{\partial x_j}\right)=3K_a\left[I_o-\frac{\sigma}{\pi}T^4\right]$$	(6)
Species transport equation	$$\frac{\partial}{\partial x_j}\left(\rho u_j m_i\right)=\frac{\partial}{\partial x_j}\left(\frac{\mu_e}{\sigma_{m_i}}\frac{\partial m_i}{\partial x_j}\right)+S_{m_i}$$	(7)
with:	$\mu_e=\mu+\mu_t$ (8) ; $\qquad G_k=\frac{\partial u_i}{\partial x_j}\left[\mu_t\left(\frac{\partial u_j}{\partial x_i}+\frac{\partial u_i}{\partial x_j}\right)\right]$	(9)

C_j: model constants

h: mixture specific enthalpy

I_0: radiation intensity

k: turbulence kinetic energy

K_a: mixture apsorption coefficient

m_j: mass fraction of species j

p: mixture static pressure

S_{m_j}: source term od spec. trasp. eq.

T: mixture static temperature

u_j: mixture velocity vector

ε: dissipation rate of k

μ: molecular viscosity

μ_t: turbulent viscosity

μ_e: effective viscosity

ρ: mixture density

σ: Stefan-Bolzmann constant

σ_j: mixture effective Prantdtl-Schmidt number for variable j

Table 1: Governing equations.

viscous dissipation terms are neglected, and the equality of Prandtl and Schmidt numbers is assumed. The radiative heat transfer is modelled adopting the moment method [11], which solves a differential transport equation for the direction independent part of the radiation intensity (I_o), and assuming a gray gas radiation, which leads to a definition of the mixture absorption coefficient (K_a) as a sum of gas-phase and particle phase contributions ($K_a = K_{a,G} + K_{a,P}$) [12]. The turbulent viscosity (μ_t) of the mixture is modelled as

$$\mu_t = g\, C_D \rho k^2 / \varepsilon \tag{10}$$

where g denotes a correction function [13] introduced for modelling the effect of the particle phase on turbulence, reading as

$$g = \left[(1 - m_p) + m_p \left(1 + C_P \frac{d_p^2 \rho_P}{\mu} \frac{\varepsilon}{k} \right)^{-1} \right] (1 - m_p)^{1/2} \tag{11}$$

the variables m_P, d_P, ρ_P, C_P denoting the particle mass fraction, particle diameter, particle density and a model constant, respectively. For modelling the two-phase turbulence, there are more sophisticated approaches in the literature proposing an improved set of transport equations for k and ε [14], which are not attempted here.

Assuming an "ideal" mixture, the mixture denisty can be derived to be

$$\rho = (1 - \theta) \frac{p}{RT \sum\limits_{j,G} \dfrac{m_j}{M_j}} \tag{12}$$

where R, M_j denote the universal gas constant and the molecular weight of species j, respectively. The index G under the summation sign indicates that the summation is to be performed over the gaseous species only. The θ in (12) is the particle volume fraction.

2.3   Pulverized Coal Combustion Model

2.3.1   The Reaction Mechanism

The following four-step global reaction mechanism has been employed

Pyrolysis:

$$v_{1,RC}\, RC \rightarrow \sum_{j=1}^{NC} v_{1,C_j} C_j + v_{1,C_x H_y}\, C_x H_y + v_{1,CO}\, CO + v_{1,O_2}\, O_2 \qquad (13a)$$

Char burn-out:

$$v_{2,C_j}\, C_j + v_{2,O_2}\, O_2 \rightarrow v_{2,CO}\, CO \qquad (13b)$$

Hydrocarbon burn-out:

$$v_{3,C_x H_y}\, C_x H_y + v_{3,O_2}\, O_2 \rightarrow v_{3,CO} CO + v_{3,H_2 O}\, H_2 O \qquad (13c)$$

CO burn-out:

$$v_{4,co}\, CO + v_{4,O_2}\, O_2 \rightarrow v_{4,CO_2}\, CO_2 \qquad (13d)$$

where $v_{i,j}$ denote the stoichiometric coefficient of the species j in the reaction i and $C_x H_y$ a general hydrocarbon. The species RC denotes the "raw coal" which produces gas-phase volatile matter ($C_x H_y, CO$) and solid phase Char (C) as a result of the pyrolysis.

In the above equations, C_j denotes the char in the size class j. Since the combustion rates of char particles are size dependent, the particle size distribution of the coal dust can be taken into account by discretizing it into several size classes, each of which having a constant particle size, for being able to model the char combustion rate more accurately. NC denotes the number of particle size classes. Since the pyrolysis rate is principally not size dependent, an assignment of different size classes to the "species" raw coal (RC) is not necessary.

2.6.2 The Combustion Model

2.6.2.1 Gas-Phase Reactions

The combustion model is concerned with the modelling of the mean reaction rates. The key problem in obtaining the time averaged reaction rates in turbulent combustion is the strong nonlinearity of the instantaneous reaction rate terms. In the present reaction mechanism, the last two reactions (13c,d) are purely gas-phase reactions. The first and second (13a,b) reactions are heterogeneous, where species in solid and gas phase are both involved.

Let us consider first the reactions in the gas phase. According to reaction kinetics, utilizing an Arrhenius approach, the molar rate of destruction of a reactant i in an irreversible reaction k is given by [15]

$$R_{i,k} = -v'_{i,k}\left\{A_k T^{\beta_k}\exp(-E_k/RT)\prod_{j=1}^{N}\left[\rho\frac{m_j}{M_j}\right]^{\eta'_{j,k}}\right\}\tag{14}$$

where $N,\eta'_{j,k}$ denote the number of species and the rate exponent for reactant j, whereas A_k,β_k,E_k are the pre-exponential factor, temperature exponent, and the activation energy of the Arrhenius rate expression for the reaction k. Since the time-averaged equations for the turbulent flow are solved, one would normally express the instantenous values of the dependent variables in (12) as the well-known sum of the averaged and fluctuational components (e.g. $m_j = \overline{m_j} + m'_j, T = \overline{T} + T',...$) and time average for obtaining the time averaged source terms of the species transport equations (7). However, due to the strong non-linearity of (14), the following inequality is obvious

$$\overline{R(T,m_j,...)} \neq R(\overline{T},\overline{m_j},..)\tag{15}$$

Thus, the expression (14) gives rise to severe closure problems due to the appearance of many correlation terms such as $\overline{T'm'_j}$, etc. Fortunately, it is observed that the turbulent combustion in many technical systems are in the high Damköhler number regime [15] and the mean fuel consumption rate is majorly limited not by the "reaction kinetics" but, by the "mixing". This observation has lead to the Spalding's well-known Eddy Break Up combustion model [16], where the mean fuel consumption rate was set proportional to the rate of small scale turbulent mixing, which, in turn, was modelled through to the dissipation rate of turbulent eddies. In the present work, we apply the Eddy Dissipation Concept (EDC) of Magnussen and Hjertager [17], which is principally based on this idea. According to EDC, the time averaged reaction rate is obtined by (instead of Eq.(14))

$$\overline{R_{EDC,i,k}} = -v'_{i,k}M_i A\rho\frac{\varepsilon}{k}\min\left\{\min\left[\frac{m_{R1}}{v'_{R1,k}M_{R1}},\frac{m_{R2}}{v'_{R2,k}M_{R2}},...\right], B\frac{\sum\limits_{P}m_P}{\sum\limits_{j}^{N}v''_{j,k}M_j}\right\}\tag{16}$$

where the coefficients A and B are the empirical model constants and the indices R and P indicate reactants and products of the reaction, respectively. The purely gas-phase reactions, i.e. the hydocarbon burn-out (13c) and CO burn-out (13d) reactions are modelled utilizing the EDC.

2.6.2.2 Pyrolysis

For determining the pyrolsis rate, i.e. for the rate of the first reaction (13a) of the four-step reaction mechanism, several models exist, with different levels of sophistication [18]. We adopt here a model similar to that of Badzioch and Hawskley [19] describing the pyrolysis rate by means of a single irreversible reaction employing an Arrhenius rate expression. According to the present approach, the mean rate of consumption of the Raw Coal due to pyrolysis is given by

$$\frac{dm_{RC}}{dt} = -k_p \, m_{RC} \tag{17}$$

where m_{RC} is the raw coal mass fraction rate and the pyrolysis rate k_p is modelled by the following Arrhenius rate expression:

$$k_p = A_P \exp(E_P/R\,T) \qquad (1/s) \tag{18}$$

The pre-exponential factor A_p and the activation energy E_p are coal dependent.

2.6.2.3 Char Oxidation

The oxidation rate of char (C) to carbon monoxide (CO) through the second reaction (13b) of the four-step mechanism is modelled based on the assumptions of Field [20] and Baum and Street [21]. According to these modelling assumptions, the kinetical rate (k_s) and the diffusion rate of oxygen (k_D) to the particle surface play a combined role in determining the effective burning rate of char. Assuming a pure CO formation as a result of the heterogeneous reaction of char with oxygen on the particle surface, the effective char oxidation rate for the particle size class j, per surface area of the coal particle $(k_{C,j})$ can be expressed by

$$k_{C,j} = -\frac{1}{(1/k_S + 1/k_{D,j})} p_{O_2} \qquad \left(kg/m^2 \cdot s\right) \tag{19}$$

where p_{O_2} is the partial pressure of oxygen. The kinetical and diffusional reaction rates are given by

$$k_S = A_S \exp(-E_S/RT) \qquad \left(kg/m^2 \cdot s \cdot bar\right) \tag{20}$$

$$k_{D,j} = \frac{48\,D_0}{R\,T_0}\,10^5 \cdot \frac{T^{0,75}}{d_{p,j}} \qquad \left(kg/m^2 \cdot s \cdot bar\right) \tag{21}$$

The pre-exponential factor A_S and the activation energy E_S appearing in (20) are coal dependent and need to be determined empirically. The constants Do and To in (21) have to do with the diffusion coefficient of oxygen within the boundary layer surrounding the coal particle, as given by Field [21]. The term $d_{P,j}$ is the diameter of the coal particle (which is assumed to be spherical) in the size class j.

For obtaining the local particle area per mixture volume, the "Shadow" method of Spalding [22] is employed, which was originally proposed for spray combustion. Here, the hypothetical mass fraction of the particles for the hypothetical case of "no-combustion" is defined to be the "shadow mass fraction". Thus, the burn-out of the particle phase is estimated from the ratio of the local mass fraction of the particles to their shadow mass fraction. For estimating the "shadow mass fraction", the variable "mixture fraction" (f) is utilized, which is "inert", i.e. governed by a source-free transport equation and has the value of 1 at the primary (fuel) and the vauel 0 at the secondary (air) inlets of the combustion chamber. Letting $m_{RC,j,0}, m_{A,j,0}$ denote the raw coal and ash mass fractions of the size class j at the primary (fuel) inlet, and $m_{RC,j}, m_{C,j}, m_{A,j}$ the local raw coal, char ans ash concentrations of the size class j, a burn-out parameter B_j for the size class j can be defined as

$$B_j = 1 - \frac{m_{RC,j} + m_{C,j} + m_{A,j}}{\left(m_{RC,j,0} + m_{A,j,0}\right) \cdot f} \qquad (22)$$

Based on this parameter the local specific particle surface area per mixture volume for the size class j ($a_{P,j}$) can be expressed as

$$a_{P,j} = \frac{6\rho}{\rho_{S,0}\, d_{P,j,0}} \left(1 - B_j\right)^{n_B - 2} \left(m_{RC,j} + m_{C,j} + m_{A,j}\right) \qquad (23)$$

where $\rho_{S,0}, d_{P,j,0}$ denote the material density and the diametrer of the coal particles at the primary (fuel) inlet. The exponent n_B in (23) describes the behaviour of the coal particles during char oxidation. The value $n_B = 0$ corresponds to the case, where the highly porous coal particles burn fully from inside, where their diameter remains unaltered and their density decrease. On the contrary, the value $n_B = 2/3$ describes the situation, where the particle burns completely on the surface, with unchanging density, but decreasing diameter. Different values can be assigned to n_B for describing the burn-out behaviour of a certain coal dust, if such information is available through experimental data.

Similar to the gase phase reactions, the pyrolysis and char-oxidation reaction rates are also influenced by the turbulent fluctuations (see eq. (15)). However, there are no sophisticated closure models proposed for modelling these effects. The state-of-the-art of the modelling [4-6] is the neglection of the turbulent fluctuations on these terms, which was also adopted here.

3   Application

The modelling is based on the commercial CFD code CFX-4 of AEA Technology, utilizing a second-order Finite Volume discretisation scheme, and a pressure correction procedure (SIMPLEC) for the velocity-pressure coupling. For the model constants, standard values were used. For the parameters describing the coal combustion behaviour, which may strongly depend on the specific coal used, it is rather difficult to assume "standard" values. Nevertheless, since more information did not exist, "mean" values found in the literature (provided for a general orientation) were used for these parameters, without any additional tuning. The employed model parameters are listed below:

$$C_D = 0.09; C_1 = 1.43; C_2 = 1.92; \sigma_{k,e} = 1.0; \sigma_{\varepsilon,e} = 1.3; \sigma_h = 0.7; \sigma_{m_j} = 0.7; A = 4.0; B = 4.0$$

$$A_P = 1.5 \cdot 10^5 s^{-1}; E_P = 74 \cdot 10^6 J/kmol; A_S = 204.0 s^{-1}; E_S = 79.4 \cdot 10^6 J/kmol; n_B = 0$$

3.1   Two-Dimensional Pulverized Coal Flame

3.1.1   Case Description

In this example the flame A1 of the Long Coal Flame (LCF) trials, performed in the International Flame Research Foundation (IFRF), is considered [7]. The furnace geometry is shown in Figure 1. As one can see in the figure, the furnace does not have a circular cross section. However, the problem can still be treated as a two-dimensional, axisymmetric problem, modelling the furnace geometry by a cylinder with the same cross-sectional area. Since the flame is quite apart from the walls, one can assume that the difference between the quadratic or cylindrical cross-section can be neglected. Since the flame is a horizontal one, the buoyancy could disturb the symmetry. Nevertheless, the measurements do not show a substantial deviation from the axisymmetry, implying that the problem can still be treated as a 2D-axisymmetric.

The coal used was a high volatile bituminous coal, dried to a moisture content of approx. 2%. The flame operation conditions and the coal data are given in Tables 1-3. Although some indications were provided, a detailed size distribution of the coal dust was not given in [7]. Therefore, we empoyed the particle size distribution of Wennerberg [8], which models the particle size distribution by four distinct particle size classes with the diameters and mass fractions as given in Table 4.

The grid is generated using totally about 2350 cells. In the furnace, 57 cells were used in the axial direction with a slight expansion towards the furnace end. The radial direction was discretized by 40 cells (equidistant in the region of inlets, slightly expanding towards the wall in the outer region). The thickness of the lip between the primary and secondary inlets is neglected. For discretizing the primary inlet, 5 cells were used in the radial direction, whereas the secondary inlet was discretized by 9 cells. The exhaust duct is discretized by 3x22 cells (In a subsequent run, the problem was solved using a coarser grid with approximately 30x 30 cells in the furnace, which showed practically no substantial differences in the results. Thus the present results can be considered to be grid independent).

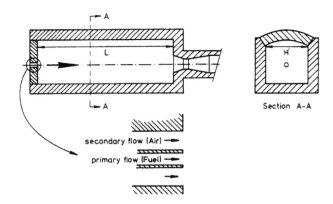

Figure 1: Two-dimensional, pulverized coal flame: Geometry of furnace and burner

Total air mass flow rate	0.6442 kg/s
Coal mass flow rate	0.0589 kg/s
Primary air mass flow rate	0.0708 kg/s
Primary air temperature	423 K
Secondary air mass flow rate	0.5733 kg/s
Secondary air temperature	763 K
Excess Air	approx. 11%

Table 2: Flame data

Char	Volatiles	Ash
60.4	32.0	7.6

Table 3: Coal composition (% weight dry)

C	H	N	S	O
80.79	5.09	1.21	0.92	11.99

Table 4: Elementary analysis (%weight, ash free)

Size Class	Diameter (μm)	Mass Fraction (%)
1	5	30
2	30	25
3	70	22
4	140	23

Table 5: Particle size distrtibution of the coal dust

3.1.2   Results

Results are compared with the measurements of Michel and Payne [7]. The present predictions are additionally compared with the available computational results of Wennerberg [8] for the same flame, which were obtained using a classical Euler-Lagrange formulation. A short description of the modelling applied by Wennerberg [8] can be given as follows: The two-phase flow modelling was achieved by applying an Euler-Lagrange modelling, solving different sets of momentum and energy equations for gas and particle phases. As the turbulence model, a high Reynolds number k-ε model was used, taking the turbulent dispersion of the particles into account. The pyrolysis was modelled according to the model of Kobayashi [18]. The applied char oxidarion model was similar to the present one. In Wennerberg's assumed reaction for the char oxidation, the *CO* oxidation was, however, neglected allowing a direct oxidation of char to CO_2. The combustion model applied for the gas phase combustion reactions was the Eddy Dissipation Concept, similar to the present approach. The same particle size distribution and similar grid resolutions were used.

Figure 2 compares the predicted radial profiles of the axial velocity with the measurements, for several axial locations. One can observe that the agreement between the present predictions and experiments is fairly good. Discrepancies are observed especially in the reverse flow region, where measurements indicate higher velocities in magnitude. However, given the difficulties in measuring velocities in recirculation zones by probes, one may assume that the uncertainies in those measured values are rather high.

Predicted radial O_2 mass fraction profiles are compared with experiments and Wennerberg's predictions at various axial locations in Figure 3. At x=0.69m, both predictions show a good agreement with experiments. However, the agreement of the present ones with experiments is somewhat better, especially near the furnace axis. At x=1.38m, both predictions show again a fair agreement with experiments, where, this time, Wennerberg's results perform better than the present ones near the furnace axis. At x=3.21m, both predictions agree quite well with each other, whereby some discrepancies in comparison to the measurements can be observed. Note that both predictions indicate an oxygen depletion region at x=3.21m, near the centerline. This could have been caused by an overprediction of the local oxygen consumption by the combustion reactions.

The temperature variation along the furnace axis is shown in Figure 4. Experiments show a steady increase of the temperature starting from the beginning of the furnace. Both computations underpredict this temperature rise in the initial part of the furnace (x < 0.75m), whereas the predictions of Wennerberg show a somewhat better agreement with the experiments. Behind x=1.0m, the present predictions show a fair agreement with the experiments, whereas the results of Wennerberg highly over-predict the measurements around x=2.0m. For the present predictions, the sudden flattening of the temperature curve around x=1.5m is probably caused by the predicted local region of oxygen depletion, which locally retards the combustion reactions. Such a tendency is not observed in Wennerberg's predictions, although they also predict a local oxygen depletion region

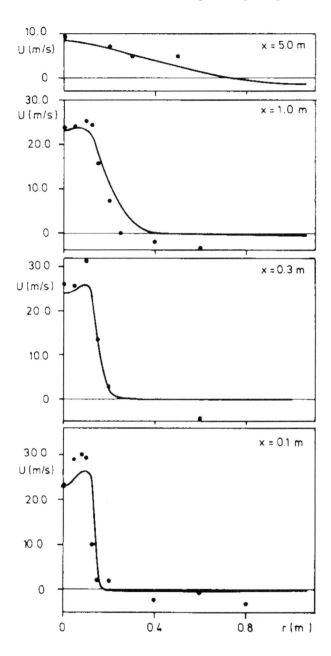

Figure 2: Radial profiles of axial velocity at x=0.1m, x=0.3m, x=1.0m, x=5.0m.
(——— present predictions, • experiments [7]).

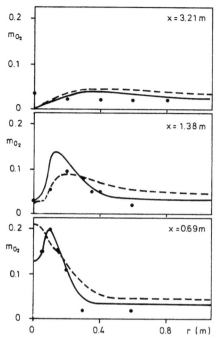

Figure 3: Radial profiles of O_2 mass fractions at x=0.69m, x=1.38m, x=3.21m.
(——— present predictions, • experiments [7], - - - predictions of Wennerberg [8])

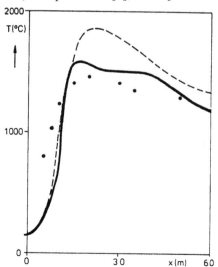

Figure 4: Variation of temperature along furnace axis.
(——— present predictions, • experiments [7], - - - predictions of Wennerberg [8])

3.2   Three-Dimensional Pulverized Coal Flame

3.2.1   Case Description

The advantage of the reduced CPU times of the present equilibrium Euler-Euler modelling (compared to a standard Euler-Lagrange modelling) becomes important especially for large scale industrial applications. In the present example, the pulverized coal flame in the furnace of a large utility boiler ($900MW_{el}$) is investigated.

The tangentially fired furnace has square cross section with a side length above 20m and a height above 80m. The furnace and the burner arrangement is sketched in Figure 5. The burners are placed in the corners of the furnace and directed to a virtual tangential circle. This creates a "vortex" in the furnace, minimizes "dead zones" and results in a rather uniform heat release on all furnace walls. There are totally six burner levels, which help to achieve a rather uniform temperature variation along the furnace height and allow additional flexibility in part load operation. Due to the applied air staging for reducing the *NOx* emissions, the burner system is quite sophisticated incorporating several primary and secondary injection nozzles for fuel and air, per burner level.

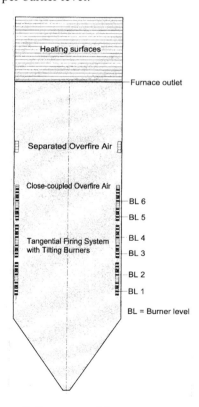

Figure 5: Furnace and burner arrangement.

3.2.2   Results

The generated block-structured finite-volume grid with about 500 000 cells is illustrated in Figure 6.

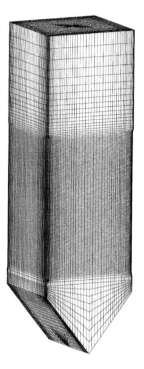

Numerical Grid System
500,000 cells

Figure 6: The grid.

Figure 7 shows the predicted velocity and temperature fields and at three horizontal cross sections along the furnace height for the full load operation. The sections are placed at the elevations of of the first, third and fifth burners levels (Fig. 5). In the present tangential firing system, the corner jets are directed to a virtual tangential circle. The penetration of the corner jets towards the middle parts of the furnace, while forming a central vortex can be observed in the figures. One can also observe that the vortex intensity increases and the penetration of the corner jets into the central parts decreases along the furnace height, due to the addition of mass and momentum at next burner levels. The temperature profile at the lowest burner level shows a minimum in the center. The reason for this is the leakage air through the hopper slots. Due to the mixing, combustion and heat exchange by radiation, the temperatures of the central core gets increased along the furnace height. One can observe that the achieved temperature patterns are rather uniform and the furnace geometry is well "filled" by the flame.

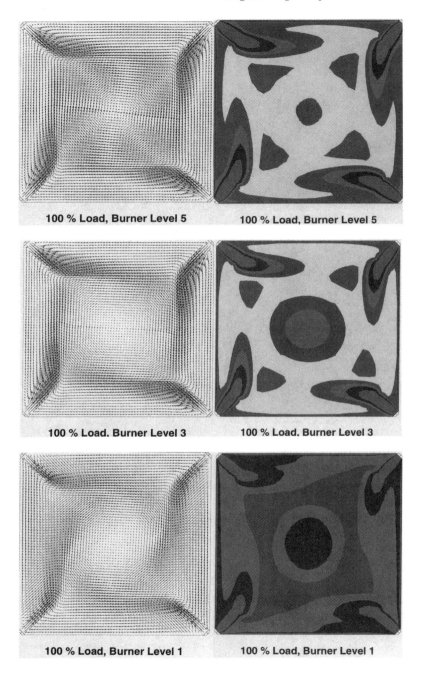

Figure 7: Velocity and temperature fields at three horizontal cross sections (100% Load).

Results for 50% load are shown in Figure 8, for the third and fifth burner levels, where the lowest burner level is shut down for this load condition. One can observe that the jet penetration and the uniformity of the temperature distribution do not deteriorate under part load conditions

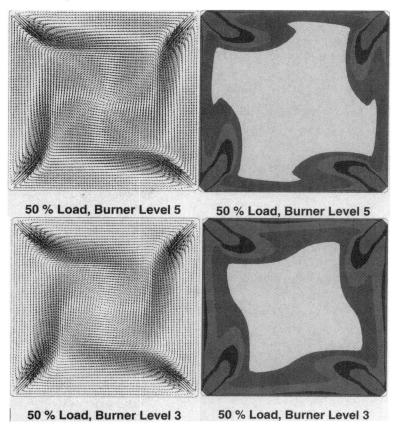

50 % Load, Burner Level 5   50 % Load, Burner Level 5

50 % Load, Burner Level 3   50 % Load, Burner Level 3

Figure 8: Velocity and temperature fields at two horizontal cross sections
(50% Load)

4   Conclusions

An equilibrium Euler-Euler model has been applied to model pulverized coal combustion. Results for a two-dimensional flame have been compared with experiments and predictions of other authors using an Euler-Lagrange approach. This comparison indicates that the degree of accuracy obtained by the present model is comparable to that of the principally more expensive Euler-Lagrange approach. The reduced computer costs by the present approach become important, especially for large scale applications. An application of the model to a large utility boiler is also demonstrated.

References

[1]   D. Migdal and D. V. Agosta, *"A Source Flow Model for Continuum Gas Particle Flow"*, Trans. ASME. J. Appl. Mech., 34 E, 860-872, 1967.
[2]   S.L . Soo, *"Fluid Dynamics of Multiphase Systems"*, Blaisdell Publishing Company Waltham, Mass.,1967.
[3]   W. Richter, *"Mathematische Modelle technischer Flammen"*, Dissertation, University of Stuttgart, 1978.
[4]   W. Zinser, *"Zur Entwicklung Mathematischer Flammenmodelle für die Verfeuerung technischer Brennstoffe"*, VDI-Berichte,Reihe 6,Nr.171, VDI-Verlag, Düsseldorf, 1985.
[5]   A.C. Benim and U. Schnell, *"Finite Element Simulation of Turbulent Reacting Flows with Emphasis on Pulverized Coal Combustion and Nitrogen Oxide Formation"*, Proceedings of the Fifth International Symposium on Numerical Methods   in   Engineering,R.Gruber,J.Periaux,R.P.Shaw(Ed.)Vol.2,5-10, Springer-Verlag,Berlin,1989.
[6]   U.   Schnell,   *"Wirkungsgradoptimierte   Kraftwerkstechnologien   zur Stromerzeugung aus fossilen Brennstoffen"*,VDI-Berichte, Reihe 6, Nr. 389, VDI-Verlag, Düsseldorf, 1998.
[7]   J.-B. Michel and R. Payne, *"Detailed Measurement of Long Pulverized Coal Flames for Characterization of Pollutant Formation"*, International Flame Research Foundation, Ijmuiden, The Netherlands, Doc. No. F09/a/23, 1980.
[8]   D. Wennerberg, *"A Mathematical Model for Pulverized Fuel Flames"*, National Swedish Board for Energy Source Development, Report No: ETU 2005/03, 1983.
[9]   O.J. Hinze, *"Turbulence"*, Mc-Graw-Hill Book Company,New York, 1985.
[10] B.E. Launder and D.B. Spalding, *"The Numerical Computation of Turbulent Flows"*, Comput. Meths. Appl. Mech. Eng., 3, 269-289, 1974.
[11] A.C. Benim, *"A Finite Element Solution of Radiative Heat Transfer in Participating Media utilizing the Moment Method"*, Comput. Meths. Appl. Mech. Engrg.,67,1-14, 1988.
[12] K.H. Hemsath, *"Zur Berechnung der Flammenstrahlung"*, Dissertation, University of Stuttgart, 1969.
[13] P.R. Owen, *"Pneumatic Transport"*, J. Fluid. Mech., 39, 407-432, 1969.
[14] S.E. Elgobashi and T.W. Abou-Arab, *"A Two-Equation Turbulence Model for Two-Phase Flows"* , Phys. Fluids, 26, 931-938, 1983.
[15] K.K-Y. Kuo, *"Principles of Combustion"*, John Wiley & Sons, New York, 1986.
[16] H.B. Mason and D.B. Spalding, *"Prediction of Reaction Rates in Turbulent Premixed Boundary Layer Flows"*, Proceedings of the Combustion Institute European Symposium, Academic Press, New York, 601-606, 1973.
[17] B.F. Magnussen and B.H. Hjertager, *"On Mathematical Modelling of Turbulent Combustion with Special Emphasis on Soot Formation and Combustion"*, Proceedings of the Sixteenth Symposium (International) on Combustion, The Combustion Institute, Pittsburgh, 719-729, 1976.

[18] L.D. Smoot, D.T. Pratt (Eds.), *"Pulverized-Coal Combustion and Gasification"*, Plenum Press, New York, 1979.

[19] S. Badzioch and P.G.W. Hawskley, *"Kinetics of Thermal Decomposition of Pulverized Coal Particles"*, Int. Eng. Chem. Proc. Des. Develop., 9, 521-530, 1970.

[20] M.A. Field, D.W. Gill, B.B. Morgan and P.G.W. Hawskley, *"Combustion of Pulverized Coal"*, The British Coal Utilization Research Assoc., Leatherhead, 1967.

[21] M.M. Baum and P.J. Street, *"Predicting the Combustion Behaviour of Coal Particles"*, Comb. Sci. Technology, 3, 231-243, 1971.

[22] D.B. Spalding, *"The 'Shadow' Method of Particle-Size Calculation in Two-Phase Combustion"*, Proceedings of the Nineteenth Symposium (International) on Combustion, The Combustion Institute, Pittsburgh, 941-952, 1982.

Chapter 12

©2002, Saxe-Coburg Publications, Stirling, Scotland
Engineering Computational Technology
B.H.V. Topping and Z. Bittnar, (Editors)
Saxe-Coburg Publications, Stirling, Scotland, 271-311.

Toward Intelligent Object-Oriented Scientific Applications

Th. Zimmermann† and P. Bommé‡
† Laboratory of Structural and Continuum Mechanics
  Faculty of Natural Environment, Architecture & Construction
  Swiss Federal Institute of Technology, Lausanne, Switzerland
‡ Consultant, Lausanne, Switzerland

Abstract

Combining rule-based expert systems and scientific applications has long been a challenge. So far, however, the idea has not had the expected success; the main reasons are the incompatibility of implementation languages used in both activities and the lack of integration methodology. But today, the object-oriented approach provides the appropriate paradigm on which an integration methodology can be built. In this paper, we outline the concept of an intelligent-object, describe its functionality, and sketch the principles of its implementation in existing applications. Details of the implementation are given in [1].

Keywords: intelligent object, object-oriented programming, rule-based expert system, scientific applications, finite elements.

1 Methodology

1.1 Introduction

1.1.1 Historical sketch

The idea of combining artificial intelligence with scientific applications is certainly not new. The work of Bennett & al. [2] demonstrated, in 1978 already, the feasibility and the great potential of rule-based expert systems interacting with scientific applications. However, all early attempts used commercial environments that introduced constraints in the setting up of scientific applications [3]. The impossibility of calling a LISP function from a C module is typical; only the opposite is feasible, because LISP is a dominant language. This leads to a master-slave relationship in which scientific applications lose part of their identity. In addition, efficiently converting numerical data from the scientific application into qualitative information processed by a knowledge-based system can be a difficult task.

Consequently, in spite of enthusiasm and research investment [4,5,6,7,8] the use of expert systems in scientific applications did not meet the engineer's expectations. The main reasons for failure lie, as already mentioned in the respective implementation languages of the two domains and also in the lack of integration methodology [9]. The high level of interoperability required between scientific applications and their associated knowledge-based systems seems impossible to achieve when different paradigms are used.

The object-oriented concept, which emerged from the artificial intelligence field [10], is receiving increasing acceptance within the scientific community nowadays. The object-oriented paradigm, recognized to help in developing better numerical simulation tools with increased maintainability, reusability, and efficiency [11,12,13,14], provides de facto a paradigm on which an integration methodology can be built.

1.1.2   Present work

The work presented herein originated in an attempt to design an expert-assistant for field engineers confronted with the problem of optimizing the proper selection of industrial equipment in a context subject to a large number of constraints of various natures: mechanical, thermal, environmental, etc. Early designs of the system used a commercial expert system to manage all constraints and it soon became evident that the resulting system was insufficient, as all decisions were extracted from a global knowledge base and did not properly exploit the specificity of the rules and the evidence that most inferences could be carried out on small subsets of the rules database. An intelligent-object concept was therefore introduced, featuring a methodology for integrating propositional rules with objects. The proposed concept achieves a good interoperability between rule-based expert systems and object-oriented scientific applications, as well as a high level of reusability and decentralization of control.

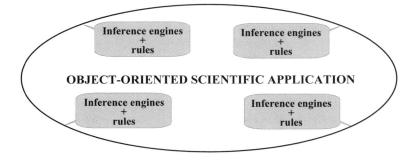

Figure 1: Decentralized object-oriented intelligence

To summarize, our goal is to design an approach, which supports implementation of multiple local reasoning systems into an existing application, and can easily be adapted to any scientific application, as illustrated in figure 1.

In section 1.1 we briefly present object-oriented programming, describe the components of rule-based expert systems and the hierarchy of implementation of the original rule-based expert system. Section 1.3 is dedicated to intelligent objects. We first define the concept, before reviewing the components of the architecture of implementation. Section 4 lists the main steps to follow for the integration of rule-based intelligence within object-oriented scientific applications; this section also proposes some guidelines in order to facilitate the formulation of rules in an object-oriented manner, to identify blackboard variables, and to select appropriate reasoning modes. An academic illustration is presented in section 1.5. In part 2, application of the concept to assistance in data preparation for a finite element code is presented.

1.1 Components of the proposed approach

1.2.1 Object-oriented programming

Much has already been written on object-oriented programming. Depending on the reader's background various references are available, [10,15,16,17] for computer scientists, [18,19,20] for engineers, and many books on object-oriented languages have an introductory chapter [21,22,23,24,25,26,27]. The authors have already given an introduction to the object-oriented approach applied to scientific applications in a previous paper [28], and the main concepts are recalled hereafter.

Class, object, inheritance and polymorphism are the main concepts of the object-oriented paradigm and encapsulation is the central principle. A class is an abstract data structure consisting of data and associated methods. When a data structure is instantiated, an object is created as an instance of the corresponding class. The data of an object can only be accessed by the object itself. Methods provide the means of interfacing an object with the environment, and objects communicate with messages. Hiding data structure and implementation of methods is called encapsulation. Classes are organized in hierarchies and can share data structures and methods with subclasses, like **Element** (C3) with subclasses **PlaneStrain** (C3.1) and **Truss** (C3.2) etc., **Load** (C2) with subclasses **BodyLoad** (C2.1) e.g.... (see Figure 2). A class may inherit data structures and methods partially or totally from a single or from multiple ancestors. This means that inheritance may be total or selective, simple or multiple. Different behaviors may result from an identical message sent to different objects: this is called polymorphism, local methods overwriting inherited ones.

Objects encapsulate their state, and if the non-anticipation principle is respected, code robustness is improved. In short, the principle requires that the implementation of an object's method should not make any assumption on the object's state [28,29].

The power of the object-oriented approach lies in the strong modularity which results from the encapsulation of data, methods, and state, which facilitates code debugging, maintenance, and reuse.

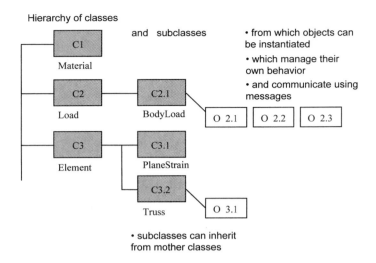

Figure 2: The object-oriented approach in two words

1.2.2   Rule-based expert system

The internal organization of a typical rule-based expert system is illustrated in Figure 3. Its three basic components are the fact base, the knowledge base, and the inference engine. All of them work on the same pieces of knowledge: facts.

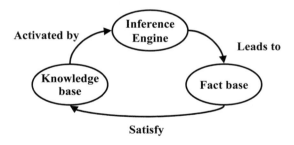

Figure 3: Components of an expert system

1.2.2.1 Facts

The formulation of facts requires a syntax, called knowledge representation language. There is no unique language, and one may find as many languages as there are expert systems. The language depends on the logic paradigm used to formulate facts. The best-known logics today are: the propositional logic (or propositional calculus) and the first order logic (or predicate calculus) [30,31,32]. Propositional logic supports declarations such as *'The stiffness matrix is symmetric'*, while first order logic allows more general statements like *'All Von-Misès material are incompressible(in the plastic range)'*.

1.2.2.2 Fact base

The fact base contains the current state of knowledge supplied by the user or obtained by inference. This is the working memory of the expert system. It is continuously updated when reasoning is in progress, and emptied when it is over.

1.2.2.3 Production rules

The knowledge base collects the know-how of an expert in a specific domain. Several paradigms exist for knowledge representation [33,34], the best-known is based on production rules. Not surprisingly, rule-based programming is one of the most popular and widely used knowledge representations, because of its usefulness in the construction of expert systems [35].

A production rule takes the following form

IF *premises* THEN *conclusions*

The IF-part defines a set of disjunctions or conjunctions of propositions that need to be satisfied for the THEN-part to be true.

Each rule represents an autonomous chunk of expertise that can be modified independently of other rules, offering flexibility and ease of maintenance. In the end, the set of rules will be yielding results, which go beyond the sum of the results of the individual rules. The main drawback of this representation is that when the number of rules increases, the processing efficiency decreases unless some structure is imposed on the rule base in order to limit the number of rules involved in the reasoning processes.

1.2.2.4 Inference engines

An inference engine is a generic process implementing a mechanism needed for the interpretation of knowledge. Usually, an inference engine interprets knowledge until a condition is satisfied. During the process, the fact base is modified. The manner in which rules are used characterizes the inference engine. Two basic reasoning strategies can be distinguished: forward chaining and backward chaining, from which mixed chaining can be constructed.

Forward chaining is a non-directed search algorithm; therefore, it usually aims at acquiring as much information as possible. This engine tries to obtain all facts that can be deduced from the rules, which have their premises satisfied by some facts of the fact base. Each time a rule is activated, the engine may use this new piece of information in order to activate a new rule. The process stops when a goal is achieved or when no additional fact can be deduced.

Backward chaining is a directed search algorithm; it usually aims at defining or proving a particular information called goal. This reasoning mode tries to prove a final goal by taking the premises of a rule (concluding on this goal) as successive subgoals. The process stops successfully when all subgoals are proved, otherwise it fails.

The above two reasoning modes can be activated together in order to obtain mixed chaining. Forward chaining may activate a backward chaining when some facts cannot be deduced, or conversely backward chaining may activate forward chaining after a goal has been proved.

1.2.2.4.1 Hierarchy of implementation

The system proposed herein is a propositional rule-based expert system, in which established facts cannot be changed once stated. This means that reasoning modes are monotonous.

ExpertSystem
KnowledgeComponent
   ExpertAttribute
   Rule
   Fact
      Premise
         PremiseWithMethod
      Conclusion
         Action
         ConclusionWithMethod
   KnowledgeBase
   Function
   ExternalFunction
   Value
      BooleanValue
      IntegerValue
      StringValue
      ...

ReasoningProcess
   ForwardReasoning
   SimpleBackwardReasoning
   BackwardReasoning
   .....

Utilities
   Parser
   LinkedList
   Stack
   ValueManager
      KnowledgeComponentManager

Blackboard

NB: class *Value*: classes in italics are abstract classes

Figure 4: Hierarchy of rule-based expert system

The system implements a forward chaining ending when no additional fact can be acquired, and a backward chaining capable of either establishing a value for a given goal, or of proving the validity of a given fact. In this presentation, there is no mixed chaining, but it will be demonstrated that customization of reasoning processes is very easily achieved.

The rule-based expert system was developed from scratch. Its original C++ hierarchy is presented in Figure 4. However, implementation principles can be found in references describing propositional expert systems [36,37,38], or in [39] presenting an integration of object-oriented programming and the principles of artificial intelligence, and also in texts dedicated to the implementation of first order expert systems [40,41]. Notice that the proposed hierarchy includes a class **Blackboard**, whose functionality will be explained in 1.3.6. The main classes, in bold characters, are described in detail in [1].

1.3 Intelligent object and architecture of implementation

1.3.1 Intelligent object

We view an intelligent object as a standard object associated with an expert system. As previously discussed and illustrated below in Figure 5, an expert system infers facts from rules using forward, backward, or mixed chaining .

The state of an object is characterized by the instantaneous values taken by its attributes (i.e. Data) and its behavior is governed by methods. Equipping an object with an inference engine and rules as additional methods and using inferred facts as additional data will allow the object to perform reasoning processes with its own set of attributes using its personal set of rules. Communication with other objects and pool-reasoning is supported by a blackboard used to share information.

1.3.2 Implementation architecture

The architecture of implementation is illustrated in Figure 6. Consider an existing scientific object-oriented application; with each class of the original hierarchy we associate a set of rules, which can be empty. As a result, a "rule hierarchy" will be created with the same inheritance path as the original hierarchy. If now an object of the original hierarchy gets access to a chaining mechanism (and we will show that this is easily feasible using simply inheritance), it will be able to infer new facts by chaining on its own rules, on rules inherited from superclasses while using information stored on the blackboard. These new facts might in turn provide new information to the blackboard or motivate messages to other objects which in turn can infer new facts.

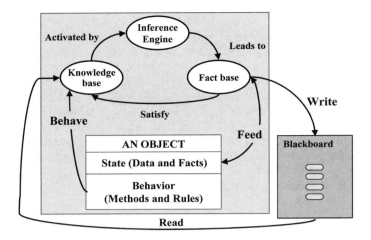

Figure 5: Integrating a rule-based expert system into an object

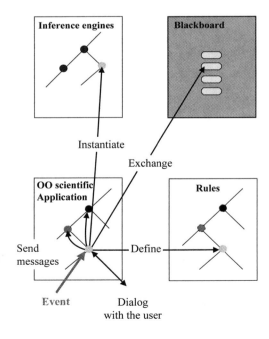

Figure 6: Implementation architecture

1.3.3   Generic intelligent object

A generic object has been designed to encapsulate the minimal basic functionalities needed for turning an object of a scientific application into an intelligent object.

Such an object possesses data in order to identify its name, its class of instantiation, and its knowledge base. Its behavior consists in (i) determining its own set of rules including the ones of its superclasses, (ii) providing some basic reasoning modes that can be redefined in subclasses if needed, (iii) asserting its state into its associated fact base, or retrieving it from the same fact base, and finally (iv) supplying minimal interpreting capabilities defining the set of methods that can be used in premises or conclusions of rules.

This generic object is implemented by class **IntelligentObject**, from which subclasses can be designed in order to fulfill customizations required by scientific applications.

1.3.4   Transforming objects into intelligent objects

Integrating basic intelligent functionalities into an existing class of a scientific application, e.g. **Material,** is achieved by defining class **Material** as a subclass of class **IntelligentObject** as Figure 7 illustrates. Either simple or multiple inheritance can be used. Thus, class **Material** is able to identify its rules among a set of rules, and is also capable of activating them within anyone of its methods. Activating a reasoning mode is done by the activation of one of the reasoning modes described in sections 1.2.2.4 and 1.3.7.

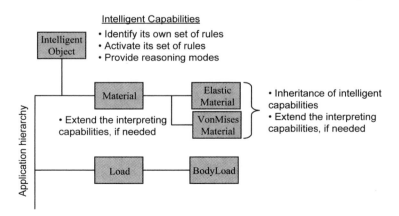

Figure 7: Transforming objects of an application into intelligent objects

Note that all subclasses of class **Material**, i.e. classes **ElasticMaterial** and **VonMisesMaterial** inherit these capabilities, and only a part of the application hierarchy can be affected by intelligent capabilities; here **Load** subhierarchy is kept unchanged. To complete the transformation, the methods of **Material** or of its subclasses employed in rules, have to be added to the interpreting capabilities of the corresponding class. We will see later that this process can be automated.

1.3.5   Rules

Rules are formulated as follows :

RULENAME: *name*
CLASS_LINK: *classname* or OBJECT_LINK: *objectname*
ACCES: *Public* or *Private*
IF *premises* [HIDDEN or NOTHIDDEN] THEN *conclusions* [SHARED or NOTSHARED]
Or
ALWAYS *conclusions* [SHARED or NOTSHARED]
COMMENT: [optional]

in which:

- a *name* (**RULENAME**) identifies the rule;
- a *membership* (**CLASS_LINK**) or   (**OBJECT_LINK**) identifies the class (*classname*) or the object (*objectname*) with which the rule is associated;
- a *keyword* (**ACCESS**) restricts the inheritance scheme applied to rules;
- A *Public* rule can be inherited by subclasses of the class to which it is attached, while a *Private* one has an influence restricted to its class or its object;
- a *body* consists either of *premises* and *conclusions,* or only of *conclusions.* The first form defines usual rules, while the second formulates initial facts always instantiated before a reasoning process starts.
  Hidden *premises* cannot be defined by external assistance (neither an object nor the user). Usually, such *premises* are used for the internal control of inferences. Shown *premises* can be determined with the assistance of other objects or of the user. By default, *premises* are not hidden.
  Shared *conclusions* are written on the blackboard when they are instantiated, as opposed to 'not shared' *conclusions* with a limited scope. The latter are mainly used as intermediate statements for the control of inferences. By default, *conclusions* are shared.
- finally, some optional comments (**COMMENT: [ ]**) may be added.

Note that rules are formulated as methods; and can therefore be considered as methods without arguments.

1.3.6   Blackboard

The architecture of the object-oriented knowledge-based system is organized around a central object called the blackboard [42,43,44]. The proposed blackboard architecture is the simplest possible that follows the concepts described in [45]. The blackboard is a common device that aims at containing a consistent state of shared knowledge in the environment, over time. Unlike usual blackboard architectures [46], it is not in charge of monitoring knowledge bases, since knowledge processing is under the control of intelligent objects.

The blackboard knows all instantiated intelligent objects and collects data. Data set on the blackboard can be accessed without restriction as they result from conclusions of rules defined as sharable variables. When an inference engine activates such conclusions, it writes them on the blackboard automatically. Notice that defining a variable as a blackboard variable is the responsibility of the designer of the application.

To summarize, blackboard variables are global variables through which intelligent objects communicate. Every intelligent object has access to its own variables and to blackboard variables. This means that blackboard variables can be viewed as an extension of the definition of intelligent objects.

1.3.7   Reasoning processes

Reasoning modes, i.e. forward chaining and backward chaining are implemented in a generic intelligent object class. As forward chaining deduces as many facts as possible, it was implemented in a method called *Deduce()*.

Backward chaining selects relevant rules among a set of rules in order to establish a value for the goal or the validity of that goal. This chaining was implemented in a method called *Define(...)*, the argument of which specifies the goal of reasoning.

1.4   Introducing rule based intelligence into a scientific application: Recipe

In this section, we establish a step-by-step procedure to be followed in order to turn an object-oriented scientific application into an intelligent application. We also define some guidelines for designing rules in an object-oriented manner in order to achieve a high level of modularity. In addition, a few hints are given for the identification of the appropriate blackboard variables and for the selection of appropriate chaining.

1.4.1   Step-by-step procedure

This procedure has two distinct phases: design and implementation. The key issue is to keep the scientific application as stable as possible, meaning that implementation changes in the application will be limited as much as possible.

❑   Design phase
    1.   Writing knowledge (rules and initial facts).
       1.1. Distinguish initial facts from rules.
       1.2. Define the blackboard variables.
       1.3. Identify the classes of the application with which rules should be associated.
    2.   Defining the most appropriate reasoning modes by answering the following questions:
       2.1. Are the needed reasoning modes already available?
       2.2. Can they be constructed with existing ones? or can they be built on existing inference engines?
       2.3. Do they require new inference engines?
    3.   Identify when and where reasoning modes should be activated.

❑   Implementation phase
    4.   Formulation of knowledge (rules and initial facts).
       4.1. Use the knowledge representation language for the formulation of rules and initial facts.
       4.2. Blackboard variables defined as sharable variables.
       4.3. Identify the methods of classes involved in rules and initial facts.
    5.   Identify the new methods used in rules and initial facts, and create a new subclass of class **IntelligentObject** for their implementation.
    6.   Implementation of reasoning modes
       6.1. If required reasoning modes are available, do nothing else.
       6.2. If a new reasoning mode needs to be built on existing reasoning ones or on existing inference engines, implement it through a method in a new subclass of class **IntelligentObject**.
       6.3. If a new inference engine is needed for the current application, implement it as an extension of the C++ expert library (see [1]).
    7.   Identify the superclass for each class that should be transformed into an intelligent class.
    8.   Transforming objects into intelligent objects
       8.1. Extend the interpreting capabilities of classes;
       8.2. Define classes of the application as subclasses inherited from class **IntelligentObject** or from any one of its created subclasses.
    9.   Modify the original methods of classes where reasoning modes should be activated.
    10. Compile and link the new implementation hierarchy of the scientific application with the C++ expert library.

1.4.2 Writing rules

The following guidelines will help generate rules in an object-oriented manner. First, state encapsulation of an object must be preserved. Rules attached to an object are designed like methods, they have to be formulated in terms (i) of variables of the owner's rules, (ii) of blackboard variables considered as global variables accessible from anywhere, (iii) of messages sent to any intelligent object including the owner. If a rule is written according to these guidelines, identifying the class of the object with which the rule has to be associated is unambiguous.

The gain of modularity resulting from the object-oriented approach at the code level should be visible at the level of rules. For that reason, premises of rules must absolutely avoid any type of enumeration describing a given state of the code development, like the hierarchy of implementation. Otherwise, it would be impossible to enhance the knowledge without revisiting the set of existing rules. Respecting this guideline will significantly improve the modularity and the extensibility of the rule set by focusing only on the extension being integrated, and by ignoring the history of the code development.

Finally, rule expressions must also be independent of the chaining that will infer on them. This can be achieved by writing rules concluding on global variables, those shared through the blackboard, or on values to be assigned to instance variables of the receiver. Therefore, both forward chaining and backward chaining could infer on these rules indifferently. But notice that rules with actions as conclusions can only be used by forward chaining.

1.4.3 Identifying blackboard variables

A blackboard variable may be a new variable or a variable attached to an object. Variables must be defined in conclusions of a rule and nowhere else, not even in an object. A blackboard variable characterizes the state of a set of objects, or a particular object, or a particular variable of an object. Setting an object's variable on the blackboard provides a shortcut for other objects without violating the encapsulation state of the object. This technique is very useful when objects of an application are influenced by a variable of a particular object. As an example, in a finite element preprocessor the options to be displayed could be closely related to the problem type, e.g. axisymmetric, plane strain, or 3D. Therefore, the variable representing the problem type is handled by an object and written on the blackboard in order to be used by objects like material, drivers [47].

1.4.5 Selecting an inference engine

Forward chaining, as a non-directed search algorithm, can be activated each time an object instantiates one of its variables in order to increase the amount of available knowledge, especially the one available on the blackboard. Most of the blackboard variables are defined this way.

As an example, consider the method in Figure 8 activated when the Poisson ratio of an elastic material is instantiated. By introducing the activation of *Deduce(...)* the rules attached to the **ElasticMaterial** are processed by a forward chaining. Therefore, the fired rule shown in Figure 9 will instantiate *Incompressibility* variable which will be put on the blackboard.

```
SetPoissonRatio(Supplied_argument) {
    PoissonRatio = Supplied_argument
    Deduce()
}
```

Figure 8: Activating a forward chaining in a method of an intelligent class

If PoissonRatio > 0.499
And PoissonRatio < 0.5
Then Incompressibility = True

Figure 9: Example of a rule instantiating a blackboard variable
by a forward chaining

Backward chaining, as a directed search algorithm, attempts to establish a value for a goal. Typically, it could be activated in order to define or verify an object variable, the object itself, or even a blackboard variable.

As an example, let us assume that the creation of the global stiffness matrix is managed by rules, which determine which type of matrix should be instantiated symmetric or non-symmetric. One of these rules is shown in Figure 10; it states that a non-symmetric matrix is instantiated when non-associative plasticity is detected. Figure 11 shows the single statement that activates a backward chaining with *stiffnessMatrix* as goal on these rules.

If Non-associativity = True
Then stiffnessMatrix =
CreateNonSymmeticMatrix()

Figure 10: Example of a rule governing the instantiation of
the global stiffness matrix

```
Matrix CreateStiffnessMatrix() {
    return Define("stiffnessMatrix")
}
```

Figure 11: Pseudo code activating a backward chaining
for the definition of *stiffnessMatrix*

1.5 A simple illustration

We illustrate here the proposed approach with a simple academic example. Suppose a scientific application depends on ten parameters but only two are independent. We wish each parameter to be equipped with a reasoning capability which allows it to identify itself automatically, if feasible. Assume also that, for each parameter, there are two methods, one getting access to its value and a second one setting up its value.

Rules will be attached to parameters so that each parameter class k will receive an associated set of rules like:

If x_i and x_j known for $(i,j) \in \{(1,\ldots,10),(1,\ldots,10)\}$ with $i \neq k$ and $j \neq k$
Then $x_k = x_k(x_i, x_j)$

The parameters will be declared as sharable variables, and automatically written on the blackboard when instantiated. These rules will be processed by a forward chaining when a parameter is instantiated. Forward chaining is provided by the method *Deduce()* inherited from class **IntelligentObject**. So, by calling this method within the method setting up the parameter's value the rules will be activated.

This reasoning mode will take full advantage of information already existing in the system. This approach could be very useful for the management of a user interface in which high a level of inter-dependency between parameters occurs.

1.6 Remarks

The proposed approach fully benefits from the features of object-oriented programming, which enables the association of rules with objects of a scientific application by using an inheritance scheme. Any object of an application can be transformed into an intelligent object. In other words, integration of rule-based intelligence may be limited to part of the application, there is no need for modifying the whole application. In addition, the proposed concept does not introduce additional constraints in the application, which keeps its own identity.

Another interesting aspect is the organization of rules, which mimics the implementation hierarchy of the application. This means that associating rules with objects automatically organizes them into a hierarchy reflecting the hierarchy of the application. Consequently, if the hierarchy of the application is modified, changes are straightforwardly propagated to the hierarchy of knowledge in order to reorganize the rules in the same manner. This hierarchical organization of knowledge automatically selects relevant rules in a reasoning process without any additional artifact.

Knowledge processing is implemented by local reasoning modes working on a limited number of rules and managed by intelligent objects. These reasoning modes are activated very easily. It is not necessary to know how they are implemented; knowing what they do is sufficient to use them adequately. In addition, the object-oriented implementation of reasoning modes provides a high flexibility for the creation of new reasoning modes, or for the customization of inference engines. The decentralization of knowledge activation allows the instantiation of multiple local reasoning modes, each of which under the control of an intelligent object.

Finally, the whole concept is implemented as an object-oriented C++ library consisting of an object-oriented propositional rule-based expert system and a generic intelligent object represented by class **IntelligentObject**. This package makes it possible to use AI techniques within a scientific application by focusing on the needs of the scientific application rather than on the AI implementation aspects. This high level of modularity leads to ease of use, genericity, portability, and efficiency of the proposed concept.

2 Application to data consistency checking in a nonlinear finite element program

2.1 Introduction

In the first part, we developed and implemented the concept of a generic intelligent object (instance of class **IntelligentObject**) with which an object-oriented propositional rule-based expert system is associated. Rules are attached to intelligent objects, behaving like methods. They respect object-oriented features of the object to which they are attached, i.e. inheritance, polymorphism, encapsulation, and the object's identity. Each intelligent object controls the activation of its own rules through local reasoning processes, and shares public information through a blackboard.

This second part of the paper aims at showing the efficiency and the ease of use of the proposed concepts on a scientific application, here an object-oriented nonlinear finite element code. Along these lines, it will be shown that the C++ package consisting of the class **IntelligentObject** and a C++ expert system library can be plugged into the application very easily and without major changes in its implementation. Required customization may be handled by subclasses of class **IntelligentObject**. Accordingly, the genericity, the modularity, and the portability of the developed concept are proved for any C++ scientific application.

This part is organized into four sections. In section 2.2.1 we discuss the needs of intelligent assistance within FE codes, then we review the use of knowledge-based systems within finite element programs, and finally point out the originality of the proposed approach. Section 2.3 presents the object-oriented nonlinear finite element code by describing its hierarchy, and some typical situations for which intelligent object are created. Section 2.4 describes the rules, and presents the corresponding formulation. Section 2.5 illustrates the implementation aspects. Concluding remarks are drawn in section 2.6.

2.2 Knowledge based systems and the finite element method

2.2.1 Need for assistance

Finite element methods have become the standard approach to the analysis of structural and mechanical systems. The availability of finite element programs with a very high level of sophistication and a wide range of modeling capabilities, operating in a graphical mode of interaction, has made the method particularly attractive for routine use. However, some software designs require that engineers fully understand every tool they use. It may take a user a year or more to learn effectively and efficiently how to use the common options and capabilities of a large FEA program. In spite of that, inexperienced engineers are often told to analyze structures without having the proper knowledge of the underlying mechanics governing their behavior. Often, this leads to inappropriate structure modeling and interpretation of results. Consequently, modeling assistants may be useful to novices who are not familiar with all modeling details, and are often not in a position to evaluate the adequacy of the various models they must build. Furthermore, finite element modeling assistants can reduce the number of tasks delegated to the user, allowing a shorter turn-around time for the analysis.

Different types of knowledge-based systems can be designed to help the user of FE packages: (i) intelligent modules helping to find particular program options, data entry procedure, suitable solutions, etc.; (ii) specification and modeling aids helping to translate appropriate element types, material model, boundary conditions, mesh density, etc.; (iii) results evaluation and design change proposals.

2.2.2 Historical sketch

The pioneer of finite element assistants, SACON [2], was designed to help the user of MARC™ code with the selection of an appropriate analysis strategy. This research showed the feasibility and the potential of using knowledge-based systems coupled with FE software. Later on, similar research projects led to the development of finite element assistants for other commercial finite element packages [48]. The literature also reports experiences to assist the user in particular problems: analysis and design of flexible elastic mechanisms [49], nonlinear finite element analysis [5,7], setting the time steps for dynamic analysis [50], finite element modeling and analysis of multichip module microelectronic devices [51], for instance.

Besides these particular applications, another research direction focuses on developing reasoning methodologies that can assist the user in setting up, interpreting and refining finite element models. Early ideas along these lines were presented in [52]. The resulting finite element assistant is based on multiple levels of abstraction through a hierarchical decomposition of the modeling problem [53]. It is also capable of reasoning about the modeling assumptions performed by the user, when data are generated. This framework maintains not only the data structure encoding the model to be analyzed, but also the assumptions used when generating the model and its history.

The aforementioned experiences illustrate the obvious need for intelligent assistance in finite element methods. Integrating knowledge-based systems in finite element modules can help accommodate CAD geometry [54,55], control input data [6,56,57], optimize meshes [58,59], and capture the structural behavior without any computation [60,61,62,63]. Reviewed works cover a wide range of applications, but none of them is directly concerned purely with computational mechanics, i.e. numerical algorithms. To the authors' knowledge, [8] is the only paper reporting on the optimization of the convergence path for a numerical algorithm using a knowledge-based system. It presents an interactive strategy of control that detects, interprets, and corrects irregular solution paths for the analysis of structural stability problems.

2.2.3 Complexity of the integration scheme

The lack of interest for intelligent control in numerical algorithms is probably due to the high level of interoperability required between the FE program and the associated knowledge-based system. For instance, controlling and setting up input data is much easier than managing numerical processes. Those tasks can easily be performed by an independent module, whereas an intelligent control of a numerical process needs to act while the process is running. Therefore, it requires a close linkage in order to efficiently convert intermediate numerical results produced by the algorithm into qualitative statements processed by the knowledge-based system. This data transfer repeated for each iteration step can adversely affect the performance of the algorithm. Therefore, integration has to be as seamless as possible to achieve reasonable efficiency.

Consequently, the complexity of integrating FE programs and knowledge-based systems depends on the nature of the tasks to be performed. The deeper the required interaction is, the more complex the integration becomes. Reported works prove the feasibility and the potential of using knowledge-based systems within FE software for some specific applications. None of them describes an integration methodology, without which efficient intelligent FE programs cannot be developed. With the intelligent-object concept, the complexity of the adopted integration scheme does not depend on the nature of the task to be performed. It remains constant regardless of the type of the object (graphical object or algorithm), and the methodology is independent of the type of application.

2.3 Object-oriented nonlinear finite element code

The object-oriented finite element code described below, is an extension of earlier work [12]. Originally, the hierarchy implemented linear elasticity. Rapidly, this code was then extended to nonlinear analysis, especially to J2-plasticity [64]. Recently, it was revisited and, with some inspiration taken from [65], modified in order to provide an optimal starting kernel for nonlinear finite element programming. Readers interested in a full description of the linear and nonlinear code should refer to [66,67].

2.3.1 Hierarchy of classes

The class hierarchy is given in Figure 12. It shows that the main objects of a finite element problem are implemented by classes **Node**, **Element**, **Material**, **Load**, and **NLSolver**, along with their subclasses. Subclasses of **Element** define various types of finite elements, subclasses of **Material** implement constitutive laws like the Von Misès model. Subclasses of **Load** define external actions and imposed boundary conditions on elements and nodes. Class **NLSolver** describes the common features of nonlinear algorithms, including linear analysis considered as a particular case solved in a single iteration. A specific nonlinear algorithm is mapped into a subclass of **NLSolver** like class **NewtonRaphson**. **LinearSystem** objects represent linear systems **Ax**=**b**, with **b** as a float array, and **A** as an instance of a subclass of **LHS** (left-hand-side). Class **Skyline** retains features applicable to both symmetric and non-symmetric matrices. Classes **TimeStep** and **LoadTimeFunction** account for the time-dependent effects. Class **TimeIntegration-Scheme** defines the time history of the problem. Its subclasses **Newmark** and **Static** deal with dynamic and static analysis respectively. Since most of those classes have some common behaviors like e.g. reading data in the input file, they are regrouped as a specialization of class **FEMComponent**. Class **Dof** implements a nodal degree of freedom. Class **Domain** describes the initial-boundary value problem. Class **GaussPoint** is clearly self-explanatory. Finally, for the sake of clarity, utility objects like matrices, dictionaries are regrouped as subclasses of an empty class **Tool**.

2.3.2 Intelligent object-oriented nonlinear finite element code

This application deals with data consistency within a nonlinear object-oriented finite element code. Keeping consistent data in a sophisticated finite element package is a difficult task. Erroneous input may be fatal to the analysis run, or generate warning messages that do not always allow the detection of errors. Consequently, debugging is time consuming, in addition to possible considerable waste of computational time. Obviously, the examples provided do not cover complete data checking for a safe use of the proposed object-oriented nonlinear code. However, they represent some typical situations one may have to face.

2.3.3 Typical situations

Typical situations that can conveniently be handled by rules and reasoning are presented now:

1. Incompressibility, dilatancy, and anisotropy may require appropriate formulations of elements (B-bar [68], Enhanced Assumed Strain (EAS) [69], mixed u-p formulation with or without stabilization [70,71,72,73]). For instance, a B-bar formulation could be used to overcome locking due to incompressibility, and EAS may be more appropriate for anisotropic media.

2. Mixture of elements is restricted to fully compatible elements. In the case of an incompressible medium discretized with B-bar elements, triangles and quadrilaterals cannot coexist.
3. Symmetry of the global stiffness matrix depends on type of plasticity (associative or non-associative), and also on the adopted formulation like stabilized mixed u-p formulation.

FEMComponent
  Node
  Element
    PlaneStrain
      Quad_U
        Quad_U_Bbar
          Triangle_U_Degen
      Quad_EAS
      Quad_U_P
        Triangle_U_P
    Truss2D
  Material
    ElasticMaterial
    VonMisesMaterial
    DruckerPragerMaterial
    MultiLaminateMaterial
  Load
    BodyLoad
      DeadWeight
    BoundaryCondition
    InitialCondition
    NodalLoad
  NLSolver
    ConstantStiffness
    ModNewtonRaphson
    NewtonRaphson
  LoadTimeFunction
    ....
  TimeIntegrationScheme
    Newmark
    Static
  TimeStep

LinearSystem

LHS
  Skyline
    SkylineSym
    SkylineUnsym
Dof
Domain
GaussPoint

Tool
  Dictionary
  FileReader
  FloatArray
  Column
  IntArray
  List
  MathUtil
  Matrix
    FloatMatrix
      DiagonalMatrix
      PolynomialMatrix
  Pair
  Polynomial
    PolynomialXY

Classes in italics are not yet implemented

Figure 12: Class hierarchy of the object-oriented nonlinear finite element code

In the following sections, we will describe how to generate rules for these situations, and how to select an appropriate chaining for their optimal use, and finally what subsequent changes in the original code are required.

2.4 Rules for finite element objects

In this section, we define rules managing the typical situations described previously. Their formulation respects the guidelines presented in part 1. After that, the corresponding implementation is described and finally reasoning modes are selected.

2.4.1 State of the hierarchy

2.4.1.1 Description of rules

The rules listed below describe the state of the hierarchy by enumerating the formulation implemented in each subclass of **Element**.
1. Displacement formulation is supported by triangles and quadrilaterals;
2. Enhanced Assumed Strains and B-bar formulations are available for quadrilaterals only;
3. Mixed u-p stabilized formulation is available for both types of element.

2.4.1.2 Formulation of rules

Each subclass of **Element** implements a specific element formulation. This means that each object of an **Element** subclass is able to instantiate a variable *elementFormulation* through an initial fact associated with the object's class. However, class **Element** provides a default instantiation. This leads to the following rules:

1. The first statement instantiates *elementFormulation* to 'Displacement' for classes **Quad_U**, and **Triangle_U** respectively;

Rulename: Set_up_element_formulation	*Rulename*: Set_up_element_formulation
Class: **Quad_U**	*Class*: **Triangle_U**
Access: Public	*Access*: Public
Always elementFormulation = 'Displacement'	Always elementFormulation = 'Displacement'

2. The second statement initializes *elementFormulation* to 'EAS' and 'Bbar' for classes **Quad_EAS** and **Quad_Bbar**;

Rulename: Set_up_element_formulation	*Rulename*: Set_up_element_formulation
Class: **Quad_EAS**	*Class*: **Quad_Bbar**
Access: Public	*Access*: Public
Always elementFormulation = 'EAS'	Always elementFormulation = 'Bbar'

3. The last statement implements a mixed u-p stabilized formulation for objects of classes **Quad_U_P** and **Triangle_U_P**;

Rulename: Set_up_element_formulation   *Rulename*: Set_up_element_formulation
Class: **Quad_U_P**   *Class*: **Triangle_U_P**
Access: Public   *Access*: Public
Always elementFormulation = 'Stabilized'   Always elementFormulation = 'Stabilized'

Default initialization is achieved by the following rule associated with class **Element**

> *Rulename*: Set_up_element_formulation
> *Class*: **Element**
> *Access*: Public
> Always elementFormulation = 'Unknown'

This last rule defined as public is inherited by subclasses of class **Element** and immediately overridden by a rule with the same name redefined in the subclass. This means that adding a new subclass of **Element** requires a single new rule defining the additional formulation. The proposed implementation only requires local changes when the hierarchy is modified, thereby achieving a good modularity.

In contrast, a bad implementation for that rule would be the following one:

> *Rulename*: Set_up_element_formulation
> *Class*: **Element**
> *Access*: Public
> If classname != 'Quad_U'
> And classname != 'Triangle_U'
> And ....
> Then elementFormulation = 'Unknown'

In that case, the premises of the rule describe the state of development of the hierarchy by enumerating the classes already implemented. This means that when a subclass is added or removed, this rule should be updated in order to account for changes.

2.4.1.3 Activation of rules

All these rules are initial facts, which are automatically activated when a reasoning process is instantiated. So, these rules can be used either by forward chaining or backward chaining.

2.4.2   Features of the material

2.4.2.1 Generation of rules

The rules listed below identify a material behavior according to the values of its paremeters.

1. An elastic material is said to be incompressible when its Poisson ratio (v) is greater than 0.499 and lower than 0.5;
2. A Von-Misès material is incompressible (in the plastic range);
3. A Drucker-Prager material is incompressible when its dilatancy angle (ψ) is null, otherwise dilatant (in the plastic range);
4. Plasticity is non-associative when the friction angle (φ) of a Drucker-Prager material is different from its dilatancy angle (ψ);
5. A multilaminate material is anisotropic by definition.

2.4.2.2 Formulation of rules

1. The first statement defines a rule attached to class **ElasticMaterial** formulated as follows:

    > *Rulename*: Incompressible_behavior
    > *Class*: **ElasticMaterial**
    > *Access*: Public
    > If give('v') > 0.499 and give('v') < 0.5
    > Then Incompressibility := True

    Notice that *give(char)* is a method of class **Material** interpreted by the local interpreter associated with each intelligent **Material** object (see section 2.5.2.2). This method gives access to a property identified by its name (here v for Poisson ratio) among all the material properties stored in a dictionary.

2. The second statement is an assertion that is true for all **VonMisesMaterial** objects. Therefore, it is formulated as the initial fact

    > *Rulename*: Incompressible_behavior
    > *Class*: **VonMisesMaterial**
    > *Access*: Public
    > Always Incompressibility := True

3. The incompressible or dilatant behavior of a **DruckerPrager** object is expressed by the two following rules

    Rulename: Incompressible_behavior | *Rulename*: Dilatant_behavior
    Class: **DruckerPragerMaterial** | *Class*: **DruckerPragerMaterial**
    Access: Public | *Access*: Public
    If give('ψ') = 0 | If give('ψ') > 0
    Then Incompressibility := True | Then Dilatancy := True

4. The next rule concludes on the non-associativity of plasticity for **DruckerPrager** objects that fulfill the rule

Rulename: Non-associative_plasticity
Class: **DruckerPragerMaterial**
Access: Public
If give('ψ') ≠ give('φ')
Then Non-associativity = True

5. Finally, the anisotropic behavior of instances of class **MultiLaminateMaterial** is defined by an initial fact formulated as follows

Rulename: Anisotropic_behavior
Class: **MultiLaminateMaterial**
Access: Public
Always Anisotropy = True

2.4.2.3 Activation of rules

These rules express their premises in term of variables of an object, and conclude on features (*Anisotropy*, *Incompressibility*...) characterizing the domain. Notice that these variables are defined as blackboard variables. These rules are perfectly suited for a forward chaining activated when **Material** objects are instantiated. More details on those implementation aspects can be found in section 2.5.1.1

2.4.3 Checking the formulation of elements

2.4.3.1 Generation of rules

♦ *Determining the best formulation*
The rules listed below define the proper formulation according to requirements set up by material definitions.
1. Incompressible media require B-bar, EAS, or mixed u-p stabilized formulations;
2. Dilatant media require either an EAS formulation or mixed u-p stabilized formulation;
3. Anisotropic media can only be simulated by an EAS formulation, so far.

♦ *Checking the consistency of the element formulation*
Every element should be compatible with the defined formulation. So, every element should be able to check its formulation with regard to the selected one in order to achieve the consistency of the mesh. This is done through the following two-steps procedure:
1. If the element formulation matches the required formulation, the appropriate element is the element itself;
2. If the element formulation does not match the required formulation, detect incompatibility, and provide a correction if possible, otherwise a general failure occurs. The correction creates a new object with the appropriate formulation that will replace the receiver.

2.4.3.2 Formulation of rules

♦ *Determining the best formulation*

A formulation is adopted for all elements discretizing the domain. So, rules defining the most suitable formulation given some media features are logically associated with an object of class **Domain**, which embodies the Initial-Boundary-Value-Problem.

A domain may consist of several types of materials meaning that incompressibility, dilatancy, and anisotropy may co-exist within the domain media. Looking at the three statements governing the formulation to be adopted, one may observe that some conditions are stronger or more restrictive than others. As an example, the third one is the most restrictive, because the formulation is established considering *Anisotropy* only, and regardless of the values of *Dilatancy* and *Incompressibility*. So the following rules will govern the selection of the most suitable formulation:

1. The third statement is formulated as:

> *Rulename*: EAS_Only
> *Class*: **Domain**
> *Access*: Public
> If Anisotropy = True
> Then Formulation = 'EAS'

2. When this rule fails, the next one is evaluated:

> *Rulename*: EAS_or_Stabilized
> *Class*: **Domain**
> *Access*: Public
> If IsNotInstantiated(Anisotropy)
> And Dilatancy = True
> Then answer := userSelectFrom('EAS', 'Stabilized')
> And Formulation := answer

Its first conclusion asks the user to select the formulation to be used, while the second instantiates *Formulation* with the user's answer. Note that the method *userSelectFrom(...)* is implemented in class **IntelligentObject** and will be inherited by its subclasses, e.g. class **Domain**.

3. Finally, when none of these rules matches, the last one is activated:

> *Rulename*: EAS_or_Stabilized_or_Bbar
> *Class*: **Domain**
> *Access*: Public
> If IsNotInstantiated(Anisotropy)
> And IsNotInstantiated(Dilatancy) = True
> And Incompressibility = True
> Then answer := userSelectFrom('Bbar', 'EAS', 'Stabilized')
> And Formulation := answer

Here, the user can select a formulation among three, and then again *Formulation* is instantiated with the user's answer.

◆   ***Checking consistency of the element formulation***
The rules that follow are used by an element in order to find out whether its formulation is compatible with the existing ones or not.

1.   The first rule expresses that any element is fully compatible when its formulation exactly matches *Formulation*.

> *Rulename*: Fully_compatible_element
> *Class*: **Element**
> *Access*: Public
> If elementFormulation = Formulation
> Then object := this

Notice that *this* is an internal variable representing the receiver. It is instantiated by an initial fact associated with class **IntelligentObject**. *Fully_compatible_element* is associated with class **Element**, and therefore inherited by objects of subclasses, which will instantiate *elementFormulation* by activating the initial fact attached to their class.

2.   The next rule detects incompatibility and recovers it by creating an object implementing *Formulation*, that will replace the receiver.

> *Rulename*: Incompatible_element
> *Class*: **Element**
> *Access*: Public
> If elementFormulation != Formulation
> Then object := newElementWith (Formulation)

These two rules are sufficient to support the verification of the consistent formulation of an element. If the returned element is null that means no correction can be supplied and this should lead to a general failure. The method *newElementWith()* provides a new object consistent with *Formulation*. Its implementation will be presented in detail later.

Finally, the last question to be solved is how to get access to *Formulation*. This variable is supposed to be on the blackboard, but it is not, it should be asked of the domain. The rule *GiveDomainFormulation* addresses the message *Define("Formulation")* to domain when *Formulation* is not yet instantiated. This message defined in class **IntelligentObject** (see [1]) activates a backward chaining on the set of rules attached to *domain* in order to find out the value of *Formulation*.

> *Rulename*: GiveDomainFormulation
> *Class*: **Element**
> *Access*: Public
> If IsNotInstantiated('Formulation') = True
> Then Formulation := SendMessage(domain, "Define", "Formulation")

GiveDomain is an initial fact attached to class **Element** that gets access to the domain through the *domain* variable within the reasoning process.

> *Rulename*: GiveDomain
> *Class*: **Element**
> *Access*: Public
> Always domain := giveDomain()

It is very important to notice that the checking process implemented by these four rules does not make any assumption on the state of the hierarchy, since the rules are attached to class **Element**, which is a higher abstraction level of elements. This means that this process is valid for any type of element, even those that are not yet implemented. The element formulation will be evaluated by the initial facts associated with each subclass of **Element**.

2.4.3.3   Activation of rules

♦   ***Determining the best formulation***
The rules associated with class **Domain** are good candidates to be activated by a forward chaining. The question is when the inference engine should be activated. The simplest answer would be when the domain is instantiated. But the **Domain** object is the first empty object to be created, this means that activating rules at that moment is totally useless. So, the solution is to delay the determination of *Formulation*, when it is needed for the fist time. This determination will result from a backward chaining activated when an element fires *GiveDomainFormulation* and sends the message *Define* to the domain. Then, a value is assigned to *Formulation,* and set on the blackboard.

♦   ***Checking consistency of the element formulation***
The rules attached to class Element define a compatible object, and are therefore efficiently used by a backward chaining as will be exposed in section 2.5.1.2.

2.4.4   Symmetry of the global stiffness matrix

2.4.4.1 Generation of rules

The rules given hereafter help in determining the shape (symmetric or non-symmetric) of the global stiffness matrix.
1. Non-associative plasticity implies non-symmetry of the global stiffness matrix;
2. Using a mixed u-p stabilized formulation generally leads to a non-symmetric global stiffness matrix;
3. Otherwise, the global stiffness matrix is symmetric.

2.4.4.2 Formulation of rules

All the above statements conclude on the instantiation class of the global stiffness matrix. This matrix is defined as an attribute of class **Domain**, so the resulting rules will be associated with class **Domain**.

1. The first statement can be written as follows:

> *Rulename*: Non-symmetry_resulting_from_non-associativity
> *Class*: **Domain**
> *Access*: Public
> If Non-associativity = True
> Then stiffnessMatrix := newObjectOfClass('UnsymSkyline')

Notice that method *newObjectOfClass(classname)* is defined and implemented in class **IntelligentObject**. The method argument represents a classname, by which a class can be identified and manipulated as a **VClass** object. This means that classes **UnsymSkyline** and **SymSkyline** will also be intelligent classes.

2. The second statement draws a conclusion upon the use of stabilized formulation, that can be expressed by:

> *Rulename*: Non-symmetry_resulting_from_stabilization
> *Class*: **Domain**
> *Access*: Public
> If Formulation = 'Stabilized'
> Then stiffnessMatrix := newObjectOfClass('UnsymSkyline')

3. The last statement covers all the other cases for which a symmetric matrix can be used:

> *Rulename*: Symmetry_for_other_situations
> *Class*: **Domain**
> *Access*: Public
> If Formulation ≠ 'Stabilized' and NotInstanciated(Anisotropy)
> Then stiffnessMatrix := newObjectOfClass('SymSkyline')

2.4.4.3 Activation of rules

The goal of these rules is the creation of an object defined as a domain variable. To avoid any anticipation of the creation of objects, a backward chaining will be activated on these rules only when the stiffness matrix is requested.

2.5 Implementation aspects [1]

This section shows how rules can be activated within methods. In other words how the C++ expert library described in the appendix is plugged into the present application, and it will be seen that required customizations are easily managed by the creation of subclasses of **IntelligentObject**

2.5.1 Activation of rules

2.5.1.1 Features of the media

Subclasses of class **Material** redefine a polymorphic method called *instantiateYourself()* that initializes the variables of the object. This is the best place for the activation a forward chaining in order to acquire as much information as possible. The implementation scheme valid for any type of material is as follows:

```
void Subclass::instantiateYourself() {
    …
    …. Initialization of the object's variables
    …
    Material::instantiateYourself() ;
}

void Material::instantiateYourself() {
    Deduce() ;   inherited from class IntelligentObject
}
```

As an illustration, consider the class **ElasticMaterial** for which *instantiateYourself* looks like:

```
void ElasticMaterial::instantiateYourself() {
    double value ;

    if (readWhetherHas("E")) value = give("E") ;
    if (readWhetherHas("n")) value = give("n") ;
    if (readWhetherHas("d")) value = give("d") ;
    if (readWhetherHas("A")) value = give("A") ;
    if (readWhetherHas("t")) value = read("t") ;
    Material::instantiateYourself() ;
}
```

2.5.1.2 Checking the formulation of an element

The creation of elements is a two-step process. The first one creates an instance of class **Element**, and the second mutates it by reading its real instantiation class in the input file. The mutation process is implemented by the method *ofType(char * aClass)* of class **Element**. Integrating the verification of the element formulation in this process is achieved by adding a third step that ensures the consistency of the mutation procedure.

Figure 13 presents the resulting code in which the method *constructCompatible-Element* takes *element* as a possible compatible element and modifies it if necessary. This method is defined in class **IntelligentFiniteElement**, the description of which is given later on. Note that encapsulating the verification process within a method allows moving this method to somewhere else very easily if needed.

```
Element*  Element :: ofType (char* aClass) {
    Element* element ;

    if (aClass =="Quad_U") element = new Quad_U(number,domain) ;
    else if (aClass == "Quad_U_BBar") element = new Quad_U_BBar(number,domain) ;
    else if (aClass == "Truss2D")  element = new Truss2D(number,domain) ;
    else if …..
    else {
        Printf ("%s : unknown element type \n",aClass) ;
        exit(0) ;
    }
    constructCompatibleElement(element) ;
    return element ;
}
```

Figure 13: Source code for mutation process

2.5.1.3 Symmetry of the global stiffness matrix

The global stiffness matrix is instantiated in the method *CreateStiffnessMatrix()* of class **Domain**. The original code is shown in figure 14. Notice that a non-symmetric matrix is instantiated without knowing if it is the optimal choice.

```
Skyline*  Domain :: CreateStiffnessMatrix() {
    return  new SkylineUnsym();
}
```

Figure 14: Source code of *CreateStiffnessMatrix()* in class Domain

Activating rules associated with class **Domain** will return the optimal object for the global stiffness matrix. These rules are activated in backward chaining. The single statement of this method will be replaced by the following code:

```
return  (Skyline *) Define('stiffnessMatrix')->GetObject()
```

2.5.2   Hierarchy of the intelligent object-oriented nonlinear finite element code

2.5.2.1 Modifying the original hierarchy

Figure 15 presents an overview of the new hierarchy of the intelligent object-oriented nonlinear finite element code.

FEMComponent   **IntelligentFiniteElement**
   Element
      PlaneStrain
         Quad_U
           Quad_Bbar
         Triangle_U_Degen
        ………

                IntelligentObject
   Material
      ElasticMaterial
      VonMisesMaterial
      DruckerPragerMaterial
      MultiLaminateMaterial

LHS   **IntelligentObject**
   Skyline
      SkylineSym
      SkylineUnsym

                IntelligentObject
   Domain

Figure 15: Hierarchy of the intelligent object-oriented

nonlinear finite element code

According to rules defined in section 2.4, class **Element** and its subclasses, class **Material** and its subclasses, class **Domain** and class **Skyline** are classes with intelligent behaviors. Class **Element** inherits a finite-element-analysis behavior from class **FEMComponent,** and an intelligent behavior from a new **class Intelligent-FiniteElement.** Class **Material** inherits a finite-element part from class **FEMComponent,** and an intelligent part from the generic class **IntelligentObject.** Finally, class **Domain**, which did not have superclass, is defined as a subclass of class **IntelligentObject.** Next, it will be seen how these classes have been modified in order to collect their set of rules and to activate them.

2.5.2.2 Class Material and its subclasses

Converting a standard C++ class into an expert class is easy, since transforming principles have been automated.
The input screen of the transforming tool is presented in Figure 16. First of all, a class is selected by opening one of its files (either its header file or its source file) through the activation of *Open* from *File* menu.
   By clicking on *Extend/Limit* of *Methods* Menu, interpreting the capabilities of class **Material** can be extended. The dialog box in Figure 17 lists the methods implemented in class **Material** on top, and the methods that can be used in rules at the bottom; here this list is emptied.

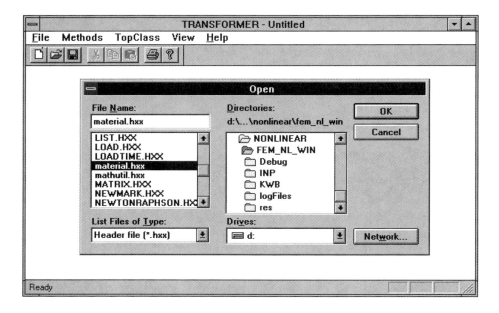

Figure 16: Opening class **Material**

The user selects methods with the mouse, and moves them down by clicking on *MoveDown* button. Once the selection is achieved, clicking on *OK* fixes the methods that can be used within rules. *Cancel* button cancels the selection. Activating the same option again makes it possible to change the list of selected methods.

In a next step, the user selects *SaveAsIntelligentObject* from *File* menu. Additional information is needed to achieve the generation of the appropriate source code. For instance, the top class beyond which rules cannot be attached to superclasses, or the superclass that brings intelligent behavior to class **Material**. Figure 18 asks the user whether class **Material** is the top class for knowledge collection; here the answer is yes, according to the hierarchy presented in the previous section. This leads to the dialog box in Figure 19 which defines the super-intelligent class of class **Material** class; here it is class **IntelligentObject.** By clicking on OK, two new files for class **Material,** are generated: a new header file and a new source code file. They include modifications that identify class **Material** as a **VClass** object, and make it possible to use *give(...)* method within rules.

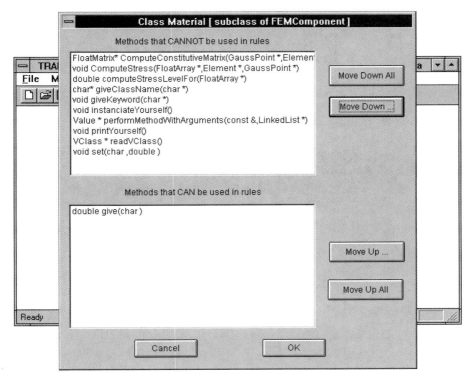

Figure 17: Extended the local interpreter of class **Material**

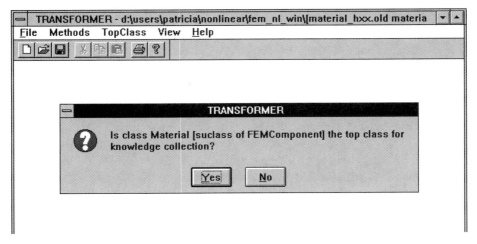

Figure 18: Definition of the top class for knowledge collection

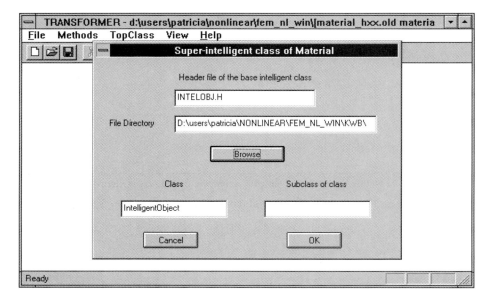

Figure 19: Defining the super-intelligent class for class **Material**

2.5.3   Class Domain and class Skyline

Transforming these two classes presents no difficulties. The process to be followed is the same as for subclasses of class **Material**. For both classes, no additional method is added to the local interpreter, and the intelligent base class is class **IntelligentObject**.

2.5.2.4 Class Element and its subclasses

As already mentioned, the rule *incompatible_element* requires the creation of a new class, class **IntelligentFiniteElement**, for the implementation of *newElementWith(…)* method. This method implements the following algorithm:

1. Collect the subclasses of class **Element**;
2. For each subclass *c* of class **Element**;
   2.1. Create a new element *e*;
   2.2. If *e* has not the same number of nodes as the receiver, go to 2.4;
   2.3. Check if the formulation of *e* matches the required formulation;
      2.3.1. Yes, return *e* as a compatible element;
      2.3.2. No, delete *e*, and go to 2.4;
   2.4. Take next subclass *c* and go to 2.1;
3. No compatible element can be created, return NIL;

with the implementation shown in Figure 20.

```
Element * IntelligentFiniteElement::newElementWith(ExpertAttribute * aFormulation) {
    subclasses = GetClassname("Element")->GetSubclasses() ;                    1
    while (c = (VClass *)sublclasses->GetElement())  {                         2
        Element * e = c->construct() ;                                         2.1
        If (e->hasSameNumberOfNodes(this)                                      2.2 /
            && e->Define(aFormulation)) return e ;                   2.3 / 2.3.1
        delete e ;                                                             2.3.2
        c = subclasses->GetNext() ;                                            2.4
    }
    return Null ;                                                              2.5
}
```

Note that the declaration of variables is omitted for the sake of clarity

Figure 20: Source code of method *newElementWith(aFormulation)*

Note that this algorithm is generic and would support any changes in the **Element** sub-hierarchy without being updated.

The second method implemented in this class is *constructCompatible-Element(Element *)* that follows the algorithm :

1. Activate a backward chaining on rules attached to *e* with *compatibleElement* as goal;
2. Retrieve the result of inference ;
3. If the inference result differs from *e*, delete *e*;
4. Assign the inference result to *e*;
5. If *e* is still null, print out an error message.
6. Return *e* ;

with the implementation shown in Figure 21.

```
void  IntelligentFiniteElement:ConstructCompatibleElement(Element * e) {
    Value * value = Define ('compatibleElement') ;                  1
    Element * result = (Element *)value->GetObject() ;              2
    If (result != e) delete e ;                                     3
    e  = result ;                                                   4
    If (e == NIL) print out an error message                        5
    return e ;                                                      6
}
```

Figure 21: Source code of method *constructCompatibleElement(Element *)*

The transforming tool is used to generate an appropriate source code for the interpretation of these methods. Like class **Material**, class **Element** is a top class for knowledge collection, but this class has a different super-intelligent class; here it is class **IntelligentFiniteElement** instead of class **IntelligentObject**. Answering the question leads to filling the dialog box shown in Figure 22. Then, the tool generates two new files accordingly. Again, in order to complete the process each subclass of class **Element** should be transformed as well.

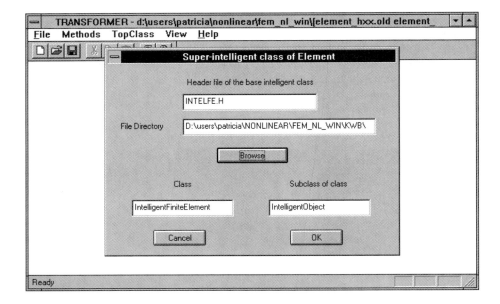

Figure 22: Defining the super-intelligent class for class **Element**

2.6   Concluding remarks

In the first part of this paper, we developed the concept of an intelligent object and described its implementation. It is based on a generic intelligent object with which an object-oriented propositional rule-based expert system is associated. Rules are attached to intelligent objects, behaving like methods. They respect object-oriented features of the object to which they are attached, i.e. inheritance, polymorphism, encapsulation, and the object's identity. Each intelligent object controls the activation of its own rules through local reasoning processes, and shares public information through a blackboard.

In the second part of the paper, we presented an application of the intelligent-object concept to data preparation for an object-oriented nonlinear finite element program. We illustrated its ease of use, emphasized the genericity and the portability of the approach. Customizations required by the application are easily managed by subtyping the generic intelligent object class. Most code changes required to plug the C++ expert library like the management of a class as an **VClass** object, and the generation of a local interpreter have been automated by the use of a transforming tool, thereby freeing the software designer from tedious tasks.

However, the modularity resulting from the integration of rules in an application highly depends on the quality of the object-orientedness embodied in that application. This quality may either facilitate or hinder the development. But simple guidelines for writing rules and selection of inference engines warrant a satisfying modularity, and genericity. It should be noticed that the proposed approach does not introduce additional constraints on the design of the host application. All code changes are performed in existing methods, and new methods are located in subclasses of class **IntelligentObject**. Another interesting aspect is that any object of an application can be transformed into an intelligent object, so the use of rules may be limited to part of the application; there is no need to modify the whole application, which keeps its own identity.

The proposed approach constitutes a step forward toward the integration of rule-based expert systems into object-oriented scientific applications. Due to its genericity, many applications could benefit from this efficient and portable approach.

Acknowledgments: This research was supported by the Swiss Commission for Technology and Innovation (CTI) under grants 2020.1 and 2516.1.

References

[1]   P.Bomme, Th.Zimmermann, *"Toward intelligent object-oriented scientific applications"*, EPFL-ENAC-LSC Rep. 2002.01, 2002.

[2]   J. Bennett, L. Creary, R. Engelmore, and R. Melosh, *"SACON: A knowledge-based consultant for structural analysis"*, Tech. Rep. STAN-CS-78-699, Stanford University, 1978.

[3]   I. C. Taig, *"Expert aids to finite element system application"*, Applications of AI in Engineering Problems, (eds) D. Sriram and R.A. Adey, Springer-Verlag, Vol. 2, 759-770, 1986.

[4]   S. J. Fenves, *"A framework for a knowledge-based finite element analysis assistant"*, Application of knowledge-based in structural design, (ed.) C. L. Dym, 1-7, 1985.

[5]   P. Breitkopf and M. Kleiber, *"Knowledge engineering enhancement of finite element analysis"*, Communications in Applied Numerical Methods 3, 359-377, 1987.

[6]   R.K. Fink, R.A. Callow, T.K. Larson and V.H. Ramson, *"An intelligent modeling environment for a large engineering analysis code: Design issues and developments experience"*, Eng. with Comp. 3, 167-179, 1988.

[7]   B.W.R. Forde and S.F. Stiemer, *"ESA: Expert Structural Analysis for engineers"*, Comp. & Struct. 29(1), 171-174, 1988.

[8]   P. Wriggers and N. Tarnow, *"Interactive control of non-linear finite element calculations by an expert system"*, AI in Computational Engineering, (ed.) M. Kleiber, Ellis Horwood, 97-119, 1990.

[9]   R. Balzer: *"AI and software engineering, Will the twain ever meet?"* Proc. of the 8th Conf. of the AAAI, Morgan Kaufmann, 1990.

[10] M. Stefik and D. Brobow, *"Object-oriented programming: Themes and variations"*, AI Magazine 6(4), 40-62, 1986.

[11] B.W.R. Forde, R.O. Foschi and S.F. Stiemer, *"Object-oriented finite element analysis"*, Comp. & Struct. 34(3), 355-374, 1990.

[12] Y. Dubois-Pèlerin and Th. Zimmermann, *"Object-oriented finite element programming (Part III): An efficient implementation in C++"*, Comp. Meth. App. Mech. and Eng. 108, 165-183, 1993.

[13] M.D. Rucki and G.R Miller, *"An algorithmic framework for flexible finite element-based structural modeling"*, Comp. Meth. App. Mech. and Eng. 136, 363-384, 1996.

[14] J. Besson and R. Foerch, *"Large scale object-oriented finite element code design"*, Comp. Meth. App. Mech. and Eng. 142, 165 –187, 1997.

[15] D. Thomas: *"Object-oriented programming: What's an object"*, Byte Magazine, March 1989.

[16} A. Snyder, *"The essence of objects: Concepts and Terms"*, IEEE Software, January, 31-42, 1993.

[17] R. Sethi, *"Programming Languages: Concepts and Constructs"*, Addison Wesley, 1996.

[18] G.R. Miller: *"An object-oriented approach to structural analysis and design"*, Comp. & Struct. 40(1), 75-82, 1991.

[19] T.J. Ross, L.R. Wagner and G.F. Luger, *"Object-oriented programming for scientific codes: I. Thoughts and Concepts"*, J. Comp. in Civil Eng. 6(4), 480-496, 1992.

[20] A. Cardona, I. Klapka and M. Géradin, *"Design of new finite element programming environment"*, Eng. Comp. 11, 365-381, 1994.

[21] W.R. Lalonde and J.R. Pugh, *"Inside Smalltalk: Volume I and II"*, Prentice-Hall, 1990.

[22] B.J. Cox, *"Object-oriented programming: An evolutionary approach"*, Addison Wesley, 1991.

[23] B. Meyer, *Eiffel: The language*, Prentice Hall, 1992.

[24] A. Paepcke, *"Object-oriented programming: A CLOS perspective"*, MIT Press, 1993.

[25] N.A.B. Gray, *"Programming with Class: A practical introduction to object-oriented programming with C++"*, J. Wiley & Sons, 1994.

[26] D. Flanagan, *"Java in a Nutshell: A desktop quick reference for Java programmer"*, O'Reilly & Associates, 1996.

[27] M.B. Feldman and E.B. Koffman, *"Ada 95: Problem solving and program design"*, (2nd ed), Addison Wesley, 1996.

[28] Th. Zimmermann, Y. Dubois-Pèlerin, and P. Bomme, "Object-oriented finite element programming (Part I): Governing principles", *Comp. Meth. App. Mech. Eng.* 98, 291-303 1992.

[29] Y. Dubois-Pèlerin, Th. Zimmermann, and P. Bomme, "Object-oriented finite element programming (Part II): A prototype program in SmallTalk", *Comp. Meth. App. Mech. Eng.* 98, 261-397, 1992.

[30] A. Galton, *"Logic for Information Technology"*, John Wiley & Sons, 1990.

[31] E. Turban, *"Expert systems and Applied Artificial Intelligence"*, Macmillan Publishing Company, 1992.

[32] T. Dean, J. Allen, and Y. Aloimonos, *"Artificial Intelligence Theory and Practice"*, Benjammin/Cummings, 1995.

[33] H. Reichgelt, *"A knowledge representation: an AI perspective"*, Ablex publishing Corporation, 1991.

[34] M.R. Klein and L.B. Methlie, *"Knowledge-based decision systems with applications in Business"*, J. Wiley & Sons, 1995.

[35] B. G. Buchanan and E. H. Shortliffe, *"Rule-based expert systems: The MYCIN experiments of the Stanford heuristics programming project"*, Addison Wesley, 1984.

[36] R. Voyer, *"Moteurs de systèmes experts"*, Eyrolles, 1987, (*in French*).

[37] J.P. Delahaye, *"Systèmes experts: organisation et programmation des bases de connaissance en calcul propositionel"*, Eyrolles Editions, 1988, (*in French*).

[38] B. Sawyer and D. Foster, *"Programming expert systems in Pascal"*, J. Wiley & Sons, 1986.

[39] K. W. Tracy and P. Bouthoorn, *"Object-oriented Artificial Intelligence using C++"*, Computer Science Press, 1996.

[40] F. Pachet, *"OOPSLA'94 Workshop on Embedded Object-Oriented Production Systems (EOOPS'94)"*, Tech. Rep. 24, LAFORIA, Université Pierre et Marie Curie, Paris VI, 1994.

[41] D. Hu, *"C/C++ for expert systems: Unleashes the power of artificial intelligence"*, Management Information Source, 1989.

[42] R. Engelmore and T. Morgan, *"Blackboard systems"*, Addison-Wesley, 1988.

[43] H. P. Nii, *"Chapter 1 Introduction, Blackboard architectures and applications"*, xix-xxix, (eds) V. Jagannathan, R. Dohiawala, and L.S. Baum, Academic Press, 1989.

[44] J. Hallam, *"Blackboard architectures and systems"*, Chapman and Hall, 1990.

[45] I. Graham, *"Migrating to object technology"*, Addison Wesley, London, 1994.

[46] N. Carver and V. Lesser, *"Evolution of blackboard control architectures"*, Expert Systems with Applications 7, 1-30, 1994.

[47] P. Bomme, *"Intelligent objects for object-oriented engineering environments"*, EPFL Ph.D Thesis no 1763, 1998.

[48] J. Mackerle and K. Osborn, *"Expert systems for finite element analysis and design optimization: a review"*, Eng. Comp. 5, 90-102, 1988.

[49] F. W. Liou and A. K. Patra, *"Development of an advisory expert system for elastic mechamisn desig'"*, Comp. Struct. 46(1), 125-132, 1993.

[50] M. R. Ramirez and T. Belytschko, *"An expert system for setting time steps in dynamic finite element program"*, Eng. with Comp. 5, 205-219, 1989.

[51] I. R. Grosse and D. D. Corkill, *"A blackboard approach to intelligent finite element modeling analysis"*, Comp. in Eng. ASME, (ed.) G. A. Gabriele, Vol 2, 61-68, 1992.

[52] G.M. Turkiyyah and S. J. Fenves, *"Getting finite element programs to reason about their analysis assumptions"*, Computer Utilization in Structural Engineering, (ed) J. K. Nelson, 51-60, 1989.

[53] G.M. Turkiyyah and S. J. Fenves, *"Knowledge-based assistance for finite element modeling"*, IEEE Special Issue AI in civil eng. and struct. eng., June, 23-32, 1996.

[54] U. Roy, *"An intelligent interface between symbolic and numeric analysis tools required for the development of an integrated CAD system"*, Comp. Industr. Eng., 30(1), 13-26, 1996.

[55] P. Dabke, *"An agent-based approach towards finite element analysis"*, Ph.D. Thesis, Stanford University, 1996.

[56] S. Nagasawa, Y. Miyata, M. Murayama, and H. Sakuta, *"Support system for finite element analysis"*, Advances in Eng. Software 27, 179-189, 1996.

[57] J. Cagan and V. Genberg, 'PLASHTRAN: *"An expert consultant on two-dimensional finite element modeling techniques"*, Eng. with Comp. 28, 199-208, 1987.

[58] E. Kang and K. Haghighi, *"Intelligent finite element mesh generation"*, Eng. with Comp. 11, 70-92, 1995.

[59] A. Dolšak, A. Jezernik and I. Bratko, *"A knowledge base for finite element mesh design"*, AI in Eng., 9, 19-27, 1994.

[60] Y. Iwasaki, S. Tessler, and K. H. Law, *"Diagrammatic reasoning for qualitative structural analysis"*, Center for Integrated Facility Engineering (CIFE), Stanford University, April 1994.

[61] S. Tessler, Y. Iwasaki, and K. H. Law, *"REDRAW: a diagrammatic reasoning system for qualitative structural analysi'"*, Advances in Eng. Software 25, 149-159, 1996.

[62] L. M. Bozzo and G. L. Fenves, *"Qualitative reasoning and the representation of fundamental principles in structural engineering'*, Research in Eng. Design 6, 61-72, 1994.

[63] L. M. Bozzo and G. L. Fenves, *"Qualitative reasoning inference strategy and its application to structural engineering "*, Research in Eng. Design 6, 73-84, 1994.

[64] Ph. Ménétrey and Th. Zimmermann, *"Object-oriented nonlinear finite element analysis: Application to J2 plasticity"*, Comp. & Struct. 49(5), 767-777, 1993.

[65] Y-D. Dubois-Pèlerin and P. Pegon, *"Object-oriented programming in nonlinear finite element analysis"*, Comp. & Struct. 67, 225-241, 1998.

[66] Y-D. Dubois-Pèlerin and Th. Zimmermann, *"Object-oriented finite element programming: Theory and C++ Implementation for FEM_Object$_{C++}$"* Elmepress, 1992.

[67] S. Commend and Th. Zimmermann, *"Object-oriented Nonlinear Finite Element Programming : a Primer"*. Advances in Engineering Software, 32:611-628, 2001.

[68] T.J.R. Hughes, *"Generalization of selective Integration Procedures to Anisotropic and Nonlinear Media"*. IJNME 15: 1413-1418, 1980.

[69] Th. Zimmermann, J.-L. Sarf, A. Truty, and S. Commend, *"Recent advances in geotechnical engineering software"*, Proc. Congress on Advances in Computer Methods in Geotechnical and Geoenvironmental Engineering, Moscow 2000. S. Yufin Ed., Balkema, 2000 .

[70] A. Truty and Th. Zimmermann, *"A robust formulation for FE-analysis of elasto-plastic media"*, Numog VI, S. Pietruszczak and G. N. Pande, 381-386, 1997.

[71] S. Commend , A. Truty , Th. Zimmermann, " *Stabilized finite elements applied to elastoplasticity: I. Mixed displacement pressure formulation,* accepted for CMAME, 2002.

[72] S. Commend, *"Stabilized finite elements in geomechanics",* PhD dissertation 2391. Swiss Federal Institute of Technology (EPFL), 2001.

[73] A. Truty , "On certain classes of mixed and stabilized mixed finite element formulations for single and two-phase geomaterials", Habilitation thesis, Cracow University of Technology, 2002.

Author Index

Keyword Index